D0041795

A WORLD *on the* WING

ALSO BY SCOTT WEIDENSAUL

Peterson Reference Guide to Owls
of North America and the Caribbean

The First Frontier:
The Forgotten History of Struggle, Savagery, and
Endurance in Early America

Of a Feather:
A Brief History of American Birding

Return to Wild America:
A Yearlong Search for the Continent's Natural Soul

The Ghost with Trembling Wings:
Science, Wishful Thinking, and the Search for Lost Species

Living on the Wind:
Across the Hemisphere with Migratory Birds

Mountains of the Heart:
A Natural History of the Appalachians

A
WORLD
on the
WING

The Global Odyssey
of Migratory Birds

SCOTT WEIDENSAUL

W. W. NORTON & COMPANY
Independent Publishers Since 1923

All maps by the author, except for National Weather Service Doppler radar map and summer tanager abundance map (latter image provided by eBird, www.ebird.org, and created Aug. 21, 2019). All photographs by the author, except as noted.

Copyright © 2021 by Scott Weidensaul

All rights reserved
Printed in the United States of America
First Edition

For information about permission to reproduce selections from this book, write to Permissions, W. W. Norton & Company, Inc., 500 Fifth Avenue, New York, NY 10110

For information about special discounts for bulk purchases, please contact W. W. Norton Special Sales at specialsales@wwnorton.com or 800-233-4830

Manufacturing by LSC Communications, Harrisonburg
Production manager: Beth Steidle

Library of Congress Cataloging-in-Publication Data

Names: Weidensaul, Scott, author.
Title: A world on the wing : the global odyssey of migratory birds / Scott Weidensaul.
Description: First edition. | New York : W. W. Norton & Company, [2021] | Includes bibliographical references and index.
Identifiers: LCCN 2020028997 | ISBN 9780393608908 (hardcover) | ISBN 9780393608915 (epub)
Subjects: LCSH: Birds—Migration. | Migratory birds. | Flyways.
Classification: LCC QL698.9 .W455 2021 | DDC 598.156/8—dc23
LC record available at https://lccn.loc.gov/2020028997

W. W. Norton & Company, Inc., 500 Fifth Avenue, New York, N.Y. 10110
www.wwnorton.com

W. W. Norton & Company Ltd., 15 Carlisle Street, London W1D 3BS

1 2 3 4 5 6 7 8 9 0

For Amy, as always (but even more than usual)

CONTENTS

A WORLD *on the* WING

PROLOGUE

Tundra may be the most gloriously comfortable mattress in the world.

A little damp, it's true, which is why it's a good idea to wear rain pants and a jacket, even on a clear, chilly morning like this one—the sun just touching the peaks of the Alaska Range with pink-orange light, the glacier-wrapped bulk of Denali a vast, rosy monolith 70 miles to our west, uncharacteristically free of clouds.

My three companions and I flopped down with happy sighs, legs outstretched and hands laced behind our heads, onto the soft, spongy cushion of sphagnum moss, dwarf cranberries, reindeer lichen, and other Lilliputian tundra plants. The break felt good. We'd risen at two in the morning, in the bright twilight that passes for the middle of the subarctic night in the interior of Alaska. By three, keeping an eye out for moose or grizzlies, we were headed west along the 90-mile gravel road that bisects the six-million-acre wilderness of Denali National Park and Preserve. We never knew what we'd see. The day before, a large male wolf had trotted warily around our National Park Service truck before sniffing nervously at the rear fender, just a few feet from my open window.

There were no such interruptions today. By four o'clock, 30 miles inside the park, we'd shouldered our packs and bundles of aluminum net poles, then trudged down a long slope to a sinuous willow thicket that snaked through a mile-long draw. As luxurious as spongy tundra is to lie on, it is a tiring chore to hike across, with every footstep sink-ing deep or rolling on some hidden tussock, while shin-high birches and willows claw at your feet and legs.

"Hey! Hey!" we yelled, to alert any moose or grizzly bear that might be hidden in the dense, 10-foot-high brush ahead. "Blah blah blah blah!" I shouted nonsensically; it doesn't matter what you bellow, just so you don't surprise a protective cow moose with a calf, or startle a grizzly whose first reaction might be to charge. Unlike many hikers, one thing we never did was yell, "Hey bear!" Those words, old Alaskan hands will tell you, should be reserved solely for the gut-twisting moment when a grizzly pops up at close range—a warning to the bear, but more importantly to everyone else in earshot.

As it was, all we alerted was a family of willow ptarmigan, half a dozen rotund, brown fledglings that boomed off in as many directions while the mother grouse barked her displeasure. We shucked our loads, and I followed Laura Phillips, the park's avian ecologist, as she wormed her way into the seemingly impenetrable tangle of willows. Somehow, the moose had no trouble maneuvering in there— the wet ground was pocked with their saucer-sized tracks and piles of oblong droppings. But in the middle we found a slender lozenge-shaped meadow just a few yards wide, blue with the stately flowers of monkshood and larkspur, its margins purple with spires of fireweed.

We weren't looking for ptarmigan or wildflowers, though, but for thrushes—and not to watch, but to catch. After more than three decades of visiting Denali, I was helping to launch a new research project there to better understand the lives of the park's birds, which every year fan out across three-quarters of the earth's surface on their migrations.

Soon, we had three 40-foot-long mist nets radiating out into the brush. David Tomeo, with Alaska Geographic, and seabird biologist Iain Stenhouse—a transplanted Scot now living in Maine, who was once Audubon's director of bird conservation in Alaska—secured the net poles with guy lines of bright red parachute cord. I jammed a long wooden dowel into the ground mid-net, and perched on its tip a painted, life-sized wooden thrush decoy. Then I thumbed the controls of a battered old MP3 player, from which emerged the buzzy, ethereal song of a gray-cheeked thrush. Our work finished for the moment, the four of us walked 10 or 15 yards up the hill, out of the willows and into the open tundra, and sank down to relax for a few minutes. Our hope was that a male thrush—hearing what sounded like an intruder in his jealously defended territory—would come barreling down through the shrubs and collide harmlessly with our delicate nets. Then we could carefully attach a tiny device called a geolocator, weighing barely half a gram, to the small of his back. For the next year, it would record the bird's location as it flew to South America and back, giving us the first glimpse anyone's had into the specifics of this bird's epic migration.

For the better part of a century, the only means scientists had of figuring out where birds traveled was by putting lightweight num- bered bands on their legs, and hoping to hear if the banded bird was ever encountered again. Banding is still a critical element of migra- tion research—some 7 million mallard ducks have been banded in the past century, for example, and 1.2 million of them recovered (mostly by hunters), providing data that help underpin our very suc- cessful management of waterfowl populations. But it's a long, slow slog if you're studying a rarely banded bird in a remote area—a bird that, unlike mallards, isn't legally hunted. In the past century, roughly 82,000 gray-cheeked thrushes have been banded in North America as a whole, but only 4,312 of them were in Alaska—and of those banded Alaskan thrushes, only three have ever been encoun- tered again. One was caught close to where it was banded, one on its spring migration north through Illinois, and one heading south in the fall in Georgia. That's not much to go on.

What banding data and observations we do have show that gray- cheeked thrushes are exceptionally long-distance migrants. Even though they weigh only about 30 grams—a shade more than an ounce—they travel from conifer forests and thickets in northern Alaska and the Canadian subarctic to South America and back each year. At least some of them cross the Gulf of Mexico in a 600-mile nonstop leap, while others may follow the long finger of Florida and then overfly the Caribbean. In winter, they disappear into the rain forests of northern South America, but we have only the sketchiest notion of where they go within that vast continent.

But where banding struggles to fill in the blanks, newly miniaturized technology is opening exciting horizons in the study of bird migration. The geolocators we were using are just one example of tiny, relatively inexpensive tracking devices that are revolutionizing migration research. Instead of depending on satellite transmitters that cost $4,000–$5,000 each (and which are, in any case, far too heavy for small songbirds), our geolocators weigh a fraction of a gram and cost just a few hundred dollars each. Our team, headed by National Park Service ecologist Carol McIntyre, was starting a multiyear project to trace the migratory links between Denali and the far corners of the globe to which the park's birds fly. Our geolocators would give us the first opportunity anyone's ever had to track the actual route and destinations of the park's thrushes.

But first we had to catch some. We'd had easy success the previous week tagging Swainson's thrushes, which are abundant in the spruce forests of Denali. The closely related gray-cheeked thrushes, on the other hand, were proving to be a little more challenging, and we hoped that the use of a few extra nets might make a difference that morning.

The tundra was almost *too* comfortable, and after about 15 minutes of waiting and dozing, I levered myself off the ground and trotted down the hill to the willows to see what we had caught. In one net, a male blackpoll warbler hung head-down in the cushioning mesh—another bird that makes an extraordinary migration, from Alaska to the Atlantic coast of Canada and the northeastern United States, then south in a nonstop 90-hour flight over the western Atlantic to South America. The next net held a male Wilson's warbler, tinier even than the blackpoll, weighing just nine grams, less than a third of an ounce. Those Wilson's warblers that breed in central Alaska migrate (we think) to the Gulf Coast of Texas and eastern Mexico and south into Central America. Many of them may commute to the Yucatán Peninsula across the Gulf of Mexico—but no one really knows. Only a single Wilson's banded in the Alaskan interior has ever been recovered away from the breeding grounds, and that one was found in Idaho on its way south.

We would be tagging blackpolls and Wilson's warblers another time, but for the moment, that mystery would have to wait, and I released them quickly. Our focus this morning was thrushes, and to my disappointment, there were none in our nets. I turned to trudge back up the hill—and in that moment, the quiet of the morning became a terrifying chaos.

"Hey bear! *HEY BEAR!*" Laura and David's voices had the rough edge of panic, their arms raised and waving wildly against the pale dawn sky. I couldn't see Iain, hidden from me by the willows.

I heard a huffing, staccato roar, and an explosive wooden sound like someone pounding two-by-fours together, which I realized were the clashing jaws of an angry grizzly "popping" its teeth in rage. Time, as often happens in moments of extremity, seemed to slow. I couldn't see the charging bear, but assumed it was coming out of the willows where I was standing. I froze.

"*HEY BEAR!*" The roaring and popping sounds were much closer now, and the thicket was filled with the crashing of a large animal, very close and moving very fast. David bellowed, "Scott, get the *hell* out of there!"

I bolted from the willows as the bear passed a few yards away, so close that I could hear its ragged, woofing breaths and smell its pungent odor, but invisible behind the screen of brush. In seconds I scrambled back up the hill to my friends. Turning, we saw the bear— a big female with a dark yearling cub in tow—burst out the far side of the willows and race away from us with the horse-like speed for which grizzlies are famous. The sow's straw-blonde fur rippled as she pounded up the far tundra slope, vanishing over the crest.

The story emerged in shaky, disjointed pieces. Everyone was still lying down when the bear emerged from a hidden draw, just 50 or 60 feet away and a little behind them. "I looked over to say something to Iain," Laura said, "and I saw this grizzly head beyond him. I said, 'Oh shit.' We started to stand up, and she just charged."

Iain was closest. "I heard you and David yell, but I couldn't move," he said in his Glasgow accent, shaking his head. "I was just— I couldn't move." The grizzly crossed the distance in seconds. Only a few feet from Iain, the bear changed its mind; Laura and Iain both said they could see the fraction of a moment when the sow decided not to maul them—and turned instead to race down the hill, directly toward me.

"It's ironic," David said, "that the one person who didn't see the bear coming is the one who was probably in the greatest danger of getting mauled." It took me a second to realize he meant me. Even for an angry grizzly, three people together is a lot to tackle. But alone and hemmed in by the willows, I would have been helpless if she'd spotted me just a few yards from her in the thicket, and had decided to vent her frustration and fear.

Laura drew a long, uneven breath and looked around. "You guys think we have any nets left?"

The bears' path had been right through the middle of our array, but somehow the 400-pound sow and her cub had missed them. And whether because of all the commotion, or because they'd fallen to the lure of the recorded song in spite of it, there were three gray-cheeked thrushes hanging in the mesh. Knowing the bears were safely gone—and with a sense of relief that we had something else to think about—we set to work.

Placing the birds in lightweight cloth holding bags, we spread a small tarp on the damp ground and laid out our tools—banding pliers, clipboard, a spring scale, a small camera, and the first geo-locator. The device was maybe a third of an inch long, with a short plastic stalk poking out its rear that carried a light sensor. Small elastic loops stuck out to either side like rabbit ears. Laura removed the first thrush, gently caging it in her hand with its neck between her first two fingers. Gray-cheekeds are two-thirds the size of robins, lovely in their subtlety. Their upperparts are a cool olive-gray, their off-white chests covered in brownish spots that look like watercolor gently seeping into thick paper. Attaching the geolocator took less than a minute. Iain worked an elastic loop high up one leg, to the top of the bird's thigh. With her thumb, Laura steadied the geolocator in the small of the thrush's back while Iain slid the other loop up the opposite leg; thus secured, the tracker rode snugly just above the bird's rump, all but the light stalk hidden beneath its back feathers.

With practiced moves, Laura banded the thrush—a standard metal band on the right leg, and two colored plastic bands, yellow above orange, on the left. When Denali's migrants returned the next spring, the color bands would make it easier to relocate this and the other tagged thrushes so we could recapture them, remove the geolocators, and download their data. One by one, we processed and released the thrushes, each of which flew back into the sheltering willows with nasal, scolding *jee-eer* notes. We packed up the gear, but as we rose to go, I realized Iain was staring out over the hills, in the direction where the bears had gone.

"You know what?" he said, his bright smile conveying a sense of cheerful discovery. "I didn't think my sphincter muscle was that strong!"

For almost six years in the 1990s, I followed birds up and down the Western Hemisphere, exploring the phenomenon of migration for a book called *Living on the Wind*. I'd come to the subject largely as a deeply interested observer—a lifelong birder who had, a decade or so earlier, become obsessed with banding raptors. Initially, I must admit, the attraction to banding was largely the adrenaline-surging thrill of luring a goshawk or golden eagle out of the sky and into my nets—fly-fishing in the air, on an epic scale, for prey with talons and a regal mastery of the wind. But with each hawk or falcon on whose leg I placed a band—and with each time one of those marked birds was recaptured or found dead in some distant place, adding a little more to our understanding of their migrations—I became more fascinated with the natural forces that push not just powerful birds of prey but even the tiniest and seemingly most fragile warbler to cross immensities of space with a speed and physical tenacity that beggars human imagination.

In the past two decades, science's understanding of migration—of the mechanics that allow a bird, alone and on its first journey, to find its way across the globe in the face of crosswinds, storms, and exhaustion—has exploded. To take just one especially mind-bending example, we've known since the 1950s that birds use the earth's magnetic field to orient themselves. Ornithologists long assumed this ability was a sort of biological compass, and the presence of magnetic iron crystals in the heads of many birds seemed to bear this out—except that those magnetite deposits actually appear to play little role in orientation. Vision, quite unexpectedly, does. Expose a bird to red wavelengths instead of natural white light, and it loses its ability to orient magnetically, regardless of what minute lumps of iron might be in its head. But just why this should be has baffled ornithologists since at least the 1970s.

It now appears that birds may visualize the earth's magnetic field through a form of quantum entanglement, which is just as bizarre as it sounds. Quantum mechanics dictates that two particles, created at the same instant, are linked at the most profound level—that they are, in essence, one *thing*, and remain "entangled" with each other so that regardless of distance, what affects one instantly affects the other. No wonder the technical term in physics for this effect is "spooky action." Even Einstein was unsettled by the implications.

Theoretically, entanglement occurs even across millions of light-years of space, but what happens within the much smaller scale of a bird's eye may produce that mysterious ability to use the planetary magnetic field. Scientists now believe that wavelengths of blue light strike a migratory bird's eye, exciting the entangled electrons in a chemical called cryptochrome. The energy from an incoming photon splits an entangled pair of electrons, knocking one into an adjacent cryptochrome molecule—yet the two particles remain entangled. However minute, the distance between them means the electrons react to the planet's magnetic field in subtly different ways, creating slightly different chemical reactions in the molecules. Microsecond by microsecond, this palette of varying chemical signals, spread across countless entangled pairs of electrons, apparently builds a map in the bird's eye of the geomagnetic fields through which it is traveling.

That's by no means the only gee-whiz discovery. Researchers have found that in advance of their flights, migrant birds can bulk up with new muscle mass without really exercising, something humans would love to copy. Because a bird's muscle tissue is all but identical to a human's, the trigger must be biochemical, but remains a tantalizing mystery. They also put on so much fat (in many cases more than doubling their weight in a few weeks) that they are, by any measure, grossly obese, and their blood chemistry at such times resembles that of diabetics and coronary patients—except that they suffer no harm. Nor do birds flying nonstop for days suffer from the effects of sleep deprivation; they can shut down one hemisphere of the brain (along with that side's eye) for a second or two at a time, switching back and forth as they fly through the night; during the day, they take thousands of little micronaps lasting just a few seconds. Researchers have found dozens of similarly extraordinary ways in which a bird's body copes with and overcomes the stress of long-distance travel.

And as science's grasp of the mechanics of migration has improved, so too has our understanding of the gritty, life-and-death challenges that increasingly face these travelers, and the almost inconceivable feats they accomplish twice each year to reach their destinations. In the past two decades we've realized how badly we have underestimated the simple physical abilities of birds.

Until recently, the acknowledged long-distance migration champion was the Arctic tern, a ghostly gray seabird the size of a dove, which breeds in the highest latitudes of the Northern Hemisphere and winters in the southern oceans between Africa, South America, and Antarctica. Draw lines on a map between those waypoints,

scratch a few calculations on a table napkin, and you reach the conclusion that generations of ornithologists had—that Arctic terns
migrate some 22,000 to 25,000 miles each year. It was a guess,
because tracking technology wasn't nearly small enough for a delicate
creature like the tern to carry. But as transmitters and data-loggers
began to grow smaller, they could be deployed on other, somewhat
bigger seabirds—which soon left the Arctic tern's assumed record in
the dust.

In 2006, scientists using geolocators announced they had successfully tracked 19 sooty shearwaters from their breeding colonies in
New Zealand. Even a "local" feeding run during the breeding season, when the parents forage for squid and fish to bring back to their
nest burrows for their chicks, carried these plump, dark gray birds
from New Zealand down into the frigid sub-Antarctic waters thousands of miles away, and back. Once the chicks fledged, however,
they and the adults all headed north, crossing the equator to reach
"winter" feeding grounds in the boreal summer off Japan, Alaska, or
California. By following wind and ocean currents in looping curli-
cues across the Pacific, the birds (in the words of the researchers)
enjoyed "an endless summer." It's a helluva road trip, since the routes
taken by some shearwaters exceed 46,000 miles a year.

Finally, by 2007, geolocators had grown small enough that my
Scottish friend Iain and several of his colleagues were able to attach
them to the legs of Arctic terns in Greenland and Iceland. A year later,
the returning birds were recaptured, and the story that unspooled
from their stored data was astonishing.

The first surprise was that the terns took one of two dramatically
different tracks south, regardless of their colony of origin. Some
veered east to the northwestern bulge of Africa, then angled back
across the narrowest part of the Atlantic to the coast of Brazil before
continuing south to the Weddell Sea along the Antarctic Peninsula.
In spring they migrated to the waters off southern Africa, then across
the Atlantic again to northern South America, and finally on to the
North Atlantic—a figure eight, inscribed on the planet by endlessly

beating wings. For some reason, other terns from the same colonies instead shadowed the coast of Africa almost to the Cape of Good Hope, then either crossed the Southern Ocean to the Antarctic coast, or followed the screaming gales of those high, storm-raked latitudes for thousands of miles farther east, south of the Indian Ocean.

In all, Iain and his colleagues found that even the least ambitious of their terns migrated at least 37,000 miles a year, though some traveled almost 51,000 miles a year—a new long-distance record, and more than twice what scientists had once assumed was possible for this species. And just to cap that, three years later researchers who had tagged Arctic terns in the Netherlands found that those birds were traveling up to *57,000* miles a year, reaching the waters off Australia and using staging areas in the Indian Ocean (where, it turns out, tagged terns from the coast of Maine also gather). Any seabird biologist will admit, especially after a beer or two, that no one really has a clue what the true limits of tern migration might be.

Many other assumptions about migration have been turned on their heads in recent years. It's the nature of the beast; ecology is an almost perversely complicated subject, and every layer of the onion that we peel back just reveals further complexities.

Twenty years ago, North American ornithologists who had assumed the biggest challenge for migratory songbirds lay in the loss of wintering habitat from tropical deforestation were coming to grips with a problem much closer to home. A growing body of research showed that forest fragmentation—the endless slicing of large, intact tracts of woodland into smaller and smaller scrubby shards, bisected by roads, utility corridors, developments, and fields—posed a serious danger to many of the most prized and lovely migrant songbirds, like tanagers and thrushes, which evolved to nest in unbroken woodland. Fragmentation, it turns out, brings a host of evils. They include so-called edge predators that thrive in disturbed habitats, creatures like raccoons, skunks, opossums, grackles, crows, jays, and rat snakes— all adept nest predators that are rare or absent from deep woods. Fragments also invite brown-headed cowbirds, grassland birds that parasitize the nests of other songbirds (and which were originally restricted to the Great Plains). What's more, fragmentation dries out the very forest itself, reducing insect abundance and creating other environmental challenges for the nesting birds.

Scientists have tracked the nesting success of so-called forest-interior songbirds like wood thrushes, monitoring their nests to see which ones produce the most eggs, and how many eggs successfully grow into fledglings that fly off on their own to form the next generation. Decades of such study confirm that when big expanses of woodland are chopped into smaller fragments, nesting success drops in lockstep with the splintering of the forest.

So, to save the bird, save the forest. While preventing fragmentation is challenging in practice, it's a simple target to articulate and to aim for, and one that has guided important elements of bird conservation since the 1980s. But—and in ecology, there is usually a *but* lurking in the underbrush—more recent research has uncovered a real surprise. This one came when scientists took the next step. Instead of just monitoring breeding success in those safe, intact forests, they began the much more arduous task of tracking fledgling thrushes after they left the nest and scattered to the four winds. When they put tiny radio transmitters on those adolescent thrushes and followed them until they were ready to migrate, the researchers found that many of the juveniles abandon the mature, expansive woods where their parents nested—the intact forests we've come to assume were of singular importance to their survival, and whose preservation has been a major focus of migrant conservation.

For a month or more leading up to migration, when the young songbirds must rapidly gain weight so they can make the exhausting flights to Latin America or the Caribbean that lie ahead, the fledglings congregate instead in scrubby, brushy, early successional thickets—the kind of habitat created, say, after a clear-cut has begun to regenerate, a clear-cut that might otherwise be seen as destroying habitat for forest-interior birds.

It's not that these birds don't need contiguous forests—they do.

But that's not *all* they need. Time and again, science has underestimated the complexity of migratory ecology.

This isn't willful ignorance; studying small, active creatures whose annual migrations cover tens of thousands of miles is inherently, extraordinarily difficult. But in a lot of ways that are not uncommon in science, ornithology has always been a victim of blinkered vision and the path of least resistance. For the better part of two centuries, ornithologists were almost exclusively North American or European—and because it's easiest to study something close to where you live and work, for a long while we therefore knew mostly about the lives of migratory birds during the few months when they were on their temperate breeding grounds. In the 1970s and '80s that began to change, and the emerging research from the tropical wintering grounds upended many comfortable assumptions about migrant ecology. Once thought to be adaptable, go-along-to-get-along types that could fit into any vacant slot in the tropics, many migrants proved to be every bit as specialized as the resident birds with which they shared the landscape, tightly bound to specific, often narrow ecological niches. Even within the same species, scientists found, different age and sex classes often had dramatically different needs, and used very different regions or habitats—adult males preferring dense rain forest, for example, and juvenile females drier, scrubbier habitat.

This new understanding came as alarm bells were sounding over rampant tropical deforestation, which quickly came to be seen as the greatest threat to neotropical songbirds. Conversely, neotropical migrants like warblers and tanagers simultaneously became poster children for the campaigns of the 1980s and '90s to save the rain forest, the most direct (and emotionally resonant) link between a distant and threatened ecosystem and American backyards.

The loss of tropical habitat was and is very real, but it was hardly the only threat. There's the degradation of habitat on the temperate breeding grounds, and the loss of stopover sites that make long-distance travel possible. You can't divvy up the lives of wild creatures, especially those with the wind at their command, into seasonal slices or geographically discrete segments. We may finally be looking at migratory birds the way they should be viewed—not as residents of any one place, but of the whole. These are creatures whose entire life cycles must be understood if we're to have a prayer of preserving them against the onslaught they face at every moment, and at every step, of their migratory journey.

We still have much to learn. For example, we still know almost nothing about the precise routes that most migratory birds take, and we have only the sketchiest notion of which sites along the way are critical for rest and refueling. We've belatedly realized—though it should have come as no shock—that regional breeding populations within the same species, even those within fairly close proximity to each other, often have dramatically different migratory paths and wintering areas. Most of the wood thrushes from New York and New England, for example, head to a narrow swath of eastern Honduras and northern Nicaragua for the winter, while those from the mid-Atlantic crowd into the jungles of the Yucatán Peninsula. Geolocators and banding records show that ovenbirds from the Philadelphia suburbs mostly migrate to the Caribbean, especially the island of Hispaniola, while those from just across the Alleghenies near Pittsburgh fly straight across the Gulf of Mexico to northern Central America.

That's of more than academic interest. Lose one part of the wintering range, or one critical way station in between, and you may lose an entire regional population. If you want to keep wood thrushes or ovenbirds—or any of hundreds of other species of migrants—healthy and abundant across their entire range, you may need to take a far more expansive, muscular approach to land protection than has been the case until now.

The first step is knowledge, and a new generation of researchers is doing the grueling, difficult field work that's necessary to tease out all the strands of a bird's full life cycle, stretching across 12 months of the year and often thousands of miles between distant corners of the world. It's a field known as migratory connectivity—in a sense, the maturation of a process that began more than 200 or more years ago, when John James Audubon tied silver wire to the legs of phoe-bes on his Pennsylvania estate to learn if the same birds came back each year to nest. Fortunately, we have more sophisticated tools than Audubon's silver wire. Mapping migratory connectivity was why we risk grizzly bears in the Alaskan interior, to better understand *exactly* where the park's birds spend the winter. It's no longer enough to say that gray-cheeked thrushes go to "northern South America." As the world changes and warms, the hurdles migrants face grow rapidly steeper—and conservationists need this information if we're to shep-herd the birds through what is already a rapidly narrowing bottleneck.

This has become a very personal crusade for me, as it has for many of the men and women who study and protect migratory birds; the idea of a world without epic migrations is simply too poor and mel-ancholy to contemplate. As with many of them, migration has cap-tivated me all my life—an obsession that began in childhood and crystallized on a windy ridgetop in Pennsylvania, and which has led me from being an eager observer to an increasingly passionate partic-ipant; from a recreational birder to someone in the trenches of migra-tion science.

I did not grow up among birders, but my parents loved the out-doors and offered encouragement (though at times bemusedly so) to their slightly strange son. My mother in particular paid attention to the rhythms of the seasons, and bird migration was central to this. She jotted down in her garden journal when the first juncos and white-throated sparrows of the fall appeared at the feeders, and when the first spring migrants returned to our yard in the mountains of eastern Pennsylvania. We paid special heed to the autumn and spring passages of Canada geese, which in the 1960s and early '70s (before nonmigratory flocks overspread every suburban office campus, city lake, and farm pond in the East) were still an electrifying benchmark of the changing seasons.

Most years there would be a single morning—exactly when depended on the severity of the winter, but it usually fell in early March—when we would wake to the sound of geese. We'd bundle

into coats and unlaced boots, racing outside to the first truly mild
morning of the year, craning our necks up at a sky layered with chevrons of geese plowing north against the bleached-denim sky. It was,
and remains, one of the most thrilling moments of the natural year
for me, and each winter as the days lengthened and the snow melted,
we looked forward to "Big Goose Day" as the singular pivot in the
seasonal round. And we still do; the phone will ring early, as the
sun is just rising and my wife is having her first cup of coffee, and
my mother will say, "Have you heard them? Have you been outside? It's Big Goose Day!" And out we go, bootlaces trailing, to soak
up the show all over again. (Many years ago I wrote about our odd
little family observance in a state wildlife magazine, and someone
who read it asked one of my sisters—who is definitely not a birder—
whether it could possibly be true. C'mon, he said, you guys didn't
really celebrate this goose-day thing, did you? "Yes," Jill replied with
an exasperated sigh, "but it's not like we baked a *cake* or anything.")

So in some ways I was primed to spend a life chasing migration,
but the pivotal moment came when I was 12. On an October day of
blustery wind and ragged clouds, we climbed to the top of the Kit-
tatinny Ridge, the southern lip of the Appalachian ridge-and-valley
system, an hour or so from our home—a highway for migrating raptors, which ride its updrafts as they coast down the long, sinuous
mountain range on their journey south.

It was, by luck, exactly the right conditions for a big flight—
a powerful cold front the night before had dragged strong, northwest
winds across the state—and the skies over Hawk Mountain Sanctuary's North Lookout were peppered with sleek, predatory shapes. My
family forgotten, I tucked myself in among the gray boulders, sheltering from the wind as best I could, eyes wide and excited. The silhouettes in the air looked nothing at all like the tiny drawings I had
studied in my field guide. But it did not matter. Hundreds of raptors
glided down the ridge that day, surfing the invisible waves of air, and
I stared hungrily through my cheap binoculars, as each passing hawk
dragged my eyes along with it.

Adults around me called out identifications and landmarks: "Sharpie on the slope of Five!" "Two redtails over left Hunter's Field." One hawk, plunging down at a plastic owl decoy (placed, for just such a purpose, on a limbless sapling jammed in the rocks near where I sat) looked for long, heart-pounding seconds as though it was going to fly right through my binoculars. The spectacle was easily the most intoxicating thing I had ever witnessed, and the memory remains almost painfully intense today.

I didn't have the words, at the time, to articulate why I was so moved, why I found the sight so spellbinding. The hawks and falcons were beautiful, of course, their passage majestic; it was enthralling to watch the way they would, with subtle corrections of their wings and tails, counterbalance the gusting wind and yoke its energy. But it wasn't until I got home that evening, and dragged out my bird books and an old *National Geographic* map, that another, even more powerful reaction manifested itself. Tracing my finger down the curving spine of the Appalachians, I thought for the first time about where those hawks had been coming from, and where they were going. I'd been fuzzy on the details before, but now I read that some of these birds—the very ones I'd seen—may have come from far-off places like Greenland and Labrador, and were heading to destinations like Mexico or Colombia or Patagonia that seemed impossibly exotic to a kid growing up on the edge of Pennsylvania coal country.

I slept very little that night, and my dreams were full of wings. Half a century on, I am still captivated by migration.

What has changed is my involvement. That electrifying day on Hawk Mountain cemented my passion for birding, and especially hawk-watching, but while I became a serious bird geek as a teenager, it was all in fun. Birding was a hobby. Then a college course in ornithology that I hadn't planned to take, snagging the last remaining opening in the class of a wise and generous professor teaching his final semester before retirement, fully opened my eyes for the first time to the fascinating *science* of birds.

Life turns on such flukes and happy accidents. As a young newspa-

per reporter, I pitched an assignment to my editor—Hawk Mountain
had hired its first director of research, a newly minted PhD named
Jim Bednarz, who was putting tiny radio transmitters on migrating
hawks. The editor nibbled, and I wrangled an invitation to spend a
day with Jim in the trapping blind, notebook poised. The first time a
red-tailed hawk dropped from the sky on folded wings, talons flexed,
hurtling down into our nets like the messenger of some pagan god, I
knew with the same lightning-bolt clarity as when I was 12 that the
ground had shifted under me again. Apprenticing with Jim, within
a few years I had a federal banding permit; when he left the organization, I ran Hawk Mountain's banding program for a time before
taking over one of the sites myself. Soon I was also banding songbirds, then owls, then hummingbirds, always driven by a curiosity
that verged on mania about migration.

Without really meaning to, I slid further and further from observer
to participant. While my day job was (and remains) writing about the
natural world, field research has occupied an ever larger and increasingly satisfying part of my life, even though I lack an academic degree
in science. Fortunately, ornithology has a long tradition of welcom-
ing experienced amateurs like me into the fold.

When I wrote *Living on the Wind*, I was very much still an outsider
looking in on the world of migration science and conservation, but in
the years since then, I've become ever more deeply enmeshed in the
research itself—not just interpreting the work of others, but adding
my own small contributions. Perhaps if research were my workaday
job, some of the shine would have worn off, but it's more fulfilling
than ever. For example, for more than 20 years I've overseen what
has grown into one of the largest studies anywhere into the movements of the northern saw-whet owl—a fetching little raptor about
the size of a robin, round-headed and enchantingly wide-eyed. Over
the years, with a crew of about 100 volunteers, we've banded more
than 12,000 of these elfin birds in the mountains of Pennsylvania,
and used a variety of technologies—geolocators, radio transmitters, forward-looking infrared, and marine radar, among others—to

track their wanderings. I also help coordinate a continental network of more than 125 owl-banding stations, all cooperating in the same kind of research.

Intrigued by evidence that western hummingbirds were evolving a new migration route to the East instead of Mexico, I spent several years learning to catch and safely band them, until I qualified as one of fewer than 200 licensed hummingbird banders in the world. Every autumn, now, I chase hardy vagrant hummingbirds, coming from Alaska or the Pacific Northwest, which show up in the mid-Atlantic and New England during the chill winds of autumn and which often linger through the snowstorms and subzero temperatures of January, confounding all of our expectations about how fragile we assume these tiny birds must be.

Those same winter winds bring snowy owls down from the Arctic, and a few years ago—when the East experienced the largest such invasion in close to a century—several colleagues and I started something we dubbed Project SNOWstorm. Working in biting cold and snow, we lay nets for the great raptors, then fit them with transmitters that log incredibly precise GPS locations every few minutes, sending us the data through the cell-phone network—a marriage of two cutting-edge technologies that allow us to track the owls' movements in stupefying, three-dimensional detail. With a few keystrokes, we can follow our tagged owls as they hunt waterfowl at night over the open Atlantic, cruise the farmland of Michigan or Ontario for rodents, or ride summer icebergs pushed by wind and tide on Hudson Bay. Some of those same colleagues and I have also installed more than 100 automated receiver stations across the Northeast that detect the signals from radio transmitters tiny enough to let us track the smallest bird—and even migratory insects like dragonflies and monarch butterflies.

The project that brought me to Denali—and our hair-raising encounter with the grizzly—was another such collaborative venture, one that had been hatched out of a chance meeting years earlier. Carol McIntyre has been studying the birdlife of Alaska's national parks for 30 years, and she is widely known for her groundbreaking studies of golden eagles in Denali—a place near to my heart, and to which I've been returning almost annually for more than three decades. The plan we concocted, at a raptor conference in Minnesota a few years ago, was a little breathtaking (and maybe a little crazy) in its audacity. We decided to launch an open-ended research program to map the migratory connectivity of an ever-changing suite of Denali's birds, shifting with time among songbirds, raptors, shorebirds, inland-nesting seabirds, and other groups. As we're proving its success in Denali, we and our colleagues are beginning to expand the study to other parks, with an ultimate goal of encompassing much of the 54 million acres of Alaska's national park lands. When you're studying a global phenomenon like migration, it pays to think big.

For that same reason, this book takes the wide view as it explores the fascinating state of migration research and conservation today. Like the birds, it required a lot of miles, and more than a little stamina. With seabird experts, I sailed through the storm-wracked waters of the Bering Sea, and to the edge of the continental shelf off the Outer Banks, the better to understand one of the least-known frontiers of migration. I talked to white-coated scientists in high-tech labs—people working at the subatomic level to understand the mechanics of navigation—and with ornithologists who work on the dusty, dangerous southern fringe of the Sahara, keeping one eye on the birds they study, and another peeled for Islamic insurgents who would gladly kill or kidnap them. I dodged gunners and trappers in the Mediterranean, where a guerilla war to stop the illegal slaughter of millions of songbirds goes on largely out of sight, and I visited China, where rampant coastal development and a hunger for wild birds in the pot are causing a conservation catastrophe, but where unlooked-for hope still glimmers. And I traveled to one of the most remote parts of Asia, to a forgotten corner of India where former headhunters turned one of the grimmest stories of migratory crisis into an unprecedented conservation success.

Nor are the scientists and conservationists who people these pages strangers; many of them have become my friends and colleagues over the years, part of the tight-knit global community working to know and save migration. Some have been mentors of mine, others collaborators, a few of them former protégés who have gone on to do remarkable work in their own right. It is a privilege to work with them, and to share the stories of their discoveries and insights.

And so, as I set out once again to follow the myriad threads of bird migration for this book, I found myself coming to the subject from a very different and, in many ways, much more intimate perspective than I had 20 years ago—not as an enthusiastic outsider, but as someone directly involved in the hard, exciting work of puzzling out how and why birds crisscross the planet, and how to make sure they always do.

Yet however much I may like to think otherwise, in truth I remain very much an outsider—as is every human who tries to penetrate the inner workings of this phenomenon. The best we can do is scratch at the margins of this majestic global pageant, to try to comprehend the sheer physicality of the migratory feats going on all around us, and to understand the natural systems on which they depend. The world is changing around us, in ways that we barely understand and show little ability to control, and birds—especially migratory birds—are our best and most compelling window into those changes. The news is often grim; by one measure, North America has lost fully a third of its birds, some three billion individuals, since the day I had that childhood epiphany on Hawk Mountain. That tells us, with frightening clarity, how very badly we've mangled our shared world. Birds are sentinels and bellwethers, the victims of our follies—but also, if we are heedful of their needs, guides to a more sustainable future for ourselves as well.

And they are everywhere, whether we know it or not. Last night, before I went to bed, I pulled up a Doppler radar image of the Northeast—not to look for rain, but for birds. On the computer screen, the whole region was occluded by immense blobs of pale blue and green, the radar signature of millions of songbirds aloft in the clear night sky on their way south. Night after night, from the muggy depths of August to the frosty weeks before Thanksgiving, they stream south in numbers that would leave those of us over whose homes they fly mute with awe, if only we could see them.

On such nights (as I knew from work we had done a few years ago in Pennsylvania, using specialized radar) migrants may pass at a rate of a couple *million* an hour. It is arguably the world's greatest natural spectacle, and a nearly universal one, playing out twice a year over every landmass except Antarctica (where the migrant penguins shamble on foot), but one hidden from our sight by the anonymity of darkness. We sleep, unaware of the marvel above our heads.

This morning I slipped outside just after daybreak, careful not to wake Amy. The air was bracing; overnight, autumn had clearly taken command, and I pushed my hands deep into the warm pockets of my fleece. The trees and thickets were trembling with movement and the flickering of wings. Tired from a night of flight, the birds were snatching a few quick bites, then moving on, looking for a safe place to nap for a few hours. Catbirds, slender and sooty gray, gobbled the blue-black berries of a dogwood. A common yellowthroat—small, plump, its short tail cocked like a wren's—eyed me from a stalk of goldenrod that matched the color of its chin. Several red-eyed vireos moved methodically through the leafy branches of a crab apple, plucking cold-numbed insects from their hiding places.

In the dim shade of the pines, where night seemed to linger, I saw a cautious movement near the ground, and raised my binoculars. The wet-on-wet watercolor breast and umber plumage of a gray-cheeked thrush came into view. The bird eyed me suspiciously a few yards away and gave a quiet alarm call, but necessity drove it. Apparently deciding I was the lesser of evils, it turned its back to scuff in the needles, looking for its first meal after 12 hours of exhausting flight. Pale tips on the covert feathers of its wings told me this thrush was a juvenile, on its first migration. It was probably born in the spruce woods of Newfoundland or northern Labrador, a continent away from those we'd tagged in Alaska. But I was gripped by the same urgent desire to know it as we would come to know those Denali thrushes—not as a here-and-gone distraction, one among a multitude of migrants on a busy morning, but as an individual, a singular creature with a singular and extraordinary life.

It was an utterly ordinary, *extra*ordinary bird—as is every migrant that makes the leap into the void, guided by instinct, shaped by millions of generations of toil and savage selection, crossing the vaults of space through dangers we cannot comprehend, by lucky chance and near-calamity and great endurance, on the strength of its own muscle and wings. For eons uncounted, that has always been enough. But no longer. Now their future, for good or ill, lies in our hands.

One

SPOONIES

The world was precisely equal halves of gray, divided by the flat line of the horizon—the smoky silver of an overcast sky, unmarked and smooth, and the darker, mottled granite and charcoal of a mudflat that stretched to every side, paper-thin sheets of water lying on its surface reflecting the clouds or ruffled by the breeze. There was a salty sharpness in the air, but the ocean was invisible many, many miles to our east. When the tide turned, the water would surge back across these flats, advancing faster than a person could easily move, but for now the Yellow Sea was only a rumor, carried on a damp and chilly wind.

I expected my rubber boots to sink, but the mud felt like concrete; the local name, here on the coast of China's Jiangsu Province, translates to "steel plate," appropriate for both the firmness and the gunmetal hue of the sediment. Even the big tractor and wagon that had hauled us out here, far beyond the seawall, barely made a dent in it. Nothing grew here, nothing broke the smooth sweep of tide-rippled mud other than a few pieces of driftwood and odd bits of plastic junk. One could hardly picture a more lifeless landscape. Except for my half-dozen or so companions, huddled inside our rain gear against the breeze and blowing mist, the only signs of any living thing were a few scrawled trails where some mollusks or worms had left sinuous paths when the fast-retreating tide had fallen away an hour earlier.

Jing Li slid her spotting scope off her shoulder, flicking open its tripod legs and starting to scan with a single, practiced motion. Zhang

Lin did the same with his scope, aiming in a different direction, while the rest of us simply picked random spots on the horizon and glassed with our binoculars, seeing basically nothing. But as I lowered my arms and swept my gaze to the left, I heard a sharp, rolling whistle behind me—and turned to find we were engulfed in birds.

The flocks came from the south, dense layers of small bodies that undulated and folded into themselves, creating sheets, splitting into tendrils, forming separate tributaries that reunited into great rivers of wings, all moving with tremendous speed. The first washed over and around us within seconds, thousands of small, fleet bodies sweeping past in a rush of thin, whispery sound very different—higher, more urgent—than the wind. I spun with them, turning on my heel like a weathervane buffeted by a changing gust, but they were already past me, receding, even as the next waves flashed to my right and left. Most were red-necked stints, the common, sparrow-sized "peep" sandpipers of Asia that are very much the size and shape of the semipalmated sandpipers I'm familiar with from home, but with a deep chestnut wash over the head and throat in this, their breeding plumage. Some of the birds were dunlin, with curved bills and black bellies, or ruddy turnstones, piebald with patches of rust and black and white like an Italian harlequin. Not that I could see such details while they flew; it was all just a mass of motion and blurred shapes and wings—now brown-gray, now flashing white as thousands of racing birds flipped in instant, eerie unison and wheeled about with their pale undersides showing.

Glancing back I saw, miles away, dense clouds of birds lifting off from wherever they had been sitting, invisible below the subtle curve of the horizon, forming amoebic masses that billowed and contracted, joined together, and then stretched out like bulbous fingers toward us. The initial vanguard had now reversed course, circling back around and passing at right angles beneath the continued flow of birds from the south, settling all around us in carpets of brownish bodies stretching for hundreds of yards in every direction. Without preamble, the moment the small sandpipers landed they plunged their beaks into the mud, creating pulsations of frantic feeding motion as though they had not a second to spare.

And in truth, they did not. Most of these birds had already come thousands of miles, from places as far to the south as Eighty Mile Beach in northwestern Australia or the Firth of Thames in New Zealand. In a week or two, they would be leaving for Kamchatka in the Russian Far East, or the Yukon Delta in western Alaska, or the islands of Ostrova Anzhu in the Siberian Arctic. Every year, some eight million migratory shorebirds pass through the Yellow Sea, on tidal flats and marshes like this one at Dongling, using them to rest and refuel. What appeared to me to be empty mud was, just below the surface, a biological stew of bristle worms, clams, snails, tiny crustaceans, and myriad other marine invertebrates—a smorgasbord for hungry birds. To migration scientists, these critical way stations are known as stopover sites,* places where a tired and hungry bird can linger and recoup its strength. The foundational importance of preserving stopover sites has become fully clear to conservationists only in recent decades, though it should have been obvious to anyone who has ever planned a cross-country road trip and tried to figure out where and when to stop for fuel, food, and sleep.

Stopover sites can range in size and quality; somewhat playfully, ornithologists have categorized them as fire escapes, convenience stores, and five-star hotels, though their importance to migrant birds is in dead earnest. Like a crowded truck stop serving as its own best advertisement, the premiere stopover locations—those offering the richest bounty of food, abundant at just the right season, with safety from danger and plenty of elbow room—are crowded with migrants that have evolved to depend on these often widely scattered places.

---

* Some biologists, especially shorebird specialists, use the term "stopover" only for places or periods in which migratory birds simply rest, reserving the term "staging sites" for those locations that provide both rest and abundant food. For simplicity's sake (and because those who study other groups, like songbirds, do not necessarily draw this distinction) I use "stopover" to refer to both.

Such sites often lie at either end of formidable geographic barriers
that test the physical limits of a migratory bird—the southern fringe
of the Sahara, for example, which is the last pit stop for northbound
songbirds that must cross first that immense desert and then the Mediterranean en route to Europe; or the thickets and coastal marshes of
New England for songbirds and shorebirds that will fly 1,000 miles
out into the western Atlantic, until the northeasterly trade winds
nudge them back another 1,000 miles toward landfall on the shores
of Venezuela or Suriname. Every flyway, every migration route has
such bottlenecks and choke points, but there is arguably no stopover
site quite as critical, as fundamentally important on a global scale for
more birds of more species, than the Yellow Sea.

Take out a map of the Eastern Hemisphere, and plant the point
of a pencil near New Zealand. Draw a line west, below Tasmania,
then northwest 5,000 miles to catch both sides of the Bay of Ben-
gal in India and Myanmar (Burma). Move your line now east across
southern China to Taiwan, then southeast to enclose the Philippines,
Indonesia, the island of New Guinea, and the archipelagos of the
southwestern Pacific, like the Solomons and Fiji. This is where the
shorebirds of the Yellow Sea spend their off-season—"winter" only
in a northern sense, which is why ornithologists prefer to say "nonbreeding" season instead. Now draw another line, starting this time
at the mouth of Canada's Mackenzie River, on the Beaufort Sea in
the Northwest Territories. Push your line west along the North Slope
of Alaska, across the Bering Sea to include all of Siberia west to the
Taymyr Peninsula, then south across Russia, Mongolia, and western China to the Tibetan Plateau. Turn your pencil east to encircle
North Korea and Japan, then northeast to include Kamchakta, the
volcanic arc of the Aleutian Islands, and most of western Alaska. This
is the vast range to which those migrants return to mate and nest.

These two great cartographic blobs cover roughly 27 million square
miles, and where they overlap, just a little, lies the Yellow Sea, separating eastern China from North and South Korea. It is the extraordinarily narrow waist of an hourglass-shaped hemispheric migration

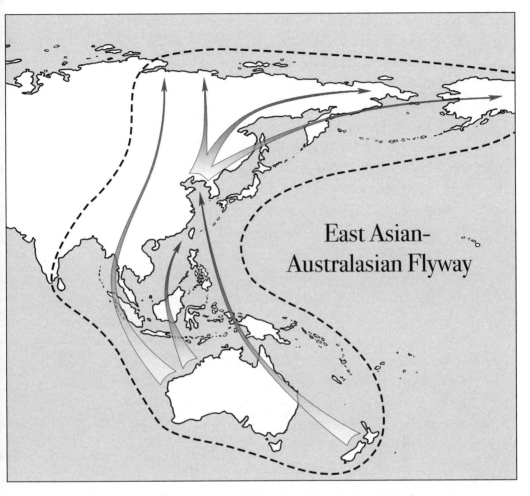

An estimated 8 million migratory shorebirds—and countless hundreds of
millions of migratory songbirds, raptors, and other species—use the East
Asian–Australasian Flyway every year.

system known as the East Asian–Australasian Flyway, or EAAF, and
its importance is more than a matter of geographic happenstance.
The Yellow Sea, especially on the Chinese side and in the northern
gulf known as Bohai Bay or the Bohai Sea, is exceptionally shallow;
during the past ice age, when global sea levels were hundreds of feet
lower, it was largely dry land, with the channel of the Yangtze River
running through the middle. The combination of a shallow coastal

margin and a tidal range that during certain lunar phases may exceed 25 or 30 feet means that when the tide goes out, it goes *out*—for miles and miles and miles, exposing the most expansive natural mud-flats in the world. They were nourished by the prodigious load of sediments washed into the Yellow Sea—thus giving it its name—by the Yangtze, Huang He (Yellow), and other major rivers flowing out of eastern China. The Huang He carried as much as 57 pounds of silt for every cubic yard of water—although "water" is probably not the right word for such a grit-filled slurry.

Historically, mudflats covered some 2.7 million acres of the Yellow Sea coast, providing the ritziest of all five-star accommodations for migratory shorebirds. But in the past 50 years, and with accelerated speed in the last decade, more than two-thirds of those coastal wetlands have been destroyed, mostly through the euphemistically named process of "reclamation"—dredging up enormous walls of mud to seal out the tide, then pumping millions of tons of seabed sediment into the artificial impoundment to create dry land for industry, agriculture, or other purposes. The flow of fresh mud down the rivers was also choked off. The Yangtze alone now has some 50,000 dams along its main stem and tributaries, which had already reduced the movement of sediment into the sea by about 90 percent even before the controversial Three Gorges Dam began operation in 2003, cutting what was left of the silt flow by a further 70 percent. What scraps remain of the Yellow Sea's tidal flats are, to the weary birds that seek them out, precious beyond expression, and probably the single most threatened linchpin among the world's many tangled migratory paths.

I'd been traveling along the Yellow Sea for several weeks, in the company of conservationists and researchers from China, Europe, Australia, and the United States, all of whom recognize the global importance of the region's coastal wetlands to migratory birds, and the existential crisis facing this ecosystem. Some days earlier—staring in shock at a mammoth complex of five new steel mills, several miles in length and a mile wide, rising on what had been tidelands—one of them had said, "There is no more buffer. There is no more 'somewhere else' for these migrants. Every new hectare gone means lost birds." The International Union for the Conservation of Nature (IUCN) says the destruction of the Yellow Sea coast and the tailspin of shorebirds that depend upon it (along with the catastrophe it represents for millions of humans who, like fishermen and shellfishers, depend on a healthy sea) is among the worst environmental crises on the planet. Certainly, it is as grim a situation as faces wildlife anywhere in the world—but more or less by coincidence, I found myself on the Chinese coast at what might also be a transformational moment for these imperiled birds and their battered ecosystem. Not long before my visit, the Chinese government had done what authoritarian governments can do—issued a sweeping edict without dissent or delay, banning in one stroke the kind of rampant, locally driven coastal ruin that has defined the Yellow Sea for decades. One conservationist described her reaction as "stunned joy," and while many others leavened hope with some hard-learned cynicism, at this bleak eleventh hour there were actually good reasons for cautious optimism. Critical areas were being short-listed for international protection; the tide of obliteration that had seemed so implacable just a few months before appeared to be weakening. And strangely enough, this possible salvation was due, in very large measure, to a dumpy little sandpiper with a bizarre beak, rock-star charisma, and one small foot in the grave.

The mist had stopped, and the rising wind now shredded the clouds into long tatters. "Mostly red-necked stints," Jing Li said, straightening from her scope and tucking loose strands of her long hair back under her cap; Jing's chartreuse pack was the only bright color for miles. "Maybe a third of them are dunlin, some grey plover, a few knots and godwits, but almost all stints and dunlin." Zhang Lin was still hunched over his scope, a clicker in his hand going as fast as his thumb could push the button, counting birds. It was an impressive sight, but Jing was looking for one specific bird, the poster child for Yellow Sea conservation—the spoon-billed sandpiper. A spoonbill has an almost cartoon cuteness; from the side, it looks like any ordi-

nary, plump sandpiper, with the same russet head as the red-necked
stint. But head-on you see that its beak is flattened at the tip into a
wide spatulate shape, as though it had been whacked with a hammer
while soft. No one really knows why the spoonie (as birders often call
it) has evolved such an absurdly weird bill; it must have something to
do with feeding, but its exact function is a mystery, like much about
this bird. The effect, though, is of a comical plush toy.

And a critically endangered one; the other part of the spoonbill's
cachet is its extreme rarity. Likely never common, the birds breed
in a few locations within a narrow band that stretches along the
northeast Russian coast of the Bering and East Siberian Seas, never
more than a couple of miles from the ocean and usually on treeless,
crowberry-covered spits of tundra that stick out into the frigid waters.
In 1977 Soviet scientists surveyed the breeding range and estimated
the global population at 2,000 to 2,800 pairs; no one looked again
for almost a quarter century, by which time half of the spoonies had
vanished. With greater urgency, intensive searches were organized in
the subsequent nine years; in areas where researchers expected to find
up to 65 pairs, they found eight. Experts came to the grim conclusion that the population was between 300 and 600, "although this
is thought to be an optimistic estimate," the IUCN admitted at the
time. More sober assessments set the upper limit at about 400 birds,
including just 120 breeding pairs. Charting the population trend for
strongholds (if such a term can be applied to so rare a species) like
Meinypil'gyno, in the Far Eastern district of Chukotka, shows an
essentially vertical drop, from 90 pairs in the mid-2000s to fewer
than 10 a few years later.

The reasons for the spoonie's nosedive toward extinction were not
at first clear, but it soon became apparent that the problem wasn't in
the Arctic. The birds were producing chicks—almost cosmically cute
balls of mottled brown fluff with their parents' signature beak—but
almost all of them disappeared each year after leaving Russia. No one
really knew where the spoonbills went, so birders and ornithologists
began combing Southeast Asia, discovering small concentrations in

Myanmar and Thailand, as well as Bangladesh, Vietnam, and southern China. It is here that juvenile birds, not yet old enough to breed, remain through their second year of life—and here that they collide with the two biggest risks that shorebirds in Asia face, heavy illegal netting and shooting for the pot, and the loss of critical habitat, especially on the Yellow Sea.

Jing Li and Zhang Lin run a small nongovernmental organization called Spoon-billed Sandpiper in China, part of a frantic international effort to save the spoonie and, through the umbrella effect of protecting its coastal habitat, the millions of other birds that rely on the Yellow Sea. Although the spoonbill is the most famous of the bunch, others are teetering on the edge as well. Nordmann's greenshank, with barely 1,000 individuals left in the world, is only slightly further from extinction, and many of the other shorebirds that use the Yellow Sea—red knots and great knots, black-tailed and bar-tailed godwits, curlew and Terek sandpipers—are declining at up to 25 percent in some years.

"What should I be looking for?" I asked, trying to make visual sense of the melee of feeding birds, but with no notion (other than carefully studied field guide illustrations) of what search image for a spoonbill I should hold in my mind.

"Look for a bird that's paler than the stints," Wendy Paulson told me. Wendy and her husband, Henry M. Paulson Jr., the former CEO and chairman of Goldman Sachs and US Treasury secretary, formed the Paulson Institute in 2011, what they describe as a "think and do tank" focusing on sustainable development and environmental protection in China. I'd gotten to know the Paulsons years earlier through their conservation work, including Hank's time as chairman of the Nature Conservancy; both have used their connections in China, and the institute's resources, to encourage the government to protect coastal wetlands, particularly along the Yellow Sea. In 2015, the Paulson Institute published an influential blueprint for coastal conservation in China, which identified the mudflats along the Jiangsu Province coast, including these at Dongling, as among the sites needing the most urgent protection—and perhaps as significantly, made the case for the economic importance of coastal wetlands for local livelihoods like shellfisheries, as a buffer against sea level rise, and for water purification, among other ecological services.

"Spoonbills are a little bigger and a little paler than the stints, but look for different behavior," Jing said. The stints, she said, probe like rapid little sewing machines, up and down, but spoonbills tend to feed in tight circles. "And they do a kind of back and forth sweep, swishing their bills through the mud," Wendy added. The previous year, she and Hank had seen seven spoonbills here, despite lashing rain and wind.

"There's a paler bird," Jing said, "I think maybe it's a—" At that moment, every bird around us, maybe 5,000 of them, roared into flight, wheeling in one densely packed squadron. Like the Red Sea parting for Moses, they split into a giant U-shape to make way for a peregrine falcon that flashed through the middle of the flock. The falcon wasn't hunting, just loafing along, but even after it had vanished up the coast the smaller birds remained nervous, settling briefly only to flush again and again, eliciting a frustrated sigh from Jing each time they did so.

In the past few years the number of spoon-billed sandpipers has, if not detectably increased, at least ceased its gut-churning plummet. The spoonie's life history, which was largely an enigma just a decade ago, has begun to take shape. Zhang Lin, Jing Li, and other conservationists found the largest concentration of spoonbills here, on the Jiangsu coast, sometimes more than 100 that rest and feed there for two months in autumn as they molt their worn feathers after breeding. Other equally dedicated researchers and amateurs have combed the tidal deltas of southern Asia to locate previously unknown wintering sites—and to stop the rampant hunting of shorebirds in general in many such locations. A small and still struggling captive population has been established against a calamity in the wild, and scientists have tried a very successful Hail Mary approach to increase the number of chicks on the breeding grounds.

The Chinese government's ban on uninhibited coastal development, if it translates into action, may have come just in the nick of time for spoonies and other migrants. But good news is thin on the land. Indeed, it is almost impossible to overstate the gravity—the sheer desperation—of the crisis facing migratory shorebirds worldwide. The collapse of shorebird numbers is on par with that of the passenger pigeon more than a century ago; in fact, the clouds of waders that once filled the sky were often compared with the flying rivers of wild pigeons. But where the passenger pigeon's extinction involved just a single kind of bird on one continent, this time the world faces the loss of entire suites of species, as dozens of kinds of shorebirds tumble toward the abyss. Globally, most shorebird species are in decline, some at terrifying rates. In North America, long-running surveys have shown that shorebird numbers in general have fallen by half since 1974, with the steepest drops among long-distance, Arctic-nesting species like ruddy turnstones, red knots, and Hudsonian godwits. Counts along the mid-Atlantic coast of whimbrels—among the largest shorebirds, the size of small ducks with long, down-curved bills, their plumage finely checked in nut-brown and white—have fallen 4 percent a year over the past 35 years, a depressing rate of erosion. A 2006 survey of the world's shorebirds, comparing them with numbers from the 1980s, found that only 12 out of 66 regional populations were stable or increasing, and the situation has only deteriorated since then. It isn't as though conservationists aren't trying. They've made huge efforts to protect networks of critical habitat—the Western Hemisphere Shorebird Reserve Network (WHSRN), for example, now encompasses more than 100 sites totaling 38 million acres in 16 countries from Canada to Argentina. But while designation as a WHSRN site brings attention to its importance, it does not confer protection; control remains in the hands of local and national governments, which do not always make the best management decisions for birds. For example, the Fraser River estuary in Vancouver, British Columbia, where 95 percent of the world's western sandpipers stage each year (along with more than a 100,000 dunlins and large concentrations of black-bellied plovers) is designated as a hemispheric-level WSHRN site, a globally significant Important Bird Area, and is listed as a wetland of international significance under the multinational Ramsar Convention on Wetlands. Yet it is threatened by plans to double the size of an existing offshore port and causeway complex that has already altered the ecology of the estuary.

Migration is the most dangerous time of the year for most birds, but it has become an especially deadly gauntlet for many shorebirds. They face poaching and subsistence hunting in much of Asia, Africa, the Caribbean, northern South America, and parts of the Mediterranean. Coastal wetlands are disappearing, while humans—faced with rising sea levels—have armored much of what remains with rock and concrete, leaving few soft tidal shorelines where waders can feed. The loss of safe, resource-rich places to rest, refuel, and winter means that increasing numbers of breeding-age adult shorebirds simply don't survive the journey, or arrive on the nesting grounds late and in such poor condition that they haven't the time or energy to breed. Once there, they may find that intensive agriculture has created landscapes unable to support their chicks, or discover they are so out of synch with a rapidly changing climate that their nesting attempts fail year after year. Those with the longest, most dramatic migrations—already delicately balanced among time, distance, weather, food, and physiological ability—are at the greatest and most immediate peril. The hazards are global, but few places tie together all the existential threats faced by shorebirds quite like the Yellow Sea.

A week earlier, I stood on a seawall at Nanpu, about 100 miles southeast of Beijing. It is hard to imagine a more altered, manipulated landscape than the one here, stretching south from the industrial city of Tangshan to the Yellow Sea. More than 100 square miles of former tidal flats have been converted into an endless expanse of salt-making ponds, known as the Nanpu Saltpans—an industry with roots here that stretch back into antiquity, but only recently on such an immense scale, said to be the largest in Asia. An equal area of "reclaimed" land holds a haphazard mix of smoke-stacked chemical factories, power plant cooling towers, manufacturing complexes, half a dozen prisons, and an oil drilling and storage facility, interspersed with immense mountains of gray-white salt, all part of a massive, half-finished industrial and port complex called Caofeidian New Area. A partially constructed highway, six lanes of gravel and incomplete paving, runs through the middle; although it was barely five in the morning, it was already rumbling with heavy machinery.

The sun was rising through a low pall of smog and dust, which stung the back of my throat and burned my eyes, but Theunis Piersma was looking away from Nanpu and into Bohai Bay, where the high tide had begun to pull back from the seawall. The receding water exposed gray mudflats where shorebirds were now gathering, first by the hundreds, then by the thousands, and soon by the tens of thousands—wave after wave of knots and stints, Terek and curlew sandpipers, godwits and turnstones, flying in from their high-tide roosts among the salt ponds at Piersma's back, clouds of birds arriving to feed and preen.

These flats and the multitudes of shorebirds they attract have drawn Piersma for more than a decade. The 60-year-old Dutch scientist from the University of Groningen, a nimbus of gray, curly hair blowing in the breeze, is a legend in the shorebird world; in fact, some of the red knots feeding on the mud below us were a richly colored subspecies named in his honor: *Calidris canutus piersmai*, which winters in Australia and nests on a few islands in the Russian Arctic.

Piersma's research has helped to show that shorebirds are, in many ways, the ultimate athletes, not only among migrant birds but across all classes of vertebrates. Seabirds like albatrosses and petrels fly farther, crossing tens of thousands of miles of open ocean, but for them the waves hold no terror. They can rest or sleep on the surface when tired, drink seawater when thirsty (filtering out the salt with special glands between their eyes), and feed on squid or fish when hungry. For a red-necked stint roughly six inches long, which cannot rest on

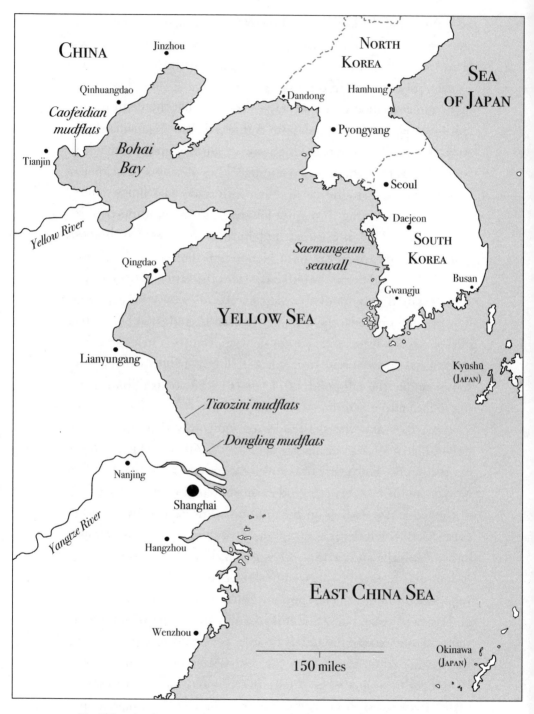

The Yellow Sea, between China and the Koreas, is one of the world's most critical stopover sites for migrating shorebirds.

the water—a bird that must make an uninterrupted, multiday journey across Indonesia, the Philippines, and the East China Sea, then on to the edge of the Arctic—places like Dongling or Nanpu are simply and critically irreplaceable.

The global scale and physiological extremes that shorebird migration represent almost beggar comprehension. Even the smallest birds on the Yellow Sea flats make epic journeys. If the stints I saw had wintered at the northern extent of their species' nonbreeding range, along the coasts of India or Vietnam, they would have flown almost 2,000 miles to get here. But many winter as far south as Tasmania and northern New Zealand, meaning that just to reach this way station they had already flown more than 6,000 miles. And from coastal China they would have further thousands to go before reaching their nesting grounds—as much as 3,400 miles more for those breeding in eastern Siberia, where a large percentage nest. This from a bird that weighs less than an ounce.

In the rarified world of shorebird migration, however, there is nothing especially notable about what the stints accomplish; of the more than 320 species worldwide, most make long-distance migrations, while at least 19 are known to make nonstop flights of more than 3,000 miles. And that's just what we know of. The limitations of telemetry technology had, for years, restricted scientists to tracking only the largest species, like godwits, which are strong enough to carry small satellite transmitters; Piersma was a pioneer in this research. Miniaturization has changed that, however. Semipalmated sandpipers, the New World analogue of the stints and almost exactly the same size, move between the High Arctic and northern South America, a migration finally revealed in detail by geolocators like the ones we used on thrushes in Alaska. One semipalm, tagged on Coats Island in the Canadian Arctic, flew down to James Bay in late summer to fuel up over the course of several weeks, then made a single flight to the Orinoco River delta in Venezuela, a distance of 3,300 miles, before continuing along the coast to the mouth of the Amazon in Brazil, where it spent the northern winter. The exertion of such a trip stresses almost every physical system in a shorebird's body. Theunis and several colleagues studying great knots, robin-sized shorebirds related to the slightly smaller red knots, have found that those on their northward journey run their metabolic tanks down to empty, arriving on the physical equivalent of fumes. Leaving northwestern Australia, the knots make a single nonstop flight of more than 3,400 miles to China and the Koreas, during which they burn through their entire fat reserves, while cannibalizing their own muscle and organ tissue to feed the enormous energy demands of their continually pumping flight muscles. By the time they arrive on the Yellow Sea, almost all of their internal organs have withered from exertion as the body catabolizes itself; only the brain and lungs appear to be unaffected by the marathon, while organs like the intestines and salt glands, which are of limited use during a long flight anyway, shrink the most.

Actually, referring to the flight as a marathon does the birds a considerable disservice. Theunis has noted that an elite human athlete, performing at maximum exertion—a male Tour de France cyclist in mid-race is a good example, he says—is operating at about five times his base metabolism. That seems to be the upper limit for sustained exercise by even the fittest, most highly trained human. A shorebird, on the other hand, is working at a rate *eight or nine* times its base metabolism, and is doing so for days at a stretch without food, water, or rest. When in 2019 an Ethiopian athlete set a new record by running a marathon in less than two hours, it was routinely described as "superhuman." Perhaps, but also subavian. That semipalmated sandpiper that flew from subarctic Canada to the jungle delta of the Orinoco managed the equivalent of running 126 consecutive marathons, and at a metabolic rate several times greater than even an elite human runner can achieve. Again: six inches long, and not even an ounce in weight.

Still more birds were arriving at Nanpu. I was having a hard time estimating numbers, but Piersma looked down the shore as far as he could see, squinted a bit, and said, "Fifteen or sixteen thousand—

that's just the knots." Both red knots and great knots formed a dense line right at the edge of the receding water, the former rusty orange, the latter densely checkered in black and white, with splashes of chestnut on their backs. He pointed out a pale gray Nordmann's greenshank, literally one of 1,000 left in the world, pirouetting and lunging like an egret as it hunted invertebrates in the shallow water, stabbing down with its long, thin bill. "Everyone gets excited about spoonbilled sandpipers, of course, but Nordmann's greenshank is almost as rare," he reminded me. A flock of bar-tailed godwits alighted, tall and leggy and the color of sun-worn bricks. The previous autumn, some of these dove-sized birds would have made the longest nonstop flight of any landbird we know of—7,200 miles across the widest part of the Pacific Ocean, from their nesting grounds in western Alaska to New Zealand, an uninterrupted flight of continuous, high-intensity exertion lasting between seven and nine days.

And staggering as such feats obviously are, these long-distance migrants are balanced on an even sharper physiological knife's edge. Like most of these shorebirds, the red knots we watched would arrive in the Arctic in late May or early June, when the land is still locked in ice and snow, so in addition to the fuel they need to make an essentially unbroken flight of more than 3,000 miles from China, they must also pack on enough fat and protein at Nanpu to last them through the initial weeks of defending a territory, finding a mate, and beginning to nest (and, for the females, producing eggs), until the tundra melts and they can find early insects and the last, withered berries of the previous autumn. Chip away at their ability to bulk up en route by destroying habitat, and not only are they unlikely to breed, they may not survive at all. The usual term for what's happening to the Yellow Sea coast is "reclamation," which suggests humanity taking back something that had been stolen, when in fact the opposite is true. In 2006, South Korea completed a 21-mile-long seawall at Saemangeum, severing the tide from two major estuaries that encompassed more than 150 square miles of once-fertile wetland, on which 20,000 people depended for a living catching shellfish, and hundreds of thousands of migratory shorebirds for their very survival. The result: more than 70,000 great knots, fully a fifth of the world's population, immediately vanished—not coincidentally, almost precisely the same number that had stopped annually at Saemangeum.

"It is almost a birds-per-hectare loss at this point," Theunis said, as he peered through his scope. He wasn't just admiring the spectacle; he was scanning for the data points that allow him to make that stark claim with real authority. "There's one—engraved yellow flag ZHT—one of your hometown birds, Chris." This was directed to Chris Hassell, a grizzled expatriate Brit who moved years ago to Australia and fell in with Piersma's Global Flyway Network, a worldwide collaboration of shorebird researchers. Each year Hassell and his team trap thousands of wintering shorebirds along the remote beaches of northwestern Australia, marking them with unique combinations of plastic color leg bands, including "flags"—bands with protruding tabs into which alphanumeric codes can be engraved. The markers allow survey crews to spot the birds later and identify them at a distance all along the migration route, not only refining our knowledge of travel routes and timing, but more importantly allowing the researchers (through some fairly sophisticated statistical analyses) to calculate population size and annual survival rates. This particular knot was one that Hassell's crew had marked back near his home in Broome, Australia.

"There's another—flag at position three, blue/yellow, red/white, 75 *rog*," Piersma said, reading off the color code from left leg to right and top to bottom, while noting that the bird had molted in 75 percent of its breeding plumage, and was of the subspecies *C.c. rogersi* that breeds in eastern Siberia, somewhat grayer above than the *piersmai* knots that nest on Siberian islands farther west.

As though on command, every shorebird before us leapt into flight, their wings making a sound like an explosive sigh that slowly died to breathy silence as they flew away to our right, down the coast. Hassell and Piersma grabbed their scopes and trotted off in pursuit; they'd warned me that because the tide goes out at Nanpu at an angle to the seawall, the survey team must leapfrog its way with the falling water, covering seven or eight kilometers over the course of a few hours, staying close enough to the feeding flocks to read their bands. In a couple hundred meters we met Katherine Leung, Matt Slaymaker, and Adrian Boyle, already scoping the now-settled birds; I remained with them while Piersma and Hassell kept moving, ready for the next shift in feeding grounds.

Leung, formerly with the World Wildlife Fund in Hong Kong, is Chinese, while Slaymaker is British and Boyle Australian. Along with Hassell, they form the core of the Global Flyway Network team in Nanpu, working with a large group of Chinese grad students overseen by Professor Zhengwang Zhang, the former deputy dean of the college of life sciences at Beijing Normal University and vice chair of the China Ornithological Society. Slaymaker and Boyle have been coming to China for 10 years, conducting these annual surveys from early April through the end of June, while Leung was helping out for a week before heading to the United States to join an international shorebird team on the Delaware Bay; like the birds they study, shorebird experts rack up a lot of travel miles.

Slaymaker—very tall and slender, a scruff of dark beard on his narrow face and his long hair pulled back in a bun—was rapidly jotting down color band combinations in his notebook, while I squinted and peered and fiddled with the focus and magnification knobs on my scope until I was forced, reluctantly, to admit I couldn't see a damned thing. "It's a knack," said Boyle, shorter and ruddier than his colleague, with close-cropped hair under a cap. "Wait until the bird turns to face you, so you can see both its legs and get all the bands at once." I realized he'd misunderstood me; at this range, maybe 300 meters in the hazy light, I hadn't spotted a single band yet, but I decided to keep my mouth shut—the distances at which the three were reading colors and engraved codes were obviously well beyond my ability, and I suspected I'd be of little help.

On the horizon, a few oil tankers were visible through the smog; some days, Hassell later told me, more than 100 queue up at the

Jidong-Nanpu offshore oil facility that forms much of the horizon here—part of an immense oil field discovered in 2005, and a chronic source of anxiety for conservationists, since a spill during migration would be catastrophic. (There have already been some major spills and well blowouts on the Yellow Sea, and less dramatic leaks are commonplace.) When the wind shifted, the refinery smell from the oil field was replaced by an outhouse reek from hog farms built on reclaimed land to our north, pumping their waste into some of the salt ponds; only occasionally did the breeze bring in a fresher, salty breath from the sea. We would race ahead to catch the retreating tide and its attendant birds, scan as quickly as possible until they moved, then scurry farther down the seawall, passing Chinese fishermen in chest waders, dragging behind them baskets floating on inner tubes, sloshing out onto the mudflats to collect clams. By mid-morning we reached the end of the seawall, where lumpish walls of hardened mud, about 10 feet high, enclosed several big, net-lined impoundments in which jellyfish were being raised for food. The birds now passed beyond even the range of Piersma and his team to see bands, so we headed back to Nanpu to get some lunch, passing again through the moonscape of salt ponds and canals, where a few motorboats pulled long trains of skifflike barges piled high with salt.

The mudflats are a predictably transitory habitat; twice a day the tide comes racing back, moving across the flats at surprising speed and eventually inundating them. At some point the water becomes too deep, first for the short-legged peeps like the stints, and eventually even for the long-limbed godwits and curlews. Once, the birds would have flown a short distance inland to brackish wetlands and over-wash flats hugging the coast, but these places are long gone, developed out of existence. So ironically the salt ponds—although created on what had once been rich mud—have become vitally important to the shorebirds as high-tide roosts. In the salt-making process, an increasingly dense brine solution is moved from initial evaporation ponds to the final crystallization stage—and the shallow brine pools used in this process, ranging from a few hundred to several thousand acres in size, provide safe harbor where the birds can rest, preen, and sleep. The numbers using such roosts can be truly eye-popping; a few years earlier, Chris Hassell told me, they'd counted 95,000 birds resting in a single salt pond, of which 62,000 were curlew sandpipers— a third of all of that species in the EAAF. A few days later, the neighboring pond held 34,000 red knots, again a full third of the flyway population, and more than half of the entire *piersmai* subspecies.

We saw no such multitudes; the ponds were uniformly much deeper than usual during my visit, though no one seemed to know exactly why. More of the impoundments seemed to be in use for shrimp farming, a lucrative business along the coast, while one of the Chinese grad students said he'd heard the water was being stored for eventual use in the massive complex of steel mills rising on the horizon miles away. Whatever the reason, instead of gathering in mammoth flocks in the middle of shallow pools, the birds were forced to crowd the edges of the mud-walled dikes, spread over a much wider area.

One afternoon, Theunis and I joined Bingrun Zhu, a tall, slender man working on a PhD focused on black-tailed godwits—a species dear to Piersma, since they nest in the meadows and pastures near his home in Friesland. Zhu (who goes by the Anglicized nickname Drew with the survey team) had gone to the Netherlands in previous years to study godwits with Theunis, but the population he was studying here breeds on the grassy plains of Outer Mongolia after stopping off along the Yellow Sea. Along the way, we passed hundreds of acres of inland fish ponds covered with solar panel arrays, shading every square yard of their surface while allowing aquaculture to continue below. I thought it was a clever way to double up land use, but Theunis had a very different reaction.

"That is the horror scenario," he said. "It's very neat and clever, yes, great for energy, but not if you're a shorebird. These freshwater ponds are used by green and wood sandpipers, ruffs, sharp-tailed sandpipers, black-tailed godwits, many more. Along the coast in the salt ponds it's godwits, knots, curlews. If the birds lose the ponds as high-tide roosts, especially the salt ponds, saving the mudflats for feeding won't really matter."

Zhu's main study area lay near Hangu, at the northwestern extreme of Bohai Bay. Getting there was a dispiriting process, looping back on ourselves through a tangle of crowded toll roads to avoid traffic, then driving through endless commercial districts jammed up against wheat fields, housing complexes, and salt ponds on what was once all mudflats. We drove through the remains of the village of Dashentang, which the local government had demolished five or six years earlier to make way for a tourist attraction before running out of money, leaving the rubble covered with acres of blue-green plastic mesh that's ubiquitous in this part of China, holding down the trash and brick as hardy weeds scrabbled out a foothold where they could. It looked like nothing so much as a bombed-out war zone or a post-apocalyptic movie set.

At the ocean, we had to shout to be heard over the din of a six-lane highway jammed with trucks hurtling a few yards from us, over which spun dozens of wind turbines framing the cooling towers of a huge coal-fired power plant a mile away, and the skyline of Tianjin's outskirts miles beyond that. The tide was wrong for waders, so Zhu navigated inland among thousands of small salt ponds, where we found flocks gathered along the downwind shorelines—many fluffed up and sleeping, but most picking at the surface of the water. "Brine flies," Zhu said. "Trillions of them." I could see that the water along the edge of the pond was in furious motion with swarms of small, blackish flies the size of rice grains; when we stepped out, they buzzed lazily in the lee of our bodies, landing harmlessly on our arms and legs. Along with tiny brine shrimp that are cultivated in some of the pools, the flies are a prime shorebird food. The mix of species here was quite different from on the flats; no knots, but rather black-tailed godwits, ruddy turnstones, common greenshanks, spotted red-shanks, and marsh sandpipers, among others. Pied avocets—stylishly lovely white birds, with inky black caps and black slashes across their

wings—waded about on blue-gray legs, their beaks tapering to needles and gently curving up like Mona Lisa smiles.

Theunis was looking hard at the godwits, however. Scientists have
always classified the black-tailed godwits that use the East Asian–
Australasian Flyway as a single, pan-Asian subspecies, *Limosa limosa
melanuroides*, but it was clear even to my untrained eye that we had
a real range of shapes, sizes, and colorations in just this one pond.
That's partly due to sexual differences, because females are bigger
and males more darkly colored, but that aside, most of the godwits
were large and bulky, with noticeably long beaks and fairly pale
plumage. But Theunis pointed out one male that was barely two-
thirds the size of the others and very darkly marked, his head and
chest saturated in deep chestnut tones, his bill shorter and straighter
than the other godwits as he daintily picked up brine flies. "That's
one of the *melanuroides* birds," Theunis said of the dark, runty godwit. The other, larger godwits, Zhu and Theunis are convinced, are
a mystery—they likely represent at least one undescribed subspecies
of black-tailed godwit using the Yellow Sea, likely with its own distinct wintering and breeding areas, another example of how little we
understand of the complexity of this migratory network. For example, while the black-taileds that Chris Hassell and his crew mark in
northwestern Australia are similarly small and brightly colored, with
the typical genetic signature of *melanuroides*, in a decade of searching the crew has never spotted one such marked bird around Nanpu;
they seem to have a separate migratory route, suggesting yet another
cryptic population. It's been devilishly hard for Zhu to net, measure,
and mark godwits in order to build up the data set he'll need to
prove their case, but if they're right, the conservation implications are
serious. Experts estimate the black-tailed godwit population in the
EAAF at about 160,000 and falling, with up to half the birds using
the Yellow Sea coast in migration. If those birds actually represent
multiple distinct populations, each with its own travel routes and
suite of threats and dangers, any one of them could be on the verge
of disappearing without anyone realizing it.

As we chatted about godwit taxonomy a very black, very shiny Audi bounced down the rutted road between two of the ponds and parked not far from us. From it emerged a heavily muscled young man in a polo shirt and sunglasses, followed by an equally fashionable young woman, both looking very badly out of place. While she waited near the car, he picked his way across the uneven, dried mud that formed one wall of the impoundment, flushing the birds, and adjusted an intake valve on the water control structure that rose from the bank. When he returned she snapped a couple of selfies as they posed together, then they got back in their car and drove past us. Zhu smiled and waved, and from inside the tinted window I could just make out an answering hand.

"That was one of the owner's guys—they're going to be stocking this pond with shrimp later today. The owner, he's a gangster, but I get along with him pretty well," Zhu said.

"When you say 'gangster,' exactly what do you mean?" I asked.

"I mean 'gangster.' Like, kill-people-and-chop-them-up gangster. Every year people are killed for control of these shrimp farms— they're worth a *lot* of money, and there's a lot of fighting to get control of them," Zhu said. "But I like this guy. He really likes birds and animals. He won't use firecrackers to scare off the birds, like a lot of the shrimp farms do. And he really likes me."

"Keep it that way," Theunis said, his eyes a little wide. "We don't want you to get chopped up."

The next morning, as usual, I waited in the cool predawn in Nanpu for the crew to pick me up, trying not to breathe too deeply; there was a perpetual sand-colored haze to the sky from the air pollution, and I'd developed a chronic sore throat shortly after arriving in China. At 4:30 in the morning the streets were softly lit with dawn, and the first workers puttered to their jobs on motorcycles and scooters wearing face masks; some women were clad in full-body fabric covers with sleeves and mittens, like biker snuggies, to protect their work clothes from road grime and soot. As they passed, more than a few openly gaped at the big westerner with a backpack and spotting scope at his side; Nanpu is not on the tourist circuit, and Americans are a stop-and-stare novelty. In fact, the only hotel in town authorized to host foreigners was actually closed for renovations during my visit, but Professor Zhang had pulled some strings, and I'd arrived a week earlier to find I was the sole guest in the sprawling, empty facility. The first morning I was served breakfast alone in an echoing banquet hall, with most of the huge, round dining tables draped in dust cloths. Since then I'd been out and gone each morning long before the skeleton staff was up, forced to scramble the perimeter wall as I left like some escaping burglar, since the wide courtyard was gated off from the street and locked tight for the night.

I heard a tooting horn, and jammed myself into a small van with the rest of the Global Flyway Network crew. Katherine passed me a bag of pork dumplings, the dough hot and slightly tacky against my fingers, as we drove south toward the bay, anxious not to be late because we were expecting visitors. Soon after we arrived on the seawall, and just as low, orange light blossomed across the water, a small convoy of vehicles rattled down the gravel road, disgorging two dozen people—Professor Zhang and a bunch of his students, along with Wendy and Hank Paulson with several Paulson Institute staff, and a British birder and environmental lawyer named Terry Townsend, who lives in Beijing and works closely with Chinese conservationists. As introductions were made and hands shaken, the tide began to pull back from the seawall, exposing the first mud, and Adrian shouted and pointed behind us. Thousands of birds—red knots glowing a deep copper in the morning light, great knots looking blacker against the pale sky—were pouring in from their high-tide roosts back in the salt ponds, just clearing the utility poles along the seawall as they glided down to feed.

Katherine, Adrian, and Matt jogged up the shoreline, scopes on their shoulders, all business; VIPs notwithstanding, they had work to do, and with the peak of the migration upon us, every banded bird they could record was critical. Professor Zhang and Theunis, however, worked the crowd, thanking the Paulsons for the institute's lobbying on behalf of coastal conservation, which both men felt was a significant factor in the government's recently announced reclamation ban. Wendy, who is the more serious birder of the couple, was picking out broad-billed sandpipers and some of the other less common species from the rapidly swelling ranks of birds on the beach. Hank, tall and lean in aviator sunglasses and a green ballcap, was immersed in an intense discussion of policy with Theunis, Terry, and Zhang, but Wendy kept pulling him back to the scope from time to time, making sure he didn't miss the birds they'd come to see. "Hank, you need to see this—a Nordmann's greenshank," she said, prying him away for a moment.

In 2016, armed with a decade's worth of data from the Global Flyway Network surveys and Zhang's grad students, the Paulson Institute and the World Wildlife Fund had negotiated a five-year agreement with the Hebei provincial forestry department and the Luannan county government to protect these Nanpu flats by setting up a nature reserve. But looking over the latest map of the proposed reserve and comparing it with the windswept landscape around them, the Paulsons, Theunis, and Zhang all expressed frustration. Not only were significant portions of the mudflats outside the core of the main reserve, almost none of the salt ponds—from which still more roosting birds were flying every minute—would have any protection at all. They are under control of the local government, which wanted nothing to interfere with its ability to develop the ponds for industry or aquaculture. Hank was hoping that meetings he had scheduled with the head of China's National Forestry and Grassland Administration might move the needle.

For quite a while, the good news for shorebirds on the Yellow Sea had been more theoretical than concrete. Early in 2017, the Chinese government added 14 sites on the Yellow Sea and Bohai Bay—including the mudflats near Nanpu, and parts of the Jiangsu coast where in a few days I would be searching for spoon-billed sandpipers—to a tentative list for potential UNESCO World Heritage recognition. It was a potent symbolic victory for conservationists, but one up to that point without teeth; while any site receiving World Heritage status must, by international treaty, receive fairly stringent protection from its home country, this interim measure carried no protective guarantees. Still, it was an important step forward, and the State Oceanic Administration's (SOA) stunning announcement, a few months before my visit, banning most reclamation along the Yellow Sea was even bigger news.

"Twelve months ago it would have been hard to imagine this amount of progress," Terry Townsend told me as we stood by the bay, watching buntings, pipits, and stonechats flitting past us in the waist-high grass along the seawall, part of a heavy flight of passerines also migrating up the coast on this cool spring morning. "But we'll have to see how it's carried out. The SOA decree bans what it calls 'business-related' development and puts the decision-making about reclamation at the national level instead of locally—that should be huge. Almost all of the coastal reclamation was being driven by commercial projects approved by local officials, often without permits. Some illegally reclaimed land is already being restored down at Yancheng, places where the walled areas were never filled with sediment—they're opening up the seawalls and letting in the tide." On the other hand, the ban leaves a loophole for projects in keeping with "the national economy and people's livelihoods," and that morning, no one could say exactly what that might mean. In the weeks ahead, they would learn that one such project would be a huge port expansion to support the complex of five steel mills already under construction on previously reclaimed land, midway between Nanpu and Zhu's godwit site at Hangu. The port would gobble up another 21 square miles of remaining mudflat, presumably in keeping with "the national economy and people's livelihoods."

Still, the freshet of good news, coupled with the spectacular show that the shorebirds, now tens of thousands strong, were putting on for our visitors, had everyone that morning in a fairly cheerful mood.

On the Yellow Sea during migration, that alone was something of a minor miracle.

Nanpu is not the only critical site along the Yellow Sea where protection is far less than it should be. Three days later I was almost 500 miles south, just above of the mouth of the Yangtze River at the southern edge of the Yellow Sea. While Hank Paulson had remained in Beijing for meetings, Wendy and I joined spoon-billed sandpiper expert Jing Li and some of her companions to visit Tiaozini, part of a mudflat complex along the coast of Jiangsu Province that is often referred to as one of the most important areas on the entire East Asian–Australasian Flyway.

"These are the largest remaining mudflats on the Yellow Sea," Jing said as we carefully walked down the steeply sloped face of the seawall, scattering sand martins and red-rumped swallows that were resting on the concrete. "At low tide, in a straight line, it is 20 kilometers to the sea."

I was sure I'd misheard her. "I'm sorry—how far?"

"Yes, 20 kilometers," she said. While one might think that such a vast area of flats would provide more than enough room and food for as many shorebirds as could possibly use them, not all mud is created equal, Jing explained—as on land, marine resources are not distributed evenly, and many parts of the mudflats may be barren of what hungry birds need. Conditions also change with the seasons; a fertile zone in spring may be less so in autumn, or vice versa, and different species of shorebirds specialize in different foods, captured in different ways in varying habitats. A godwit with its stiltlike legs and a nearly four-inch bill can wade into rather deep water and reach far below the surface, using touch to locate mollusks and worms, while a red-necked stint with its stubby legs and a dinky bill must forage where the tide is just departing, visually spotting prey that has yet to scuttle below the surface. Red knots, as Theunis Piersma and his colleagues discovered years ago, use a sixth sense unique (so far as is known) among any animals to locate clams far below the surface. The knot's rapidly probing bill sets up compression waves in the water between sand grains, which "echo" off the hard shell of a mollusk and are detected by a network of densely packed sensory organs in the tip of the beak. Each species thus has its own niche on the flats, and we would see such resource partitioning, as scientists call this phenomenon, as the morning went on; the area within a mile or so of shore, which we could reach on foot, was alive with stints and other small shorebirds, while the larger species that had dominated the scene at Nanpu, like knots and godwits, were only occasionally glimpsed, far out and in flight—at Tiaozini they feed many miles farther out on the flats, where the estuarine environment is right for their prey.

Close to shore were dozens of room-sized lozenges of smooth cordgrass, *Spartina alterniflora*, a North American tidelands plant that has become a massive problem along the Yellow Sea, where it was (and in some places, still is) planted to "stabilize" natural mudflats. While it is the foundation of fertile, biodiverse salt marshes in places like Maryland or Georgia, here in China cordgrass chokes out the natural flats with sterile monocultures in which little can survive. The Paulson Institute has been advocating for the Chinese to take action against cordgrass at places like Tiaozini and Nanpu where they still can, putting government scientists in touch with Americans grappling with *Spartina* infestations on the West Coast of the United States, where it is also a serious invasive.

Nearby, I heard a high, long whistling note that fell and repeated. Chen Tengyi—a stocky young man in a brown camouflaged jacket—was blowing on one of the cluster of small bamboo whistles around his neck, his eyes locked on an incoming flock of stints. He whistled again, and the birds instantly altered their flight path and rocketed toward us, surging past in a hurried storm. Tengteng, as he's known, grew up on Chongming Island just down the coast, learning to use homemade whistles on which local hunters mimic shorebird calls to lure the flocks into their nets. An ardent conservationist and photographer, Tengteng uses those traditional skills to help scientists and

visitors alike see the migrants that gather here by the tens of thousands. Now he switched to a rolling trill, as still more birds swept
past, splitting and curling like something out of a traditional Chinese
ribbon dance, as though the streams of birds were swirling at the
ends of invisible rods.

Mixed with the multitudes of red-necked stints was a potpourri of
other small waders—sharp-tailed and broad-billed sandpipers, dunlin, a few Terek sandpipers with their weird upturned beaks, and
lesser sand plovers. The plovers were by far the snazziest birds on
the flats, about eight inches long, intense cinnamon on the crown
fading to brown on the back, a black robber's mask, a white throat
with just the thinnest black line encircling it, and two white spots on
the forehead like headlights. The only larger shorebirds I saw were
a few gray-tailed tattlers, a rapidly declining species endemic to the
EAAF, hugging the edges of tidal creeks and occasionally snagging
food from the turbid water. Everywhere I looked the mud, rippled by
the departing tide, was embroidered with the delicate tracks of shorebirds, filigrees interwoven and branching, marked with tiny holes
where the birds had, in endless, hopeful repetition, probed the surface for a meal. At Nanpu, the flocks preyed mostly on tiny clams;
here, Jing said, she thinks the birds are taking worms and tiny, almost
transparent crabs, but no one is really sure.

We were half a mile or so from shore, threading our way around
a capillary network of tidal creeks. A large, abandoned fishing boat
lay up to its gunwales in mud, its bow facing the vanished sea. The
farther out we ventured, the more careful we had to be about our
footing as the mud became progressively looser and gooier. I'd almost
lost a boot twice, but we were following a friend of Jing's, Dongming
Li, an avid local bird photographer who is on the flats frequently and
knows them well—though not well enough, apparently, for shortly
thereafter he plunged waist-deep in a sinkhole, barely keeping his
camera dry. He came out in his socks, his boots lost for the ages.

"There has to be a spoonbill in here somewhere," Wendy muttered, hunched over her scope as we methodically sorted through the masses of shorebirds all around us. Tiaozini is one of the best places on the Yellow Sea coast to find spoon-billed sandpipers; Jing and her crew have counted as many as 100 of the endangered birds here during the prime season in autumn, roughly a quarter of the global population, which on the face of it would argue for urgent protection for this place. A significant part of the Tiaozini wetlands has already been lost to reclamation, but part of this coast was declared a nature reserve—although local officials later excised the remaining mudflats from its boundaries, all except a distant offshore island, using the patently laughable argument that no birds use the site. "The officials did their surveys in August, not taking the tides into account, and some of their reports included winter records, when the migrants are not here. They worked in an office, not in the field," Jing said with evident disgust. "They asked me to prioritize [what to save]—I told them everything that's left is a priority, because the birds have already been squeezed into what remains, and any loss will put multiple pressures on them. The whole Jiangsu coast should be a nature reserve."

Saving the coast for birds would also save it for humans; almost everywhere that natural mudflats remain in Asia, shorebirds share them with people, millions of whom in China alone depend on the flats for collecting crabs or shellfish, or as nurseries for finfish that mature farther out to sea. The remaining flats at Tiaozini, once slated for reclamation, have been returned to local control and are seeded each spring with hard shell clams by fishermen from the local village (one of whom, distrustful of our presence and angry that we were walking near his clam beds, came storming out toward us, bellowing and brandishing a stick until Dongming intercepted him). Such human uses are not entirely incompatible with the mudflats' role as hemispheric stopover sites, and although commercial shellfish culture has reduced the mollusk diversity along the Yellow Sea coast, it's probably better than another "reclamation" project. Tiaozini may

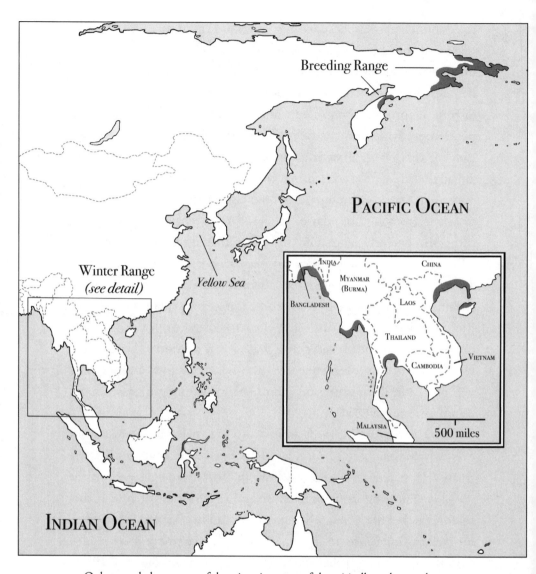

Only recently have some of the wintering areas of the critically endangered
spoon-billed sandpiper been found in southern Asia, though migration surveys
suggest other important wintering locations remain to be discovered.

have been cut out of the nature reserve, but it was one of the 14 proposed UNESCO World Heritage sites—a designation that would, if
it went through, preserve the area for clam fishermen as well as birds.

For spoon-billed sandpipers, there is no other one place as crucial as Tiaozini, no other place where so many spoonies are known to gather for weeks or months at a stretch. As recently as a decade ago, the spoon-billed sandpiper appeared to be doomed, but heroic efforts have clawed it back from the edge of extinction—helped, it must be said, by the magnetic charm of this small, peculiar bird and the enthusiasm and affection it engenders in people. Overall, increased law enforcement has reduced the risk of illegal netting and hunting on the spoon-billed's nonbreeding grounds in southern Asia. At Sonadia Island in Bangladesh, for example, where in addition to increased poaching patrols conservationists offered microcredit loans to allow former hunters to take up fishing, shopkeeping, or other alternate employment, there has been a complete cessation of shore-bird killing of any sort. In 2017 part of the Gulf of Mottama in Myanmar, where half the spoonie population winters, was declared a Ramsar conservation site, with the remainder proposed for a similar designation. The government of Chukotka in Russia, where most of the spoonbills nest, announced it would set aside nearly 200 miles of coastal tundra in a new "Land of the Spoon-billed Sandpiper" nature park. Advances in satellite tags, small and light enough for this roughly one-ounce bird, have allowed scientists with the multinational Spoon-billed Sandpiper Task Force to track them for the first time, further illuminating the routes the sandpipers take and revealing places no one knew they visited—although adding up all the spoonies counted each year at known wintering sites yields only about half the estimated population, suggesting that other, potentially unprotected wintering and stopover sites remain to be found.

Thanks to color-flagging and resighting work along the EAAF, experts like Jing now suspect the spoonie's population may be a bit larger than once thought, between 220 and 340 pairs, rather than 80 to 120—not reflecting an actual increase so much as better statistical methods, but nevertheless nudging the species a bit further back from the void than we once believed. Still, with only a few hundred individuals in the world, the risk of an apocalyptic loss—say, from a cyclone or oil spill—can't be discounted. For that reason, the task force and its partners established a small captive breeding population in England, though it saw more failures than successes until 2019, when two chicks were fledged. On the other hand, task force biologists in Chukotka have taken the innovative step of "head-starting" spoonies—removing about 30 eggs each summer from wild nests where predation and loss from flooding or bad weather are serious threats, raising the chicks to fledging in a large, outdoor aviary, then releasing them in late summer to join the wild flocks so they all migrate together. (Because shorebird chicks can run and feed themselves right out of the egg, this system works very well—and the adults whose eggs are taken usually lay a replacement clutch, thus increasing the overall number of chicks each year.) This technique has quickly paid big dividends, with more than 140 chicks produced in just a few years, some of which have since returned themselves to raise babies.

For all our searching at Tiaozini, we came up dry on finding a spoonie ourselves. My last chance would be the next day, on the gray "steel plate" mudflats down the coast at Dongling, where we were joined by Jing Li's colleague Zhang Lin, one of China's best-known bird guides and the man who first discovered the importance of the Rudong coast for spoon-billed sandpipers. Unlike the peaceful emptiness of Tiaozini, Dongling was a beehive, the site of a bustling commercial clam farm, with hundreds of scurrying employees and jostling vehicles. Huge swaths of the flats have been reclaimed in recent years, and we drove for more than seven miles through a monotony of aquaculture ponds, each a couple of acres in size. "This used to be a favorite roost spot for waders," Jing said sadly, showing me a photograph of the same area in 2012, with thousands of birds roosting quietly in wide expanses of shallow water. Closer to the seawall, we found active fish ponds roofed over with solar panels—the "horror scenario" Theunis had warned of, since it renders even this marginal habitat useless as a high-tide roost. (Once home, where my access to Google Earth was no longer blocked by the Chinese government, I was able to measure the extent of the aquaculture area from satellite images, on which the ponds look like densely packed cells under a microscope—more than 12,000 acres, parceled out into thousands of impoundments, along with a 1,000-acre reservoir Jing said would be used for watersport recreation.)

The remaining flats at Dongling are, technically, protected as a local nature reserve for the clam fishery, which is run by a company managed by the local village. As the tide drew away from land, tractors pulling big wagons piled high with bulging mesh bags full of golf ball–sized seed clams headed for the shellfish beds offshore. We climbed into the back of one of the empty wagons, a roughly welded, wholly utilitarian conveyance, and tried to arrange ourselves on two hard benches as the tractor stammered to life and lurched down the seawall. Once on level ground, the tractor slewed through deep wallows as we pitched and rolled, bouncing painfully on the steel seats. Thousands of acres closest to the seawall were jammed with invasive cordgrass—a loss for both birds and humans, since the dense vegetation excludes clams, too.

A white cardboard shipping crate banged up against my shin; moving it, I saw that it was emblazoned with the name and logo of a Massachusetts seafood company, to which the harvested clams were destined. Zhang scowled, pushing the crate with his toe. "We are only 30 kilometers from the big chemical factories up the coast— all that pollution washes into the Yellow Sea. But these guys don't care—they're not going to eat them, they figure they can eat the clams in the United States."

A tooth-rattling mile or more from the seawall, the tractor finally slowed and stopped, and we untangled our limbs and spotting-scope tripods to climb warily down a rickety metal ladder. Other tractors passed us, with workers in rain slickers and rubber boots perched atop the clam bags, following lines of wooden stakes tipped with plastic flagging that marked the clam beds. To our backs, on land, stood ranks of towering wind turbines spinning slowly in the sea breeze. But looking seaward, there was nothing but gray, the clean divide of the horizon line separating mud from sky—a monochrome backdrop across which tens of thousands of birds flew. Jing and Wendy were on high alert, trying to find me a spoon-billed, while Zhang unfolded a small camp chair, sat down, and began making methodical counts, his hand clicker going like a buzz saw. When I tried to talk with him, I found him curt and fairly dour—and probably with good reason. He and Jing formed their NGO, Spoon-billed Sandpiper in China, in 2006, and they've made big strides, especially demonstrating the importance of the Rudong coast for spoon-billed sandpipers and other waders. But it's been a pretty disheartening battle so far; all they have to do is look around to see how much of what they were trying to save has already been lost, and the remorseless pressure of raw economics that they're up against. Even when unalloyed good news comes along, like the tentative UNESCO listing for the Rudong coast, or the government decree promising an end to much of the Yellow Sea reclamation, they've learned to be skeptical. Eventually, I realized the question wasn't so much why Zhang Lin was grumpy, but how on earth Jing Li maintains her perennially sunny disposition. Yet even she feels weighed down by the immensity of the forces that have been destroying the Yellow Sea.

"Every two months or so I ask myself if I should quit this job," Jing admitted. "If we're not making any progress it's very depressing, but there are always some encouraging solutions. And the birds are still here. But it's hard."

We scoured the Dongling flats for hours and reveled in the great clouds of shorebirds that moved restlessly around us. But rarely has a haystack looked so large, nor a needle so elusive. Eventually the tide moved far enough offshore to draw the last flocks beyond our reach, and I realized I would be leaving China without having seen the one bird that, more than any other, I had hoped to find. But that may have been appropriate, for how better to emphasize the thin edge on which the spoonbill—indeed, all the migrants of this flyway—still exist, suspended between a legacy of loss and nascent hope?

The day before, we'd stopped on the outskirts of Rudong city, with apartment blocks and China's ubiquitous construction cranes forming a solid wall advancing into wheat fields and marshes, which themselves had been taken from the sea years earlier. Zhang wanted to see if poachers had been around; a few local men often catch pipits and buntings here to sell for food. Instead, we found gray-headed lapwings flying in raucous, annoyed circles around us, protesting our presence near their nests, and dozens of 10- or 11-year-old kids in matching navy blazers, the boys in ties, the girls with red-and-black plaid skirts—and all wearing binoculars, shepherded on this birdwatching field trip by their elementary school principal. Although the presence of a couple of Americans overshadowed the birds for several of the kids, most seemed genuinely enthusiastic about watching the noisy lapwings, a cooperative long-tailed shrike that posed for them in our scopes, and Oriental pratincoles that look like overgrown swallows the size of small falcons. The next day, we dropped in at their school to find the entire grade, including our friends from the field trip, outside on their athletic field, drawing colorful and highly creative spoon-billed sandpipers on 50-foot-long rolls of white linen. Inside, members of a print-making class were cutting wood blocks with their own spoonie designs—part of the public pride campaign that Jing and Lin's NGO has mustered in the local community on behalf of the birds.

It brought to mind a question posed by some of the scientists working feverishly to protect the spoon-billed sandpiper, recognizing how the ripples of public attention and governmental concern created by this one small bird have had a far greater impact than anyone would have guessed just a few years earlier—an impact that may (just may) have tipped the scales in time. "The spoon-billed sandpiper is one of the best examples of a species uniting conservation NGOs, science organizations, grant-giving bodies, corporate donors . . . and passionate conservation volunteers around the world to work harmoniously and with common cause," they wrote. "Can a species save a flyway? We do not yet know, but the verdict will be in quite soon."

But even those to whom the death of the Yellow Sea flyway very recently seemed a foregone conclusion are learning to accommodate a new and unfamiliar emotion: hope. After meetings with Hank

Paulson shortly after our morning together on the Nanpu mudflats, the head of China's National Forestry and Grassland Administration reached a consensus with the provincial government to designate those flats a wetlands park, one of the protected categories in China. And in July 2019, UNESCO approved the first World Heritage designations along the Yellow Sea, which will bind China by treaty to the protection of more than 188,000 hectares—more than 466,000 acres—of coastal habitat for shorebird sanctuaries, including the critical but vulnerable mudflats at Tiaozini, which China added to the nominated areas late in the process. Under UNESCO supervision, some of the artificial fish ponds at Tiaozini would be restored as high-tide roosts for spoonies and other shorebirds, and cordgrass control would become a priority. Other crucial Yellow Sea sites including the shellfish farming area near Dongling and the Nanpu mudflats, totaling a further 643,000 acres, were at that time under consideration by UNESCO for a second phase of World Heritage designations.

On one of my last days up in Nanpu, Theunis Piersma and I walked out along the seawall. It was cool, the sun coming through high cirrus and the wind blowing in from the sea, wiping away the smog. Theunis settled himself among the rocks, adjusting his scope and scanning for a while, jotting down notes on the flagged birds he saw. After a while, though, he straightened and sighed happily. "Ah, this is lovely," he said with evident contentment. "It's so much quieter than in past years. This was a coastal hell. A couple of years ago there were dredges everywhere, huge plumes of sediment being pumped inside the seawalls, the smell of oil in the air. None of that now. There used to be big banners along the highway, promoting industrial development—we still have banners, but now they have birds on them. Quite a change. But there were more birds on the mudflats in those days, too."

The flocks were several hundred meters away, but clearly across the distance came a long, two-noted wail, *pooh-weee*, repeated after a few seconds, very different from the feeding chatter we'd been hearing all morning. "You hear that?" Theunis asked excitedly. "That's the

Arctic display song of the red knot. Some years ago I wrote a paper, 'Arctic Songs on Temperate Shores,' about how knots in the Netherlands and Iceland sometimes begin to sing in migration. They do so here, too." To those accustomed to shorebirds feeding with quiet efficiency on beaches or mudflats, the notion of a sandpiper singing is jarring, like seeing a staid, gray-flannel coworker suddenly cut loose at a karaoke bar, but many species sing once they arrive in the Arctic. The male red knot rises rapidly hundreds of feet into the air and then sweeps back and forth in great figures-of-eight around his territory, all the while pouring out that weird, two- or three-noted moan we were hearing along the seawall: *pooh-weee, poor-oh-weee, poor-oh-weee.* His flight often provokes similar displays from neighboring males until the tundra is ringing with these plaintive, almost poignant songs as the birds crisscross the sky.

The knot called again, and a second answered. The Arctic was tugging at them, and their songs reminded me that this was merely a way station—that the life-and-death conclusion of their journey, the chance to reproduce that is the crux of their existence, was still thousands of miles away. *Poor-oh-weee, poor-oh-weee.* To my ears, those slurred syllables carried both melancholy and expectancy.

Theunis sighed again. "It was very difficult, you know, the first couple of years. I was very pessimistic—I thought our job here was to document the extinction of these birds. And there would have been value in that, importance in that. But now. . . ." He trailed off for a moment, listening to the knot's song. "Now, I hope I have enough life left to see some sort of a recovery."

Two

QUANTUM LEAP

The red knots Theunis and I heard singing along the Yellow Sea were males, and their songs were an aural reflection of internal changes. During the northern winter, when the knots had been in Australia, their testes were shrunken and all but functionless. But now, as they pushed north toward their breeding grounds, the organs were beginning to swell. By the time a male knot reached Siberia his testes would have ballooned to almost 1,000 times their minute winter size, pumping testosterone into the bird's bloodstream. What had been a mild, occasional itch to sing in China would become a constant hormonal compulsion in the Arctic.

Something similar was happening inside the female knots, each of whose single developed ovary (usually the left) was growing, albeit not as dramatically as the testes, to prepare for breeding. The seasonal expansion and contraction of gonads is an evolutionarily clever way of saving weight, and is common among birds. But the more scientists study migrants—especially extreme long-distance migrants like shorebirds—the more they realize that these creatures have evolved extraordinary physiological abilities that touch on every aspect of migration, from speed and endurance to memory and brain function, metabolism, disease immunity, blood chemistry, and much more. Some of these discoveries, as fascinating as they are in their own right, also hold the promise of future breakthroughs for human health.

Migratory birds can grow and jettison their internal organs on an as-needed basis, bolster their flight performance by juicing on naturally occurring performance-enhancing drugs, and enjoy perfect health despite seasonally exhibiting all the signs of morbid obesity, diabetes, and looming heart disease. A migrating bird can put alternating halves of its brain to sleep while flying for days, weeks, or even months on end, and when forced to remain fully awake has evolved defenses against the effects of sleep deprivation; in fact, birds actually seem to get mentally sharper under such conditions, the envy of any human slogging through the day after a poor night's sleep. If all that isn't sci-fi enough, we now know that they navigate using a form of quantum mechanics that made even Einstein queasy.

Just as there are many migratory strategies employed by different species of birds—long-distance or short-, diurnal or nocturnal, over land or over water—so too are researchers finding a kaleidoscope of seemingly contradictory approaches that birds use to achieve the same ends. For example, among the birds that depend upon the Yellow Sea are bar-tailed godwits, pigeon-sized shorebirds with long legs and slender, slightly upturned beaks, close relatives of the black-tailed godwits Zhu was studying on the gangster-owned shrimp farms. This species is largely an Old World bird, nesting on wet tundra from northern Scandinavia across the top of Asia to the Russian Far East, as well as in western and northern Alaska. Those from the European and central Eurasian portions of this range winter along the muddy coasts and mangrove lagoons of Africa, the Middle East, and the Indian Ocean, and across southeastern Asia. These are not insignificant migrations by any means, but those godwits that breed in eastern Asia and especially Alaska undertake a journey that on its face seems frankly inconceivable, and which is made possible by physiological changes that sound like something out of a mad genius's lab.

Twenty years ago, scientists employing some of the first miniaturized satellite transmitters were stunned to learn that many godwits make a 7,200-mile nonstop flight each autumn from western Alaska to New Zealand, a journey that takes them eight or nine days of uninterrupted flight—the longest nonstop migration known, exercising at the same metabolic rate as a human running endless four-minute miles. They accomplish this astounding feat by first larding themselves with thick layers of fat, feeding with manic energy on the rich tidal flats of the Alaskan Peninsula, eating marine worms and other invertebrates. They more than double their weight in about two weeks, so that a 1.5-pound godwit is carrying more than 10 ounces of fat under its skin and within its body cavity. So obese that they jiggle when they walk, the godwits then undergo a rapid reorganization of their internal anatomy. Digestive organs like their gizzard and intestines, which they no longer need, shrink and atrophy, while the pectoral muscles that power their long, slender wings double in mass, as does their heart muscle, and their lungs increase in capacity. (Like so many discoveries involving shorebirds, this one was made by Theunis Piersma, working with American scientist Robert Gill Jr. from the US Geological Survey.) The godwits time their departure from Alaska with the passage of autumn gales, when they get a boost from powerful tail winds speeding them along the first 500 to 1,000 miles of their journey across the Pacific. Along the way, they must overcome extreme dehydration and sleep deprivation, to say nothing of the physical exhaustion that must come from pumping their wings millions of times without the slightest pause. With time, however, they come within range of more tail winds, the austral westerlies, which hurry them along on the final 600 miles or so of the trip.

Once in Australasia, the godwits quickly regrow their digestive organs and spend the austral (or Southern Hemisphere) summer feeding normally, but as the days shorten, hormonal changes trigger another episode of binge feeding (known as hyperphagia) and explosive weight gain, followed by a similar but not as extreme atrophy of digestive organs. This time the godwits depart to the northwest, leaving New Zealand by early April and crossing more than 6,000 miles of the western Pacific to China and the Koreas in another uninterrupted eight- or nine-day flight. Landing, they repeat the cycle of organ regrowth and gluttony yet a third time before making their final, roughly five-day ocean crossing of only—only!— about 4,000 miles back to Alaska. Journeys like this have prompted

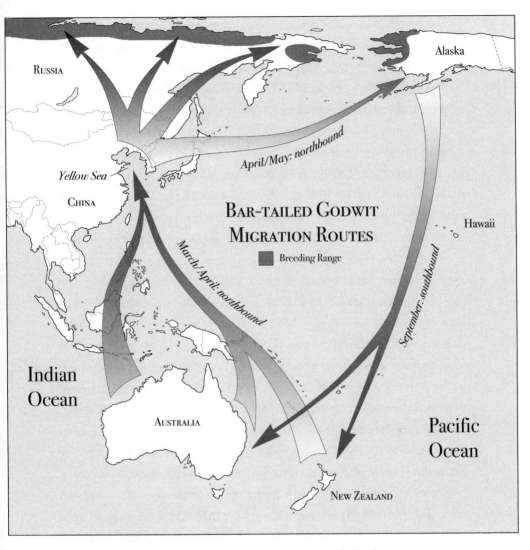

The trans-Pacific journey of Alaskan bar-tailed godwits, which takes seven to nine days of continuous flight to reach New Zealand and eastern Australia, is the longest nonstop migration of any landbird. Asian populations, which winter in western Australia and breed in northern Russia, face a formidable flight as well.

one specialist in migrant physiology to say, "The metaphor of marathon running is inadequate to fully capture the magnitude of long-distance migratory flight of birds. In some respects a journey to the moon seems more appropriate."

I've watched newly arrived bar-tailed godwits along the Keokle-vik River in remote western Alaska—a flat, treeless, waterlogged land not far from the Bering Sea, part of the 19-million-acre Yukon Delta National Wildlife Refuge, where the incessant wind lashes beds of lush grasses and sedges along meandering river channels, and scours the slightly higher benches and ridges of spongy, flower-spangled tundra. Having just completed an 18,000-mile roundtrip—something a god-wit may repeat 25 or 30 times in its long life—the birds waste little time. The male circles high overhead in a courtship display that biol-ogists call "limping flight," a stuttering rhythm in which his silvery-white wing linings flash in the sun against his rusty-red body plumage, as he sings an endlessly repeated, shrill song, *a-week a-week a-week a-week a-week.* Having mated, the female creates a small, snug nest cup in the sphagnum moss, lined with strands of lichen, for her four heavily camouflaged eggs. The chicks—should the nest evade foxes, jaegers, weasels, ravens, and other predators—will be walking and feeding themselves almost from birth. As soon as they are old enough to fly, at barely a month of age, they will be abandoned by their parents, who join the earliest waves of migrants headed back to New Zealand. Guided only by their genes, the youngsters will gang up and move to the coast as well, there to surrender to an innate command that they eat without surcease, and then to fly off into a wide and hostile ocean.

Sitting on the wet ground along the Keoklevik, watching this annual cycle unfold, I have tried to put myself into the head of a juvenile godwit, as the land falls away behind it for the first time and the Pacific, vast and deadly, rolls for days beneath its laboring wings. Is there doubt, deep in the long nights as the unfamiliar stars of the Southern Hemisphere wheel overhead? Is there fear? Is there any emotion that can penetrate what I can only assume must be absolute and numbing exhaustion? Or does the young godwit feel only cer-tainty, a sense that it is doing merely what it must at this moment, drawn toward some unseen place by a biological magnetism? There is no answer for me in the dark brown eyes of the female, sitting quietly on her nest.

The Alaskan godwits are only the most extreme example of migratory birds reordering their insides; such internal flexibility is actually common among migrants of many sorts; a thrush or catbird, feeding on the dogwood berries in a corner of the backyard, has undergone a late-summer expansion of its intestines to squeeze every calorie from that lipid-rich fruit. Even within a single species, there can be dramatically different variations on the binge-bulk-and-shrink theme. The population of red knots that winter in northwestern Europe and breed in Greenland and the eastern Canadian Arctic, for example, stop off in Iceland for about three-and-a-half weeks on their journey north. During their first week they gain little weight overall, but their hearts, stomachs, and livers increase in mass. In the next 10 days their intestines, kidneys, and leg muscles also grow. The knots are by this point slathering on fat as they gorge on small mollusks—but curiously, given their rate of feeding, their stomachs now begin to shrink again, and will have decreased in weight by one-quarter by the time the birds finally depart. On the other hand their livers mushroom, more than doubling in weight by departure.

Along with Chinese colleagues, Theunis (who was, no surprise, also part of the team studying Icelandic knots) looked as well at stopover physiology in the subspecies of red knot named for him, *Calidris canutus piersmai*, preparing to fly from the Yellow Sea to the Russian Arctic. Curiously, they found significant differences from the way the knots in Iceland responded to migration. First, and as expected, the thin and exhausted birds—having just arrived on the Yellow Sea from Australia after a 4,000-mile flight—rebuilt their organs and tissues. This included not only their digestive tracts, atrophied before takeoff, but muscle mass that was consumed for energy during the long flight. After this initial, protein-building phase, the birds appeared to flip a metabolic switch and began serious fat deposition, increasing their fat stores by a factor of almost 17, until like the godwits they more than doubled their lean weight—more than they actually needed to complete their flight to the Arctic. At this point, the metabolism flipped again, and their flight muscles built rapidly—but instead of their digestive organs shrinking, as happens with godwits and even with their cousins in Iceland, the organs in the Yellow Sea knots actually increased in size along with their breast muscles, a change that was especially pronounced in females.

Why? Theunis and his colleagues believe the birds were banking extra fat and protein in their organs because conditions in the Arctic, in the days and weeks after their arrival, would preclude them from eating. They haven't much time, for one thing; the window for breeding in the North is so short that the birds can't afford to rest and rebuild their energy stores if they hope to find a mate, set up a territory, nest, and raise a brood before winter returns. Worse, when they arrive the Arctic is still going to be locked in the last of the previous winter's ice and snow; there is simply little food to be had. The only way to breed successfully, therefore, is to carry ample stores of fat and protein from the far shores of the Yellow Sea, especially in the case of females, who must convert those energy reserves into eggs.

There is another advantage to having muscle and organ mass to spare, and it has to do with the complex chemical pathways by which birds efficiently burn fat for fuel—a task that, based on mammalian physiology, should be nearly impossible. Take water. One of most difficult hurdles for any long-distance migrant ought to be dehydration. Even though birds don't sweat, they do lose a lot of moisture through respiration, perhaps one reason why many migrants fly at night, when the air is cooler and more humid. They must also excrete waste which, however concentrated, still costs them additional water. Especially for birds crossing wide barriers like oceans or deserts, in flights that last for many days, dehydration ought to be a serious threat. Yet it appears not to be. When tested, willow warblers crossing the Sahara from Africa were found to have normal internal water balance, even those that were dying from starvation after using up their energy reserves. Theunis was part of a team that looked at the hydration rates of bar-tailed godwits migrating between western Africa and the Netherlands, a 2,700-mile nonstop flight. Luring newly arriving godwits from the sky into their nets, they injected the captured birds with a tiny amount of deuterium oxide, a nonradioactive form of water that has long been used to calculate total body water. They found no difference between the water balance of godwits that had just been migrating for days on end and those that had been resting and feeding for similar periods.

While fat is a terrifically dense, highly potent fuel, providing eight times the stored energy of protein or carbohydrates, it's challenging to burn—the reason we mammals rely mostly on carbs. Birds have adaptations that allow them to burn fat roughly 10 times more efficiently than humans, but fat contains almost no free water that would be released when it's broken down. Muscle and organ tissue does, however—protein releases up to five times more water than fat when metabolized by a bird. The beefed-up insides that red knots develop during their stopover in the Yellow Sea, then, may not only be an energy bank on which they can draw in the Arctic, but a reservoir from which their cells can, in a sense, drink on the way there. Alex Gerson, who studies the physiology of migrant birds at the University of Massachusetts, has been exploring this aspect of migration by testing Swainson's thrushes in a massive, climate-controlled wind tunnel, using a portable quantitative magnetic resonance imaging (MRI) machine to quickly and harmlessly calculate the lean mass, fat, and body water of the birds before and after exercise. He's found that by cannibalizing its muscles and organs in addition to burning fat, a bird can constantly adjust its production of metabolic water to keep up with the loss from breathing and excretion. In the process, a thrush weighing a shade more than an ounce can extend its flight range by almost 30 percent, to more than 2,000 miles, beyond what it could fly on fat alone—a critical cushion for birds that, like the Swainson's thrush, make long overwater crossings.

Scientists have really just started exploring the ways in which the harsh selective pressures of migration have fine-tuned the physiology of migrants. They've discovered major adaptations at the cellular level, including ways that birds can rapidly process lipids, increasing the proteins that speed lipid transport and goosing the cellular machinery that breaks them down into glycerol and fatty acids. The migrants have naturally high levels of the mitochondrial enzymes that oxidize fatty acids, and those levels increase even more as migration season approaches, and when they're resting at stopover sites.

Birds may also improve their muscular efficiency and performance by picking the right foods. Semipalmated sandpipers that gather on the Bay of Fundy in autumn, before making a 2,000-mile nonstop flight to the northeastern coast of South America, spend several weeks feeding selectively on a tiny marine amphipod called *Corophium*, which burrows by the trillions into the mudflats exposed by the bay's immense tides. *Corophium*, it turns out, is exceedingly rich in polyunsaturated fatty acids like omega-3, widely touted as beneficial to human health—in fact, no other marine invertebrate comes close to *Corophium*'s omega-3 levels. In birds, omega-3 not only serves as a fuel, it also primes the flight muscles and increases their aerobic capacity, what some avian physiologists have dubbed "natural doping." Experiments have shown that landbirds like white-throated sparrows—which don't have access to seafood rich in omega-3—can generate omega-6 fatty acids that also boost muscular performance, even on a controlled diet in captivity.*

Of course, people have long been warned about the dangers of yo-yo dieting, gaining and losing and gaining weight in repetitive cycles. Birds like godwits, though, oscillate between extremes of grossly fat and starving-thin that no human ever approaches, and they do so

---

* Interestingly, this case of a bird dosing itself on a natural performance-enhancing drug may be a relatively new thing. Genetic evidence now strongly suggests that the *Corophium* in the northwest Atlantic got there from Europe, likely on muddy ballast rocks carried in the holds of European vessels in the seventeenth or eighteenth centuries. Because this small amphipod burrows and energetically rakes the mud for food, using two long, leg-like antenna—and because it numbers in the countless trillions—it's considered an "ecosystem engineer," a species that, like humans and beavers, can radically reshape its environment. No one knows if *Corophium* displaced other invertebrates on which the sandpipers once depended, or if the birds have shifted their migratory strategy—as well as their bodies' physiologic approach to migration—to take advantage of this novel situation.

several times a year, sometimes for decades—yet they seem to suffer none of the consequences that would plague a human, including increased risk for hypertension, heart disease, and stroke. Their blood chemistry during migration season shows many of the same warning signs seen in a person with diabetes or coronary disease, but without the negative consequences. By any typical measure, a migratory bird ready for travel ought to head to the ER, not the skies. "By human standards, premigratory birds are obese, diabetic, and likely to drop dead of a heart attack at any moment," in the words of two researchers studying this phenomenon. How birds protect themselves is still a mystery, but researchers hope that insights from avian physiology may help unlock new treatments and preventive approaches in people.

Carrying enough energy and water are only two of the hurdles a long-distance migrant must overcome. Flapping flight also requires huge amounts of oxygen, and the fastest, most energetic fliers are operating at almost 90 percent of their maximum rate of oxygen consumption—more than twice that of a comparably sized mammal. Godwits increase the number of red blood cells in their circulatory system before migration, allowing them to extract more oxygen with every breath (and, unlike in human runners who train at high altitudes to produce the same result, this change occurs at sea level, with no exercise on the bird's part). This also helps offset the thinner atmosphere 9,000–10,000 feet up where the godwits typically fly, where they gain the benefit of cooler (and thus less dehydrating) temperatures. But what about birds that fly far higher than that? Bar-headed geese and ruddy shelducks have both been documented flying over the Himalayas at altitudes of 4.5 miles, where the effective oxygen level is only a half to a third that of sea level. (The percentage of oxygen in the atmosphere is actually constant regardless of elevation, but the decreasing air pressure as one rises makes it progressively harder for the lungs to exchange gases.) Humans without supplemental oxygen can suffer confusion from hypoxia, and seem to suffer long-term cognitive damage even after they return to lower elevations; worse, the altitude can cause cerebral edema as the brain swells, or pulmonary edema as the lungs fill with fluid, both of which can be rapidly fatal. Even elite climbers on the flanks of Everest find taking a single step a lurching, exhausting battle—but may look up and see geese, cranes, or ducks passing overhead.

Bar-headed geese, the Himalayan migrants that have been most closely studied, face a suite of daunting challenges. Although they try to thread their way through lower valleys where they can, even these routes lie at what would, anywhere else in the world, be considered ridiculously high altitudes. Not only is the available oxygen scarce, the low air pressure means they have to work considerably harder with each flap of their wings just to stay aloft—and this species, crossing the world's highest mountain range, undertakes the longest powered ascent of any bird we know of. In spring, as the flocks rise from the Indian lowlands and confront the Himalayas, they claw their way up at an average of more than 3,000 feet an hour (and in one remarkable case, more than 7,200 feet an hour) for more than three hours. That is all the more astonishing when one considers that human climbers require weeks or months to acclimate to such high altitudes.

Birds in general, and bar-headed geese in particular, have advantages even elite climbers lack. The avian respiration system is vastly more efficient than human lungs. Instead of drawing air into, and expelling it from, a dead-end system like ours, where the exchange of old air and new is as low as 5 percent, birds have a series of air sacs, distributed through the body cavity and even into the leg and wing bones, which connect with the lungs. With an inhalation, fresh air from the trachea enters not the lungs first but the air sacs in the rear of the body. The exhalation that follows moves that air into the lungs, which are far denser than mammalian lungs and have greater surface area for gas exchange. The next inhalation, which brings a new batch of fresh air into the rear sacs, also moves the depleted air from the lungs to the forward air sacs, while the final exhalation of the four-beat sequence pushes the depleted air out of the front sacs as it carries fresh air into the lungs. Not only is this unidirectional respiration setup dramatically more efficient at processing oxygen than the mammalian model, it is inherently more resistant to pulmonary edema, one of the dangers of high-altitude exercise. A bird's heart is also proportionately larger, its muscles have a greater capillary density than ours and exchange oxygen at the cellular level more efficiently, and its brain cells appear to tolerate a lack of oxygen better than a mammal's, although it's not known if they are more resistant to cerebral edema.

Scientists have recently found that in addition to these general avian advantages, bar-headed geese have evolved a suite of specific adaptations that help them on their rarified journey. Even compared with other birds, the geese can tolerate extremely low blood oxygen levels—while resting, they can function at levels equivalent to the atmosphere at almost 40,000 feet. Their lungs are even larger than in other comparably sized waterfowl, they breathe more deeply and less frequently (which increases and improves gas exchange), and their hemoglobin is better at carrying oxygen. All this means a migrating goose's body can provide more dissolved oxygen in the blood for its mitochondria, the source of cellular energy.

You can be forgiven for thinking that evolution has consistently shaped migratory birds to be tougher and harder than their nonmigratory counterparts, but in one surprising way, the opposite may be true. Because long-distance migrants, moving through many habitats across wide distances, presumably face an unusually rich variety of diseases (especially during their time in the tropics), one might expect them to have especially robust immune systems. But by comparing the diversity of genes that recognize pathogens among closely related songbirds—some of them tropical African residents like red-backed pipits, some nonmigratory northern European birds like meadow pipits, and some, like tree pipits, migrants between the two regions—Emily O'Connor and her team at Lund University in Sweden found quite the opposite. The diversity of immune-response genes was quite low in migrants (although not as low as among year-round northern residents). The scientists speculate that the cost of carrying a powerful immune system, including heightened risk from autoimmune diseases like chronic inflammation, may outweigh the benefits to migrants. Their research also lends credence to what's known as the "pathogen escape" hypothesis, which suggests bird migration evolved in part from pressure to move away from tropical regions where disease rates are high, lowering the risk to vulnerable young chicks. O'Connor and her colleagues strengthened that line of reasoning by comparing the genetics of more than 1,300 species of songbirds in Eurasia and Africa, again looking at closely related species groups; they found that African origins were 16 times more common than the reverse, and that only a few of what are now African resident species originated in the north. (This so-called northern home theory has long been thought to be the case in the Western Hemisphere as well, but recent modeling, based on a newly detailed genetic family tree of more than 800 species of New World songbirds, suggests that long-distance migrations in the Americas were twice as likely to have begun when northern birds extended their winter ranges farther and farther south, rather than the reverse. Once in the tropics, so this reasoning goes, these birds radiated into the diversity of resident tropical species found there today.)

Isla Genovesa is the northeastern-most of the Galapagos Islands, 600 miles out in the Pacific off the coast of Ecuador. It is, like all of the islands in that archipelago, volcanic in origin and is, like most, rather unforgiving in nature, its vegetation sparse and low, clinging to the island's cracked and rumpled lava rock surface. It is rimmed by sheer cliffs and pounded by heavy surf around all of its somewhat horseshoe-shaped perimeter, except for a collapsed caldera, the remnant of an ancient eruption. Breached by the ocean on one side, the caldera forms Darwin Bay, a sheltered anchorage a mile across with a small beach tucked behind an outcropping of black rock that pokes into the water, forming a natural seawall.

Genovesa is known as "Bird Island" because it hosts some of the largest nesting colonies of seabirds in the Galapagos—hundreds of thousands of red-footed, blue-footed, and Nazca boobies; thousands of ethereal red-billed tropicbirds with long, slender tail plumes that twist and shiver in the air; and immense colonies of wedge-rumped storm-petrels as big as plump swallows, which nest down among crevices in the lava rock—and which, unlike most of the world's storm-petrels, come and go by day, because their primary predators on this island are short-eared owls, nocturnal and as dark and sooty as the outcroppings among which they perch. We got an early taste of Genovesa's charms even before we arrived, as our dinghy slapped across the waves of the bay and neared shore. A single red-footed booby flew out to meet us, circled once, and folded its long, black wings to settle on the head of one of the women in our group, its crimson feet splayed on her hat. The white bird faced into the breeze with a look of nonchalance that the tourist—her eyes huge and her mouth hanging open in shock and delight—did not share. It was the group's first day in the Galapagos, and they were quickly learning that the island chain's reputation as a wildlife Eden was not promotional hype.

For the next two hours everyone wandered in a happy daze, watching charcoal-black marine iguanas clamber up from the surf like miniature Godzillas, and booby parents feeding their downy white chicks. Endemic cactus-finches, one of the 13 species of famous Darwin finches, foraged in prickly pears. But while I was charmed by all of them, the birds that fascinated me the most that morning were the great frigatebirds, which shaded their chicks in stick nests built in the low trees and shrubs, or hung effortlessly on the strong sea breezes sweeping the island.

Everything about a frigatebird is long, as though someone had stretched a normal seabird well past any common-sense proportions. In flight, the most arresting feature is its astounding wings, spanning 7.5 feet but barely a hand's-width across, crooked and bowed and wickedly tapered. The tail is long and deeply forked, the neck (though usually tucked low to the shoulders in flight) has the reach of a heron when lunging for food, which it snags in a slender, hook-tipped beak half again longer than its head. The male is black with a glossy, oily green sheen and a sliver of crimson skin down the middle of its throat that it can, when displaying, inflate into a soccer-ball-sized sac. The female has a white throat and breast, the juveniles all-white heads. But regardless of age, frigatebirds have the lowest ratio of body mass to wing surface area—what's known as wing-loading—of any bird; they are so pared-down for flight that their skeletons weigh less than their feathers. Evolution has sculpted them into consummate aerialists, unrivaled in their ability to capture rising thermals of air and soar without a flap.

A male frigatebird glided into a nest, supplanting the female, which flapped heavily into the air, then pivoted in the breeze and zoomed out of sight. The chick—about three weeks old, judging from the mantle of dark back feathers emerging from an otherwise white coat of scraggly down—made begging motions. When the male opened his mouth and leaned forward, the youngster plunged its head partway down its father's throat, gulping as the adult regurgitated again and again, pumping up the prey it carried in its stomach. Once the feeding binge finally ended, both the chick and parent sagged against one another, and both soon fell asleep.

A sleeping bird may not sound exciting, but just a few weeks earlier, an international team of scientists had published a remarkable paper about these very birds, the nesting frigatebirds of Genovesa, which had made headlines around the world. Although frigatebirds are not migratory, the foraging trips like the one from which the male had just returned (and on which his mate had just embarked) can last a week or more and cover thousands of miles over the empty ocean, during which the frigatebird, which is not waterproof, cannot land. How frigatebirds deal with the matter of sleep, it turns out, was not just cool science in its own right, but cast exciting light on one of the biggest questions involving long-distance migrants—the avian ability to forgo and reshape sleep, or avoid the consequences of missing it, in response to the demands of very long nonstop flights.

The frigatebird study was headed by Neils Rattenborg, an American scientist who directs a research program at Germany's Max Planck Institute devoted to understanding sleep by studying birds. Rattenborg and his team captured 15 nesting female frigatebirds on Genovesa and anesthetized the birds to attach EEG sensors to monitor brain activity; a small logger containing a tiny accelerometer was temporarily glued to each bird's head to collect the data, and a GPS tracker was attached to its back feathers. Although females were selected because they are larger than the males, the devices together weighed less than 1 percent of each bird's total mass. After a recovery period (during which their chicks were kept warm and safe) the adults were returned to their nests. Once their mates came back from sea, the tagged frigatebirds set out on foraging trips themselves.

The team, upon removing the devices once the frigatebirds returned, found that such hunting flights averaged about six days but might last as long as 10, during which the frigatebirds covered more than 1,800 miles, scribing a clockwise loop far out to the northeast of the Galapagos. During their time away from the island, the frigatebirds slept an average of just 42 minutes in each 24-hour period, usually dozing just after sunset, catching a rising thermal far above the water that lofted them ever higher. These were the definitions of power naps, averaging just 12 seconds each—and while the EEG readings showed that the bird's entire brain sometimes slept, more often only one-half of the brain lapsed into sleep. The other half remained awake—usually the hemisphere connected to the eye facing the direction in which the gently circling bird was heading. It was, Rattenborg realized, very similar to something he'd documented in mallards during his doctoral research. Those ducks on the edge of a flock kept one eye open, and the associated brain hemisphere active—the eye facing away from the flock, toward possible danger. In the frigatebirds, it probably wasn't the risk of a predator, or even the chance of colliding with another bird. Great frigatebirds carefully follow ocean eddies, where the hunting for flying fish and squid is best. Rattenborg and his team speculate that the mini-naps allow the birds to track such eddies through the night, to be in the best location to hunt come morning.

Unihemispheric sleep, as it's known, has been documented in marine mammals like dolphins and manatees, which must consciously take and expel each breath. Recently, a somewhat analogous condition has been found in humans as well. Most of us have experienced a poor night's sleep the first time we stay somewhere new; it's common enough that sleep scientists refer to it as the first-night effect. Scientists at Brown University and the Georgia Institute of Technology found that under such circumstances, one brain hemisphere remains, if not exactly awake, at least "less-sleeping," in their words, and more sensitive to stimuli—not fully unihemispheric sleep as birds exhibit it, but a closer match than had been realized.

The flying frigatebirds mostly exhibited what is known as slow-wave or deep sleep, but at times they entered rapid-eye movement, or REM, sleep—the form of sleep in which humans dream, and during which land mammals lose muscle tone and control. The latter would be disastrous for a flying bird, and the tagged frigatebirds were somehow able to remain in controlled flight during REM sleep, perhaps because REM lasts mere seconds in birds, compared with 20 minutes or more in a sleeping human. However they manage it, this system works for the frigatebirds, and not only on weeklong foraging trips. Great frigatebirds off the coast of Madagascar, tagged with satellite transmitters, have spent two months in continuous flight, catching turbulent updrafts within cumulus clouds that lofted them 2.5 miles up in the sky, allowing them to glide for hours looking for the next updraft. Once Rattenborg's study subjects returned to their nests on Genovesa Island, they would spend up to 13 hours a day sleeping, apparently recuperating; how long it takes a frigatebird to recover from the sleep loss on a two-month flight—or whether they even need to—is still unknown.

Frigatebirds are large enough to carry data loggers; most migrating birds are not. Consequently, we know little about how other birds handle the conflicting demands of sleep and flight, but we do know that they seem astoundingly resistant to the problems that come with limited sleep. Pectoral sandpipers, which migrate from southern and western South America to parts of the Arctic stretching from western Canada to Russia, must already be sleep-deprived when they arrive on the nesting grounds. But males then launch into a 24/7 marathon of courtship and territorial defense, mating with as many females as time and energy allow. A team including Rattenborg found that one especially ambitious male was awake 95 percent of the time for 19 straight days—and far from suffering for it, a male sandpiper's reproductive success is an almost perfect mirror of his ability to avoid sleep: less sleep, more offspring. For migratory songbirds, like white-throated sparrows and hermit thrushes, the onset of migration season brings a restlessness (known by the German term *zugunruhe*) that decreases the amount of time they sleep by two-thirds, even in captivity, and well before they start migrating. They may compensate by taking micronaps during the day, and birds that travel primarily over land may be able to more easily balance the need to fly, eat, and sleep than those crossing wide barriers like oceans or deserts. But even when such birds are experimentally prevented from sleeping, they fail to develop the loss of cognitive function that is the classic sign of sleep deprivation—but only at the right time of year. White-crowned sparrows trained to peck a lighted key for food were, if sleep-deprived outside of migration season, just as woolly-headed and bumbling as any human who just pulled an all-nighter. But during spring and autumn, those same sparrows not only maintained their ability to correctly peck for treats—their response time actually showed a dramatic improvement. Like a number of other aspects of avian physiology during migration, this has remarkable similarities to mania in humans, suggesting research avenues for human physiologists.

Scientists don't yet understand the biological mechanisms that safeguard birds, at least during certain times of the year, from the debilitating effects of sleep loss, but it may help that a bird grows more brain before it migrates—or at least, that it grows more of the neurons that store spatial information. Brain growth in birds is actu-

ally fairly common. Among male songbirds, just as their testes balloon in size with spring so, too, do the parts of the brain that produce
and respond to song. Chickadees, whose lives depend on their ability
to store and later retrieve food during the winter, experience a 30 percent increase in the size of the hippocampus, that part of the brain
that processes spatial information and memory, in autumn. A bigger
brain would seem to be an advantage to a migratory bird, given that
it must navigate immense distances, so it's surprising that migrants
actually have smaller brains, proportionate to their body size, than
do resident species that remain in the cold climates throughout the
winter. It may be that it's too much work to lug a big, heavy brain
across thousands of miles, or perhaps it's because brains are meta-
bolically expensive, consuming a lot of energy that would better fuel
flight muscles. But research suggests the disparity between migratory
and resident bird brains has more to do with resident birds evolving
bigger brains, to deal with the dramatically different challenges of
different seasons, than migrants evolving smaller ones.

Even if a migrant's brain is smaller, though, it has more horsepower
where it counts—in the hippocampus, the seat of spatial awareness.
The hippocampus of a dark-eyed junco that migrates from southern
Canada to the southeastern United States is more densely packed with
neurons than that of a nonmigratory junco that spends its life on a
peak in the southern Appalachians, and the migratory junco performs
better on spatial memory tests. Migratory birds grow fresh neurons
before autumn migration; when scientists compared the brains of reed
warblers from Europe with those of nonmigratory clamorous warblers,
the former had many more newly grown neurons. (Incidentally, neuronal growth isn't restricted to birds—despite what you likely learned
in high school biology, humans can grow new neurons as well.) That
same international team, led by Shay Barkan from Tel-Aviv University
in Israel, also found that the density of neurons was linked with how
far a species migrates—among reed warblers and turtle doves that they
studied, those individuals that migrated the farthest (as determined by
subtle chemical isotope signatures in their feathers) also had the great-

est neuronal growth. There were differences in where the new neurons appeared, however. The warblers, which migrate alone and mostly at night, grew the majority of their new brain cells in the hippocampus, as expected. In doves, on the other hand, most of the growth occurred in a different brain region, the nidopallium caudolateral, which handles executive-level thinking and may be more important for social, often daytime migrants like turtle doves, which must watch and interpret the actions of their companions.

Think of it this way. The warbler may have more neurons processing navigational and spatial data, but the dove has buddies—and in a sense, the dozens or hundreds of doves flying together in a flock can act like individual cells in a navigational brain. Each dove has an onboard compass to direct it, and each bird's directional sense is, to a greater or lesser degree, somewhat inaccurate. No one's perfect, but by flying together, the doves average out their inaccuracies and arrive at a better, more precise collective decision than any one of them could produce alone. This is known as the "many wrong" theory (as in, many wrongs make a right), and it's the same kind of wisdom-of-crowds effect that can, as was first noted in 1906, lead hundreds of English fair-goers to correctly guess the weight of an ox to within 1 percent of the actual figure, when all their guesses are averaged.

Which brings us to the emerging science on navigation, where arguably the single most mind-blowing discovery about bird migration has been made. It's probably a good thing Einstein wasn't a birder, because this one wouldn't make him happy.

A migratory bird's ability to traverse thousands of miles is perhaps the greatest physiological feat of all, especially because almost all of them do it entirely on instinct, without help from parents or other adults; only a few groups of migratory birds, including waterfowl and cranes, travel in multigenerational flocks. The rest are born with a genetic road map that compels them to fly in a certain direction, for a certain length of time, at a certain time of the year. We know that they use a variety of cues, including landscape features like mountain ridges and coastlines; stellar guideposts like the stars (not the stars' position and pattern in the night sky, but their apparent lack of rotation around Polaris that indicates north), the movement of the sun across the sky, and the concurrent shift of bands of polarized light that are invisible to us but easily seen by birds; even "odor landscapes" of volatile chemicals that remain remarkably stable across hundreds of square miles, despite wind and weather, and remain so from season after season—sniffable road signs on the migratory highway.

Perhaps the most important of these migratory cues, but for many decades also the most mysterious, is magnetic orientation. Researchers had suspected since the 1850s, and confirmed by the 1960s, that birds possess a magnetic sense. If you want to be fancy, you can demonstrate this by gluing a tiny electromagnetic coil to the head of a homing pigeon, a gizmo that looks a lot like a dunce cap, thus creating a magnetic field that overwhelms the less powerful effects of the earth's own field, disorienting the pigeon, which is unable to find its way home. (And if you lack access to one of these miniature Helmholtz coils, as they're known, a simple bar magnet, attached to the pigeon's back, will do the same thing.) The seat of this magnetic sense was long thought to be tiny deposits of magnetite, a magnetic, crystalline form of iron oxide that had been found in the upper beaks of a variety of birds. Sitting in my undergraduate ornithology class 40 years ago, reading about magnetite in the beak of a bobolink, it was easy to imagine the iron crystals acting like an onboard compass, tugging the bird's nose northward. It was a tidy image, but there were a couple of major hitches in this explanation. For one thing, experiments showed that birds weren't responding, as does a compass needle, to polarity—that is, to the north-south alignment of the planetary magnetic field. Instead, they appeared to be detecting inclination, the angle at which the magnetic field lines, emanating from the earth's core, intersect its surface at changing angles the closer to the poles or equator one goes. By the 1990s, scientists realized there was an even more inexplicable problem with the magnetite explanation.

For reasons no one could explain, a bird's magnetic compass worked just fine *except* when exposed to yellow or (especially) red light. Nor was it only birds; almost any animal with a magnetic sense—newts, fruit flies, whatever—similarly lost the ability to orient when bathed in red light.

Actually, it's not entirely true that no one could explain this phenomenon—it's just that the explanation seemed so bizarre that, when it was offered (coincidentally, the same year I was studying my ornithology text) essentially no one took it very seriously. The editors of one prominent journal to which it was submitted suggested that the author toss the manuscript in the trash. He did not, but it was another four decades before many people paid attention to Klaus Schulten's idea.

In 1975, Schulten was a young postdoc physicist at the Max Planck Institute for Biophysical Chemistry in Göttingen, Germany, working with chemical reactions that were influenced by magnetic fields. He realized the reaction he was seeing in his test tubes—one in which so-called radical pairs, two molecules linked at a quantum level, were influenced by a common bar magnet—could play a role in the still-mysterious ability of birds to orient magnetically. The right molecule, in the right place in a bird's body and activated by either light or dark (he wasn't sure which), could create a chemical compass that would be sensitive to the earth's very weak magnetic field. So along with two colleagues, Schulten wrote a mathematically dense paper explaining the hypothesis and, in 1978, submitted it to the prestigious journal *Science*.

"I got the paper back with a rejection note saying, 'A less bold scientist may have designated this idea to the waste paper basket,'" Schulten told an interviewer in 2010. "I scratched my head and thought, 'This is either a great idea or entirely stupid.' I decided it was a great idea and published it quickly in a German journal!"

The publication did not, to put it mildly, make waves. Some of the current experts in the field today ascribe the universal yawn it received back then to the thicket of equations that Schulten and his colleagues laid out, which may have deterred biologists from tackling their core idea. It may also have been due in part to the fact that no one, including Schulten, knew of a molecule that embodied the necessary properties for this kind of light-induced magnetic perception. So Schulten went back to forging an astonishingly rich and varied career melding a variety of disciplines into computational biophysics—using supercomputers to simulate the 64 million atoms in the protein shell of the HIV virus, for example. But he did not let go of his idea about magnetic orientation. In 2000, by which time he was heading a number of major research groups at the University of Illinois, Schulten circled back to the problem, because by then, someone had noted that a photoreceptor protein called cryptochrome might be a fit for his mystery molecule. Schulten coauthored a paper that laid out his argument in ways both more detailed and more easily accessible to nonphysicists. This time the scientific world paid attention, and a recent and rapidly accelerating torrent of published research makes most experts confident that Schulten had, indeed, found the grail of magnetoreception.

It's a pretty weird grail, but then, so are most things in the quantum realm. Here is the framework, as it's currently understood. A migrating bird, flapping through the night sky, glances up at the stars. A photon, having left one of those stars millions or even billions of years earlier, enters the bird's eye and strikes a molecule of a form of cryptochrome, almost certainly a specific variant known as cryptochrome 1a, or Cry1a. This encounter takes place in the retina, probably within a set of specialized vision cells known as double-cones, whose function had heretofore been a mystery. The photon knocks free one of the Cry1a's electrons, kicking that electron into a neighboring Cry1a; because they now each have an odd number of electrons, the two molecules are known as a radical pair, and are linked—entangled, in the jargon of quantum mechanics. They are also magnetic, because the electrons have a property known as spin (which isn't really spin, in the way you'd picture a top; it's actually a state known as spin angular momentum, but never mind—let's not go down too many quantum rabbit holes right now). Such entangled particles are joined regardless of distance, defying classical physics and common sense. They have become, in effect, one thing; if you measure the properties of one—even were they separated by millions of light-years—you could infer the properties of the other.

Einstein, whose own work helped spawn the concept, famously rebelled against this idea of entanglement, which he dismissed in the 1930s as "spooky action at a distance." Yet experiments have borne it out. In the eye of a migrating bird, the effect of countless radical pairs probably creates a dim shape or smudge—visible as the bird moves its head, but not opaque enough to interfere with normal vision—that shifts with the bird's position relative to the ground and to the inclination of the magnetic field lines arcing out of the planet. But if you've heard about entanglement, it's likely because of the seriously strange uses to which it's being put. In 2017, for example, Chinese scientists used entanglement to "teleport" two entangled photons—or, at least, the information in those photons—from an orbiting satellite to ground stations more than 700 miles apart. And in 2020, the same Chinese team said they used quantum entanglement to transmit an unbreakable encryption code to a satellite. These are a long way from beaming up a *Star Trek* character, but they were hailed as the first steps to an unhackable quantum internet, and possibly even faster-than-light communication. (Ironically, entanglement itself may not be essential to the process that allows birds to see a magnetic field. This strange branch of quantum theory may be, in the words of two leading researchers, "something one gets 'for free' in a cryptochrome," not essential for the molecule to act as a magnetic compass.)

I was anxious to visit Klaus Schulten at his labs at the Beckman Institute for Advanced Science and Technology at the University of Illinois Urbana-Champaign. Having been in touch with him previously, I sent him a follow-up email toward the end of 2016 to discuss possible dates—and thus was not prepared for the automated response I received, notifying me that Dr. Schulten had died a few weeks earlier at the age of 69. His obituaries justly lauded his profound impact on computational biology, pioneering techniques that allow scientists to model bogglingly complex living systems, but for a birder like me, his greatest discovery wings through the night skies twice a year.

It now seems all but certain that radical pairs, Cry1a, and quantum entanglement provide a flying bird with its magnetic compass sense. But birds possess a second magnetic ability, a map sense that allows them to navigate as well as orient, and radical pair theory doesn't explain that. What about those little deposits of magnetite in a bird's beak, the structures that I, daydreaming over my ornithology text, once pictured as a nasal compass? One team of researchers concluded that they weren't even magnetite at all—they contend the structures are iron-rich macrophages, a type of white blood cell that is part of the avian immune system, masquerading as magnetite because of the staining process used to prepare specimen slides. Macrophages play no known role in orientation. Other scientists reject that conclusion, and note that running through the upper beak is the trigeminal nerve, which seems to somehow provide the map sense that birds use. Eurasian reed warblers, caught near Kaliningrad and exposed to a magnetic field identical to that found 1,000 kilometers east across Russia, reorient themselves just fine and attempt to migrate in the correct direction to their breeding grounds in Scandinavia—but if part of their trigeminal nerve is surgically cut under anesthesia, the birds can't account for the displacement, and orient as though they were still on the shores of the Baltic Sea. Such invasive experiments, uncomfortable as they may be to read about, show how little we still know about the underpinnings of avian orientation. As the scientists who conducted the Russian study concluded, migratory birds possess "a second magnetic sense with unknown biological function." Inside each mystery, there is always a new one waiting to be found.

Sometimes—maybe most of the time—the most startling discoveries are wholly unexpected. That was the case in 2011, when ornithologists attached tiny data loggers to alpine swifts at a breeding colony in Baden, Switzerland. Swifts are the most aerial of all birds, built like blunt cigars with scimitar wings, their feet so small and almost vestigial that all they can do is cling to upright surfaces like cliffs, caves, and the insides of hollow trees; they cannot perch, they cannot walk, and swifts even mate on the wing. Alpine swifts are an unusually large species with a distinctive white throat patch and a wingspan of 22 inches, easily capable of carrying the devices, which weighed slightly more than a gram, and had two components— a geolocator that would, once it was recovered the following year, allow the scientists to reconstruct the birds' migration route to Africa in the intervening months, and a sensor that recorded each bird's wingbeat activity and the angle at which it was holding its body. That way, the team could determine the swifts' daily time budget—how many hours they flew and foraged, how many they roosted or slept.

The next spring, three of the tagged swifts returned to Baden— and the data contained in their loggers were a jaw-dropper. For the period when the swifts were in Switzerland, there was a clear day/night pattern of diurnal flights and nocturnal rest. But once they began migrating south, crossing the Mediterranean and the Sahara, down into west Africa, that pattern vanished. Day or night, for 200 days— more than six months—the swifts appeared not to have landed at all. This surely ranked as one of the most—perhaps even *the* most— extraordinary and unexpected physical feats in the natural world, but the alpine swift was soon knocked off its pedestal. Three years later, scientists announced that common swifts, similarly tagged in Sweden, spent fully 10 months on the wing in their west African wintering grounds—confirming a suspicion about that species that some naturalists had expressed dating back almost a century. More recently still, another team confirmed the same behavior among pallid swifts that breed in the Mediterranean. (North American species, like the chimney swift of the East and the black and Vaux's swifts of the West, apparently do roost and rest during the nonbreeding season.)

How do they do it? As aerial insectivores, finding food aloft is no problem for swifts, and given what we know of frigatebirds, it's likely the swifts are also practicing unihemispheric sleep. What's more, swifts surpass even the masterfully gliding and soaring frigatebirds at extreme energy conservation while aloft. A recent study of common swifts found that they glided nearly three-quarters of the time, and they were so skilled at playing off changes in the airflow from rising thermals and other currents that their overall energy expenditure was "not significantly different from [zero]." A human, eating quietly at the dinner table, loses more net energy than a swift swooping around 1,000 feet above the African plains.

Even for migratory birds, which routinely shatter our expectations, such discoveries set a new benchmark for the utterly incomprehensible. Most common swifts live five or six years, but one banded bird survived for 18 years—meaning that in the course of its life it flew some 4 million miles, most of that without ever landing. Few ornithologists think such previously unimaginable discoveries are all behind us. As we'll see, tracking technology gets ever tinier as Big Data grows ever bigger, and their roles are converging in ways that profoundly affect our ability to study migration—and to be continually awed by it.

Three

WE USED TO THINK

The discovery that common swifts fly nonstop for up to 10 months would not have come as a surprise to Ronald Lockley. The Welsh ornithologist (perhaps most famous today for his studies of wild rabbits, which became the inspiration for the novel *Watership Down*) made just that suggestion in 1969, prompted in part by his observations of large flocks of swifts at his home in Devon, ascending vertically at dusk until they passed beyond his vision. Lockley knew that a French pilot in World War I, gliding at night almost 10,000 feet above the ground with his engine off, had found himself surrounded by swifts, and that earlier naturalists all the way back to Gilbert White in the eighteenth century had suspicions about how infrequently swifts came to earth. "Perhaps other observers, suspecting this continuous aerial existence, have been less willing to risk a sweeping generalization such as I may now appear to be making," Lockley told his audience at an ornithological congress in South Africa. Lockley had no qualms about staking out a bold position.

Of course, it's one thing to suspect something and another to confirm it, and Lockley, who died in 2000 at the age of 96, did not live to see his prediction come true. In 1969, there was no practical way to track a bird as small as a 1.3-ounce swift. There were radio transmitters, some just barely small enough for the task, but they required a person to maintain constant line-of-sight contact with the signal, using a handheld receiver. That's hard enough with a relatively stationary subject, but if the bird is migrating, you need an airplane—

or, more realistically, a number of them. Even then, it is incredibly difficult to maintain contact with a constantly moving target.

Miniaturization of electronics, especially the power source in batteries and solar panels, has been a game changer for migration research, but it's just one of many advances that are reshaping the field. I've been fortunate enough to have seen, firsthand, how many of these innovations have transformed our understanding of migration, and have been lucky enough to have been directly involved with the groundbreaking use of some of these new technologies. It's a heady time, especially for those who study the movements of birds that have always been far too small to fit with any kind of traditional transmitter. For the first time we have the ability to follow individual birds, of even some of the smallest species, through their annual cycles of breeding, migration, and wintering. This new capability has uncovered previously unrecognized threats, in some cases finally explaining long-standing, once-mysterious declines, and giving conservationists a roadmap for reversing the damage. We've learned the ways that places thousands of miles apart are tightly and inextricably bound by very specific, very local migratory connections, and it's made us realize how fragmentary our understanding has been of critical life-stages—how some of our efforts at protection have been incomplete at best or counterproductive at worst. This is an exciting moment—but for many species on the edge, it remains to be seen if this rush of new information has come in the nick of time.

"You ready, Todd?" Dave Brinker yelled, shading his eyes as he peered up, sunlight shining through his white beard. Forty feet above our heads, a figure in a bucket lift gave him a thumbs-up in reply.

"All right, nice and easy," Dave said, as he and I took up the slack on a long rope that angled from the lift down to the ground, and began to pull. A nine-foot-long metal antenna arm, with short metal crosspieces every foot along its length, began to rise slowly, swaying in the hot August breeze, inching higher and higher toward the top of an old utility pole, where Todd Alleger reached out and gingerly untied it. With expert movements, the young man slid the antenna's center bracket down over a metal mast that rose another six feet above the top of the weathered pole, positioning it so the arm pointed due north, then tightened the bolts to secure it.

Over the course of the next hour, we sent three more of the long antenna arms up to the top of the pole, which stood behind a Pennsylvania Game Commission field office surrounded by cornfields and fencerows, a couple of hours northwest of Philadelphia. If any of the drivers zipping by on the busy state highway a stone's toss away noticed, they may have wondered why anyone in this age of high-def digital streaming was bothering to erect an old-fashioned television antenna. That's certainly what it looked like, but this gleaming metal array was actually part of a revolutionary effort to harness a rather old technology in an innovative way, allowing scientists for the first time to track even the smallest migrants across incredible distances.

Radiotelemetry has been around for ages; attaching VHF radio emitters to animals was the first transmitter technology used to study wildlife movements. Anyone who has watched a nature documentary understands the principle: the small radio broadcasts a signal, which a biologist with a receiver and handheld directional antenna, known as a yagi, picks up. (Those old TV antennas that used to loom over every suburban home are just collections of oversized yagis.) Point the yagi toward the radio and the beeps get louder, swing it away and they get quieter; voilà, you know the direction in which your transmitter lies. This works fine so far as it goes, but it's incredibly labor-intensive. The receiver/antenna and radio must be within line of sight of each other (because mountains, buildings, dense forest, or the like block the signal), and tracking requires the minute-to-minute involvement of a trained worker.

It's hard enough if you're tracking a deer or a bear, but following a migrant bird across the landscape adds further complications, unless you have an air force at your disposal—and even then, it's a challenge. In the late 1980s I was part of a research team tracking red-tailed hawks on their autumn migration. My job was catching the birds and affixing a transmitter to their central tail feathers, and I did not envy the chase teams, who often spent 10 or 11 days at a stretch on the road, eating junk food, catching sleep in snatches, forever being left behind by their vastly more fleet and mobile aerial targets while the humans were snarled in traffic or navigating unfamiliar roads on the ground. Many days the chase team would lose the tagged hawk's signal and call in our own air force—a retired engineer and private pilot named Frank Masters, who would on almost no notice jump into his single-engine plane in central Pennsylvania and fly to, say, western Virginia or eastern North Carolina. Using yagis mounted on the wing struts of his plane, Frank would relocate the signal from the air, land, contact the chase team with the coordinates, and fly back home before daybreak. Chances were good he'd do it all over again the very next night.

Radiotelemetry still works well for many wildlife studies; my owl research team used it to great effect for many years to study the stop-over ecology of northern saw-whet owls in the central Appalachians, locating the soda-can-sized birds on their day roosts or using a team approach to triangulate their movements around a limited territory at night. But to follow a bird in migration, scientists turned increasingly to technologies like satellite transmitters, which communicate with the orbiting Argos system and give you the location of a transmitter anywhere in the world. But all such technologies come with tradeoffs. With VHF radios, which are cheap and small, it's the labor; with satellites, it's weight and cost. For many years, satellite transmitters were too heavy to use on anything smaller than a medium-sized raptor, and even today, the smallest units are generally about five grams, meaning they can't be used on a bird smaller than about 5.5 ounces—the size of a plover or large sandpiper. That leaves many thousands of species of small birds, the vast majority of the world's migrants, out of the picture. Sat-tags are also expensive—thousands of dollars per unit, plus thousands more in annual fees for satellite time.

Recently, though, miniaturization and automation converged to give VHF radiotelemetry a new and exciting life. With highly efficient batteries, it's now possible to build radio transmitters that weigh a tiny fraction of a gram, small enough to deploy on a hummingbird and even large migratory insects like monarch butterflies and some dragonflies. Combined with automated receiver stations, it's possible to create a global tracking network that, for the first time, allows us to follow the hemispheric movements of even the smallest migrants. It's known as the Motus (from the Latin word for movement) Wildlife Tracking System, and it was the brainchild of Birds Canada (formerly Bird Studies Canada), that country's largest bird conservation organization. Biologist Stu Mackenzie and his colleagues began experimenting with tiny transmitters known as nanotags and automated receivers in 2012—the same year, by coincidence, that Dave Brinker, a biologist with Maryland's Department of Natural Resources, and I were tinkering with them in the mountains of Pennsylvania to track owl migration. At that time, Bird Studies Canada quickly recognized the enormous potential this new tech had to upend migration research. The tags are cheap, costing only a few hundred (versus a few thousand) dollars each, and all of them broadcast on the same frequency, transmitting an identification code unique to that tag, so any tag can be detected by any receiver. Each receiver station—an array of directional antennas like the one we were installing, connected to a very basic computer, a GPS receiver, and a few other instruments, with a solar power source if it's in a remote area—costs less than $5,000 and runs itself. Scientists have seen the potential, too, and in less than a decade Motus has exploded into a network of nearly 1,000 receiver stations from the Arctic to southern South America, and increasingly in Europe, Africa, Asia, and Australia, tracking tens of thousands of birds, bats, and bugs at a level of detail no one could have imagined.

A lot of those receiver stations have been erected by scientists with specific, local projects—tracking the movements of terns foraging along the New England coast, for example, or studying the roost-

ing behavior of bank swallows in southern Ontario. But because the
receivers pick up any tagged animal passing within range (which,
depending on topography and weather, can be 15 or 20 miles), any
new station, regardless of its immediate purpose, helps the overall
effort. If a semipalmated sandpiper tagged in the Canadian Arctic
flies through Ontario, the bank swallow towers may pick it up, as
may a tower along the Georgia coast erected specifically to follow
warbler migration.

Dave, Todd, and I are part of a group, known as the Northeast
Motus Collaboration, that's coming at it from a different angle.
Although we want to use the Motus network for our own research
(which varies from studying owl migration to songbird stopover ecology to whether birds that have collided with windows and survived
suffer any impairment in their ability to navigate), we also see the
value of building out the network for the general benefit of migration
science. So since 2015 we've raised private and foundation money,
along with state and federal grants, to build increasingly large regional
receiver arrays in the interior northeastern United States. That tower
we were installing on top of the old utility pole was part of our first
phase in 2017, where we erected 20 stations diagonally across Pennsylvania, every 30 or so miles from near Philadelphia to Lake Erie.
Since then we've expanded, first across the mid-Atlantic region and
New York, and more recently throughout New England.

There's a little bit of kid-at-Christmastime feel whenever I visit one
of the stations to download the accumulated data. In early December
I was back at that newly installed receiver, where I opened a green
weatherproof plastic case, connected my laptop to the small computer inside to check its status—all normal—and then pulled the
stored data from its memory card. I uploaded the data bundle to
the Motus website, and a short while later received the processed
data—a glimpse into what had been passing overhead, unheeded by
the humans sleeping below, during the preceding autumn migration.

Lots of thrushes, for one thing—several scientists in eastern Canada were using Motus technology to study that group of migrants.

One team was tagging Swainson's thrushes in Nova Scotia to learn more about large, regional movements the birds make after they finish breeding but before migrating to South America, and the receiver had picked up more than a dozen of them, passing through over a several-week period. Another researcher, this time in Quebec, was using the Motus network to see whether several species of thrushes, after breeding, were moving to shrubbier areas rich in fruit to fatten for migration; we'd detected three of their gray-cheeked thrushes passing by, as well as half a dozen Swainson's thrushes and a Tennessee warbler from a different project in Montreal investigating the importance of small habitat patches in an urban landscape. In all these cases the main focus of the research was very local, but because the network is hemispheric, most of these tagged birds would be detected repeatedly by other stations as they migrated south—and if their batteries lasted long enough, again heading north in the spring.

Let's see, what else . . . several barn swallows and cliff swallows from colonies in southern Ontario; white-rumped and least sandpipers, and lesser yellowlegs, tagged by a shorebird research project on James Bay in the Canadian subarctic; silver-haired bats and a monarch butterfly tagged along the north shore of Lake Erie; and an American woodcock tagged the previous winter on the Eastern Shore of Virginia, and heading back south again to its wintering grounds after a summer in eastern Canada. There was a Virginia rail (an incredibly secretive species about which little is known) that had been marked the previous spring in western Ohio heading north, and was now on its way south by a very different easterly route, as well as two common nighthawks tagged in southern Canada, whose tracks all the way down to Florida and into northern South America were picked up by a succession of towers after passing us. I knew about these subsequent hits and far-flung movements, incidentally, because the data that Motus collects are, for the most part, available to the public to explore on the project's website, www.motus.org. Like me, anyone can click on an individual receiver location, look up what's been detected there, and map the movements of any of the animals on the list.

Motus's impact, in just a short time, has been incredible. Much of what we know about migration has come from bird banding, which (as I know from 35 years of such work) is rewarding but laborious. Since 1960, more than 64 million birds have been ringed in North America, of which only a small percentage have been encountered again. For waterfowl, the ratio is high; for example, some 4.6 million mallards have been banded since 1960, and almost a quarter have been reencountered, most by hunters. For Swainson's thrushes, more than half a million of which have been banded, the encounter rate is a miniscule 0.4 percent, while for black-throated green warblers, one of the most common birds in northern forests, it is an even tinier 0.08 percent. By contrast, in just a few years, Motus has generated more than 1.5 billion detections from more than 17,000 tagged animals, mostly birds—an astoundingly detailed record, rich in terms of both time and space, that complements the banding data, revealing previously unrecognized migration routes and stopover sites, and forcing us to notice what we've long ignored. For example, our line of stations stretching across Pennsylvania—a functionally landlocked state with few expansive areas of wetlands—has shown that each May and early June, large numbers of migrant shorebirds leaving the Atlantic coast pour through Pennsylvania's skies, including many federally threatened red knots that were tagged on the Delaware Bay. No one ever really considered that fact, which means—to take one example—that decisions about siting industrial wind farms along high ridgelines haven't been taking into account the full measure of risk to migrant birds.

Motus also opens a window into aspects of a bird's life that were hidden before, and clues to where the biggest threats to them lie. Scientists are especially concerned about aerial insectivores, the guild of birds that includes swallows and swifts, whose numbers have plummeted worldwide. In southern Ontario, where Birds Canada has created a dense gridwork of Motus receivers evenly checkerboarding some 60,000 square kilometers of land, scientists used transmitters that each weighed just two-tenths of a gram—one-seven thousandth of an ounce—to tag more than 200 young barn swallows, then followed them for months after they left the nest, a job that would have been challenging to do with even a single swallow before Motus. Because it was nearly impossible, no one had ever really tried to find out how young swallows fared once they fledged—so no one knew, as these biologists learned, that independence is deadly. Even before the swallows' dangerous migration to Argentina began, almost 60 percent of the young birds had perished—an unsustainable rate of loss, and one that easily explains the barn swallows' crashing population. Just as importantly, it shows that at least some of the species' underlying problems lay close to home, not in distant parts of the world, prompting further work to discover and correct them.

The barn swallow study also shows how important it is to understand all of a migratory bird's life cycle, which we've rarely been able to do. Most of what we know about migrants comes through limited snapshots, the few places and times where their travels intersect with humans who take the time to notice, leaving us to try to imagine the wide landscape of their lives by peering through tiny, scattered peepholes. Only for a very few species do we have any real sense of their full, annual cycle—the routes, timing, habitats, and underlying resources that make such global journeys possible, and which support them even when they are hidden from human view. It seems that whenever we take a closer look, we make discoveries that challenge our assumptions—and even make us realize that our attempts to help may have been making things worse.

"You know what bothers me about scientists?" my mother asked some years ago. "They always say, 'We used to think, but now we *know*.'" If I remember correctly, she was ticked off over some flip-flopping research about diet and health, maybe the endless eggs-are-good/eggs-are-bad debate, but I have to admit, she had a point.

Science is a process, one where ideas are proposed, tested, and discarded if new evidence demands it. Any good researcher (if they're being true to the scientific method) ought to say, "We used to think, but now we think." Of course, human nature doesn't work that way. Even scientists tend to look for certainty, and often put their trust in the latest, sexiest research to be published, tacitly assuming it's the final word on a subject.

In the past four decades, the field of migratory bird conservation has experienced several we-used-to-think epiphanies. For more than a century, ornithology—which as a science had its roots in the northern urban centers of Europe and North America—had a decidedly temperate, breeding-season focus. Birds are easiest to study when they are tied to a nest, and breeding season is when they tend to be most colorful, vocal, and visible. It wasn't until 1977, when the Smithsonian Institution sponsored a symposium on migratory birds in the tropics, that major attention in the Western Hemisphere shifted for the first time from that blinkered focus on the breeding grounds, and a lot of eyes were opened to the fact that migrants spend the majority of their lives either in transit or on the wintering grounds. This was also the time when the degree of tropical deforestation became alarmingly clear. By the late 1980s and early '90s, North American conservationists were also growing increasingly concerned about nosedives among some groups of migrant birds, especially neotropical migrants that breed in the north but winter in Latin America and the Caribbean. Understandably, many conservationists assumed the problem lay on those besieged tropical wintering grounds. Neotrops (as they are known) were routinely enlisted as "Save the Rain Forest!" icons, used to implore the public to save the jungle in order to save the lovely birds that nest in their yards. (In fact, the loss of many kinds of tropical habitats besides lowland rain forest—mangrove swamps, cloud forests, savannahs, grassy wetlands, and especially high-elevation oak-pine forests where a majority of neotropical migrants wintered—was affecting birds.)

Meanwhile, a quietly growing body of research suggested by the

1990s that at least some of the blame lay much closer to home. Many of the neotrops in steepest decline appeared to be those known as forest-interior species—birds like wood thrushes, scarlet tanagers, and many warblers that breed deep inside large, contiguous tracts of forest, where they nest on or near the ground—a risky place if there are predators about, but the interiors of such forests have relatively few predators. Trouble was, there was little intact forest left, especially in eastern North America. Even the most heavily forested regions were tattered, moth-eaten blankets of woodland, fragmented into millions of pieces by roads, power lines, towns, clear-cuts, and developments. In such fragments, so-called edge predators—species like raccoons, skunks, opossums, house cats, crows, black snakes, blue jays, and grackles that are largely absent from contiguous forest—were abundant. So were brown-headed cowbirds, an otherwise open-country species that parasitizes the nests of other birds, tossing out the host's eggs and laying its own in their place.

Another epiphany: we used to think, but now we know. Hundreds of papers and journal articles appeared over the following decade or so, cataloging the many facets of the fragmentation problem—comparing the nesting success of ovenbirds or thrushes in small fragments versus larger forest tracts; examining how drier, warmer air infiltrates the edges of broken forests, reducing the number and diversity of leaf-litter invertebrates on which some birds feed; showing how even a small intrusion, like a dirt road, can allow cowbirds to penetrate woodland and parasitize nests. Researchers deployed automatic cameras, aimed at dummy nests with quail eggs, to learn what kinds of predators were preying on nests most heavily. (A big surprise: cute little chipmunks were among the most pervasive nest predators in the East.) Beyond the impact on eastern woodland species, scientists dug into the effects of fragmentation globally and were horrified by what they found: worldwide, 70 percent of all remaining forest lay within a kilometer of an edge, and habitat fragmentation overall reduces biodiversity by up to three-quarters. From this research came a raft of management recommendations designed to protect existing intact forests, from setting aside old-growth recovery zones where middle-aged forests would be allowed to mature without disturbance, to regulations restricting (and in some cases prohibiting) clear-cutting—or at the very least, concentrating it in fewer, larger areas to minimize the impact on the wider forest.

No one was suggesting that tropical habitat loss wasn't important in the decline of migratory birds—it just wasn't the only factor driving it. This knowledge was hard-won; when I was writing *Living on the Wind* in the mid-1990s, at a point when the focus on fragmentation was near a peak, I spent time with a team of scientists in the mountains of Pennsylvania, led by my friend Dr. Laurie Goodrich, who were trying to measure its impact on nesting ovenbirds. An ovenbird is a warbler that has largely forsaken the treetops for the forest floor, where its olivey-brown back and streaked belly allow it to disappear into the shadows—invisible but not unheard, since its explosive *tee-chur teech-CHUR tee-CHUR TEE-CHUR!* song is one of the signature sounds of an eastern hardwood forest. Laurie and her crew were in the field long before daybreak, every day from the end of April to the middle of July, lugging equipment from expansive ridgetop forests miles from the hardtop to buggy little woodlots along pasture creeks—11 study sites in all, encompassing a range of fragment sizes. In each site they set up nets, played recorded ovenbird songs, color-banded the birds they captured, looked for color-marked birds from previous years, and then followed the marked birds until they located their nests. This was no easy task, because the ovenbird is named for the fiendishly camouflaged nest, shaped like an old beehive oven, that it builds among and below the leaves on the forest floor. (One year Laurie's team enlisted a volunteer to buy an expensive bird dog and train it to search for ovenbird nests by smell; the dog instead proved incredibly adept at finding turtles.) Each of the dozens of territories they found had to be visited every four days to locate and then check on the nests. It was brutally hard, slogging work, with a goal of determining how many chicks from each nest fledged—that is, survived to leave the nest, the typical measure of breeding success.

Once the last chicks fledged, Laurie and her colleagues could breathe a sigh of relief, scratch their bug bites, and start working up their season's data; although the ovenbird parents would be tending the fledglings for weeks more, they would do so without nosey biologists prying into their lives. In fact, among the small army of biologists studying fragmentation effects and the nesting success of woodland songbirds, pretty much everyone did the same thing—exerted huge efforts studying the nestling phase, tallied up the statistics on fledging success, and called it a season once the babies bolted to the four winds, to be tended by their increasingly harried parents. But this brings us back to full-life-cycle biology, the necessity of knowing every major aspect of a bird's life in order to understand what it needs, to identify where the threats lie, and to take action to protect the species. We used to think, but now we know. There was another epiphany waiting to be revealed in that time after the nestlings fledged.

Few biologists had given much thought to the couple-of-months period between the end of nesting and the beginning of migration. We knew that it was a time of preparation, when many songbirds begin laying on fat, as well as beginning the long, energetically expensive process of molting, replacing old feathers with newly grown ones. (Many songbirds replace both their full coat of body feathers, as well as their wing and tail feathers, in late summer.) It seemed a pretty dull, placeholder time of the year, although there were hints that it was anything but. For example, ornithologists had long known that many species of waterfowl undertake "molt migrations" after they finish nesting, traveling hundreds of miles or more to reach safe havens where they can lose all of their major wing feathers at once—a habit almost unique among ducks, geese, and swans—becoming flightless for many weeks while they grow a new set. But such migrations were largely unknown for small songbirds until about 1990, when research showed that a number of birds from western North America—Bullock's orioles, Lazuli and painted buntings, warbling vireos, western tanagers—make long migrations into the Southwest and northern Mexico. They do not become flightless, but the late-summer monsoon rains that wash that region provide a flush of insect food to fuel their molt. Even more surprisingly, in 2005 we learned that some birds, including yellow-billed cuckoos, Cassin's vireos, yellow-breasted chats, and hooded and orchard orioles, not only make a late-season migration south—into western Mexico, again into the monsoon zone—but then breed there for a second time within the same summer, having just raised a brood of babies in the north. There was obviously a lot more going on in this supposed off-season than anyone had imagined.

But exactly what was happening with recently fledged juveniles remained something of a mystery, in part because it was so damnably hard to try to track babies scattering in all directions through summer-thick forests and bogs. (It must also be said that, because adding even the negligible weight of a tiny transmitter lessens, at least a little, a baby bird's already poor odds of survival, many researchers were reluctant to run the risk.) Most people assumed the youngsters stayed near their birth sites anyway until it was time to migrate, though some experts argued that the chicks began a slow migration south well before their parents. By the mid-1990s, several researchers were curious enough to tackle the thorny challenge of radio-tagging nestlings—mostly wood thrushes, a model species for fragmentation research and one that's big and strong enough to handle a tiny transmitter well—and following them as they left the nest.

To the scientists' surprise, those supposed deep-woods specialists were moving into the polar opposite habitat—dense thickets and tangled undergrowth in old clear-cuts, field margins, abandoned farms, roadsides, and the like. The juveniles were gorging there on ripe blackberries, in brambles so dense that it was hard to imagine even the most agile hawk squeezing itself through; they haunted poison ivy jungles, jumbles of wild grape vines, and sumac so thick one could barely see into them. And while they were tracking their young wood thrush subjects, the researchers were surprised to also find the juveniles—lots of them—of many other erstwhile deep-forest birds like red-eyed vireos, ovenbirds, and Kentucky, hooded, and worm-eating warblers. What was going on? If large, unfragmented woods were so vitally important to these species, why were their kids gravitating to overgrown clear-cuts, the very antithesis of a mature forest? It was almost like teenaged rebellion, avian style, turning their backs on their parents' habitat.

Food and cover seem to be the answers. Early successional habitats—thickets, shrubby copses, and other forms of dense brush and young forest that pioneer recently cleared land—are photosynthetic juggernauts, producing huge quantities of insects and, in late summer, highly nutritious fruits and berries that are ideal premigratory fuel. They are also hard to move through, as the thorn-scratched, tick- and chigger-bitten, poison-ivy-itching researchers could attest, and thus provide shelter from predators during one of the most dangerous periods of a young bird's life. Further research has confirmed the importance of young forest and shrubland habitat for species to which we once thought such places were anathema. What's more, biologists have found that there's an even more surprising two-way street of habitat cross-pollination occurring—while deep-woods species are moving into the thickets in late summer, the birds that nest in them, like golden-winged warblers, are moving out, shifting their chicks into or through mature forests.

Today, some songbird biologists are reassessing their assumptions about what constitutes good habitat for migratory forest birds. One of them is Ron Rohrbaugh, who long worked at the Cornell Lab of Ornithology and is now the forest program manager for Audubon Pennsylvania; he chairs an international alliance dedicated to wood thrush conservation and has worked for years on issues surrounding golden-winged warblers, an increasingly rare species that sports a black mask and lemon-yellow crown and shoulders. But it took a while even for him to see the thicket for the trees.

"We're finding that where young forest is available, birds like wood thrush are using it," Ron told me. "The adults are moving their fledglings into those areas to bulk up on all the energy that's provided there, during that staging period before migration." Trouble is, there's not a lot of thicket habitat left, and most of it is not where it does forest-nesting songbirds any good. In some parts of the East, only 1–2 percent of the woodland cover is young forest, mostly in the form of reverting fields and utility rights-of-way; species that depend on early successional habitat, like brown thrashers, box turtles, prairie and golden-winged warblers, and eastern towhees, have declined dramatically. That imbalance has deep roots. Americans took forests of great structural complexity in the East and Great Lakes—complexity that had developed over thousands of years—and destroyed it all, down to virtually the last stick, in an orgy of timbering in the late nineteenth and early twentieth centuries. (Such destruction continued for most of the last century in the West, where even today we keep cutting priceless old-growth stands.) Gone were different age classes of trees, known as seral stages, from the youngest saplings to the oldest Methuselahs; gone were narrow light gaps where huge trees had fallen, shrubby meadows growing on abandoned and drained beaver ponds, multilayered subcanopies and understories beneath the overarching trees, regenerating fire burns, stands of insect-killed dead timber—all of it forming intricate and ever-shifting mosaics on the land. What grew back in the past 100 years or so was a remarkably uniform forest of middle-aged trees with very little structural complexity, and far fewer of the resources that birds need.

"That forest is no longer producing the food, and the calories, and the energy necessary for migratory birds to get what they need, because there's no understory left—they stop in these forests and it's all just canopy. They fly down to the midstory and the understory and there's nothing there. The natural structure of the forest has been lost," Ron said.

It's not that the focus on fragmentation was wrong; like the earlier preoccupation with tropical deforestation, it was just incomplete.

Wildlife managers, especially those working with game species, traditionally saw edge habitat as good, promoting diversity, but it became a dirty word when fragmentation research really took off. So the message flipped, edge became the enemy, and in a wider sense the notion of cutting trees became an outrage. "Here's where I think we got off track when we started to talk about the impacts of forest fragmentation," Ron argues. "Edge can become bad when you get too much of it, and you get forest fragmentation, and nest parasitism, and higher levels of predation from edge-specific predators like raccoons and skunks and crows. But the piece that we missed in the eighties and nineties was this idea that you can have edge, and it can be beneficial for a whole suite of species without really impacting those birds that need core, interior forest. You do that by creating a landscape that's got these shifting seral stages, and making sure we're not just creating giant, square clear-cuts." It's all about location and spacing, and (maybe not surprisingly, once you think about it) creating landscapes that mimic what once occurred naturally, with maybe 10 percent of the forest in early successional stages—not in huge, geometric clear-cuts but a kaleidoscope of habitat types and tree ages, no straight lines and right-angle corners but lots of meandering, curlicue edges and ink-blot shapes; creating small thickets close to deep forest, because fledglings are weak fliers and anything more than a half-mile or so is just too far for them to travel. Ron Rohrbaugh thinks one reason that wood thrushes are declining may be that few pairs can find good nesting habitat that's also close enough to food-rich thickets into which they can safely move their chicks. "That could partially explain why these birds are not doing well," he said. "Yeah, they're breeding, they're producing young, but are those young birds energetically prepared for the long migration they're about to undertake?"

Ron isn't the only biologist who's thinking this way. I got to know Jeff Larkin when he was a grad student in 1999, helping him to trap elk in the mountains of Utah for relocation to eastern Kentucky, where he studied the newly reestablished population. Since then he got his PhD and now teaches at Indiana University of Pennsylvania, specializing in forest songbirds like golden-winged and cerulean warblers, with a particular goal of finding ways to manage woodlands to benefit these seriously troubled species. Like Ron Rohrbaugh, Jeff Larkin is all about variety and complexity in forests.

Even tree age classes that ornithologists long considered useless for birds seem to exert a powerful attraction to some key species at the right time of year, he told me. He and his students have tracked nearly 100 golden-winged warbler fledglings in Pennsylvania, and found that many moved from their young-forest nesting sites to pole-timber stands, the weedy growth stage 20 or 30 years old. "I'm talking dog-hair thickets of saplings—what most of us would have called avian deserts," he said. "But we find golden-winged warblers in mixed flocks with a variety of birds using those stands. They're looking for that sapling cover." Like Rohrbaugh, Jeff is convinced that part of the answer to conserving forest birds is acting on our new understanding of this previously overlooked part of the birds' annual cycle, and artificially restoring complexity in the monotonously even-aged forests of the East and Midwest.

"If you don't have a diversity of all forest age classes and structural conditions—whether for a golden-winged warbler, a cerulean warbler, wood thrush, you name it—you're going to have less than optimal conditions for any of those species. It's the landscape they evolved with, long before we mucked things up," he told me. Jeff's work today focuses on exactly how to unmuck the forest—for example, by designing timber cuts that specifically benefit cerulean and golden-winged warblers, two species whose populations have fallen by as much as 98 percent in recent decades. First, he said, the idea is to shave off a little of the canopy cover in a small area to allow sunlight to jump-start the understory, then come back in a couple of years and cut back a bit more, creating thick, ideal foraging habitat for cerulean warblers, a species in which the male is the color of faded blue jeans. After six or eight years, once that habitat has grown out of usefulness for ceruleans, the loggers do a heavier cut, leaving about 20–30 percent of the canopy, which creates small, irregular thickets for nesting golden-winged warblers lasting the next 10 or 15 years, and ideal brood-rearing habitat for species like wood thrushes and ovenbirds.

That's not the typical recipe for industrial forestry, but proponents like Jeff see it as an economical alternative for landowners who want to manage for timber but also create superb bird habitat. A broad coalition, from conservation NGOs like National Audubon and the Cornell Lab to game-focused advocacy groups like the Ruffed Grouse Society, along with federal and state agencies, are determined to bring young forest back to the landscape. Right now, the amount of early successional habitat varies dramatically by region. In the upper Great Lakes, where commercial forestry is still strong, it may already account for 15–25 percent of the landscape. (Perhaps not coincidentally, that's the one region where golden-winged warblers are doing well.) In the Appalachians, though, it's maybe 2–3 percent, well below the 9–10 percent experts suggest is ideal. It's also important to note that not every biologist buys the new young-forest mantra, seeing it as a turn away from allowing middle-aged woods to naturally mature into old-growth. (In fairness, Larkin, Rohrbaugh, and others point out that they're also pushing for old-growth restoration areas, and using management techniques to mimic the structural complexity of old-growth in younger stands through canopy gaps, diversified canopy layers, fallen logs and branches, and the like.) But logging has such a bad public image in many areas that it can be impossible to undertake even modest projects. Plans to create several hundred acres of early successional habitat on a 3,400-acre wildlife management area in New Jersey were stopped a few years ago by a buzz saw of ferocious public opposition.

Still, the US Fish and Wildlife Service has thrown its muscle behind the effort. In late 2016 it acquired the first parcel of land in what will become Great Thicket National Wildlife Refuge— eventually slated to encompass 115,000 acres across six states, from New York to Maine, expressly managed to create this long-neglected habitat. The restoration of thicket cover in the Northeast has already moved one early successional specialist, the New England cottontail, off the candidate list of federally endangered species, where it had been placed in 2006.

We used to think, and now we know? What are the chances that 20 years from now, we'll be backpedaling on this rush to create young forest, having realized we missed some vital piece of information? It's certainly possible, maybe even likely. But with each we-used-to-think, we peel back another layer of an impossibly complex onion, and develop a better grasp of the big picture. Each time we take the time and effort to understand the full life of a migratory bird, it seems, we learn something wondrous, and also critically important for its survival—like the weird synergy between Arctic migrants, Atlantic hurricanes, and Caribbean gunners.

The Delmarva Peninsula has always reminded me of a hand, viewed from the side, hanging down along the mid-Atlantic coast of the United States with one long finger pointed south. I was following that finger, driving down Route 13 through the slender southern tip of the Delmarva near Machipongo, Virginia. The countryside here is flat and soggy, not far above sea level, lined with drainage ditches that keep the fields, where electric-green corn shoots were ankle-high on this late-May afternoon, moderately dry. Cardinals flashed red in the thickets lining the edges of loblolly pine woods. I turned down a dead-end road, past marshes of the short cordgrass known as salthay, past a house and an open gate to a private dock on Boxtree Creek, one of a maze of twisting, meandering tidal creeks that vein the Eastern Shore of Virginia. The gut was 20 feet wide, surging with an incoming tide; to the south and east, the tidal flats stretched for miles.

Much of what was in view was part of the Nature Conservancy's Virginia Coast Reserve. Comprising 40,000 acres, including 14 barrier and marsh islands and stretching some 50 miles, the reserve is, as TNC boasts, the longest coastal wilderness left along the Atlantic seaboard. The reserve actually got its start as a strategic block-

ing maneuver in a bid to prevent wholesale destruction of the lower
Eastern Shore coastline. When the Chesapeake Bay Bridge-Tunnel
opened in 1964, linking this once-isolated area with the mainland,
it was only the first of an envisioned string of bridges and cause-
ways linking all the barrier islands along the Virginia and Maryland
coasts—and opening them to the kind of crushing development that
has ruined so much of the Atlantic coast. The Nature Conservancy
began buying up the islands at the southern end of the peninsula,
beginning with Smith Island, which it purchased from developers
who were poised to start building the first bridge. (The group also
bought up many waterfront farms, put them under permanent conservation easement and resold them, now undevelopable, as a further
bulwark against coastal ruination.) Along with land managed by the
US Fish and Wildlife Service, National Park Service, the state of Virginia, and what's in conservation easements, fully a third of the land
on Virginia's Eastern Shore is protected today. As a result, the lower
Eastern Shore is both a natural gem and a migratory stopover site of
hemispheric importance, especially for shore- and waterbirds. One of
the most dramatic of those birds had drawn me here, in the hope of
witnessing an annual ritual linking the tropics with the far Arctic.

I parked the car and walked toward a half-dozen people gathered
on the dock around tripod-mounted spotting scopes, huddled inside
fleece and windbreakers, wearing hats and gloves. The temperature
was in the upper fifties, with a biting east wind coming in off the bays
and across the flats. Around us rose a chorus of marsh bird sounds—
clapper rails whose stuttering grunts sounded like balky engines
trying to turn over; the ringing *pill-will-willet, pill-will-willet* of territorial willets chasing intruders; bubbling marsh wren songs; and all
of that against a continual backdrop of whoops and chortles from
hundreds of passing laughing gulls.

Bryan Watts detached himself from the group and shook my hand.
He was a trim man with a salt-and-pepper beard and an old, well-
worn pair of binoculars around his neck. Then he introduced me to
the small crowd—there was Alexandra (Alex) Wilke, a coastal scientist

with TNC who manages the bird nesting islands in the reserve, and Ned Brinkley, whom I knew as the longtime editor of the journal *North American Birds*. "It don't feel a whole lot like Memorial Day weekend," grumbled Barry Truitt, the coastal reserve's retired chief conservation scientist, shrugging deeper into his jacket, his flowing, gray-white beard and ponytail fluttering in the breeze. Watts is the director of the Center for Conservation Biology (CCB) at William and Mary College, and along with Truitt has, for years along the Virginia coast, been studying a shorebird that has always been among my very favorites.

I'm not sure why I love whimbrels; maybe it's because they have such physical presence. A whimbrel is among the largest of shorebirds, its gracefully tapered body bigger and heavier than a pigeon's, patterned in a fine-toothed checkerboard of warm browns. Viewed at rest, the most striking thing about a whimbrel is its boldly striped crown and the long, gently down-curved bill it uses to pursue fiddler crabs, its most important food along the Virginia coast—although elsewhere on its annual rounds, the whimbrel may shift to grasshoppers, beetles, and other insects when foraging in uplands, or to crowberry, blueberry, cloudberry, and other fruits of the tundra in late summer. The bill also gave the whimbrel and other curlews their scientific name, *Numenius*, from the Greek *neos mene*, the new moon: the bill struck the eighteenth-century scientist who named the genus as resembling the fingernail curve of a new moon.

Even at the peak of migration, when the tidal flats and salt marshes are seething with multitudes of shorebirds, I have rarely seen whimbrels mixing much with other species; they are a bit standoffish and clannish, although they were said to have once formed mixed flocks with their slightly smaller (and now almost certainly extinct) relative the Eskimo curlew. Once aloft, whimbrels fly with speed and authority. Their wings are long and tapered, like those of a peregrine falcon—another species with unrivaled command of the air, and the typical hallmark of birds that migrate tremendous distances. It was to see whimbrels in the midst of that migration that brought us on this chilly evening to Boxtree Creek, where for 10 days each spring scientists and local birders conduct a whimbrel count, tallying the birds as they head north to the Arctic.

In the early 1990s, CCB and the conservancy began conducting aerial surveys of whimbrels along the Virginia coast. The results were alarming: whimbrel numbers declined by half between 1994 and 2009, leaving Bryan, Barry, and their colleagues wondering what was happening. They knew that the whimbrels' time in Virginia is critical. Whimbrels gain about 7.5 grams of fat per day, basically doubling their weight in a couple of weeks, bulking up on the trillions of fiddler crabs that crowd the Eastern Shore marshes in such densities that at low tide the mudflats appear to be moving. "You can really see it as the season goes on, they just balloon up like footballs," Bryan said. He was fairly sure the problem, whatever it was, lay somewhere other than this secluded, well-protected, and biologically rich place. But little was known about exactly where whimbrels went, or how they spent most of their lives. Bryan and other scientists assumed those stopping in spring along the Virginia coast nested up on Hudson Bay, the easternmost breeding range for the species, and wintered in the Caribbean or South America, but as with most migratory birds, it was little more than an educated guess. Might there be something during the birds' multi-thousands-of-miles-per-year migration, they wondered, that was dragging down the whimbrel population?

Fortunately for Bryan, by the time he realized whimbrels were in trouble, satellite transmitters had shrunk enough that he could use them on the 14-ounce birds, deploying the devices starting in 2008 on whimbrels that he and his team netted along the Virginia coast. The surprises were immediate. One of their first tagged birds flew north into Canada and hooked a left past Hudson Bay, its expected destination, and continued all the way up to the Mackenzie River delta above the Arctic Circle in the Northwest Territories, some 3,000 miles in all. Many of the whimbrels stopping in Virginia take a similar route, as it proved. In autumn, Bryan and his team watched, astonished, as their tagged whimbrels left their staging grounds in the central Canadian Arctic, flew east off the coast of the Canadian

Maritimes, and then navigated directly into tropical storms or hurricanes out in the Atlantic—not once, by accident, but repeatedly and with apparent deliberation, using the slingshot tail winds of the giant storms to propel them south.

"There are two migratory routes," Bryan told me. "One group, mostly the Mackenzie birds, makes this monster flight nonstop from the Maritimes to Brazil—the birds bow way out to the east, so that at some points they're closer to Africa than South America. Then there's another group [from Hudson Bay] that takes an inshore route, maybe over the course of a month, making landfall in Venezuela. It seems like the birds heading out to sea are staying over cooler water, which lessens the risk of hitting storms, but the inshore birds are heading right through Hurricane Alley."

For all the incredible distances they fly, whimbrels have proven to be extraordinarily faithful to the four or five places, total, that they stop along the way each year. Bryan mentioned one bird they'd named Hope, easily the most famous of CCB's whimbrels, tagged in 2009 and followed for almost a decade as she migrated 18,000 miles a year. Hope always stopped here along Boxtree Creek on her way north, always nested in the same part of the Mackenzie delta, and always wintered in the same small mangrove swamp known as Great Pond on St. Croix in the US Virgin Islands.

"The tracking age is on us, and it's revealing all these patterns. Satellite tracking has really opened up our eyes to the fact that, here's a species that flies thousands of miles [a year], but it's not really using much real estate in its annual cycle," Bryan said. "Like Hope—even though she flies thousands of miles, over the course of the year that bird uses about 500 acres. She was caught on Boxtree Creek, she comes back every year to Boxtree Creek. You go up another couple of creeks, it's a different group of whimbrels." That was a shocking realization, he told me, because the implication is that if something happens to these very specific, very particular locations, the whimbrels that depend on them may not be easily able to shift to new sites. (Hope's transmitter, which had stopped functioning, was removed

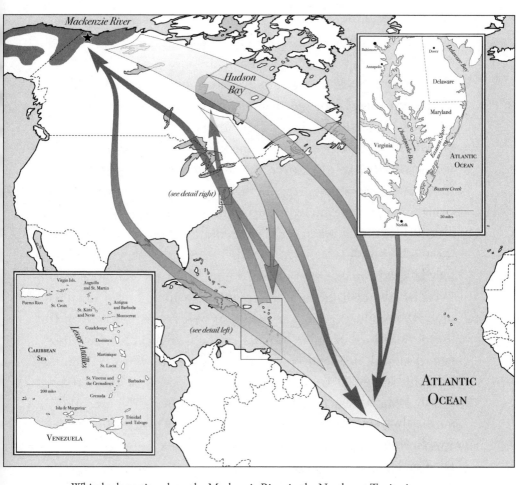

Whimbrels nesting along the Mackenzie River in the Northwest Territories perform a loop migration every year, taking advantage of seasonal winds to cross
the western Atlantic in the autumn en route to the Lesser Antilles and South
America.

in 2012, though she was still marked by distinctive leg bands that
allowed her to be identified at a distance. In September 2017, some
months after my visit to Boxtree Creek, Hurricane Maria roared
across St. Croix, and the far-traveled whimbrel vanished in the storm,
never seen again there or in Virginia.)

"It's really strange to me that they fly thousands of miles, then they pick a little patch to utilize as a stopover, then fly another couple thousand miles," Barry said. "That just blows my mind."

"It does," Bryan agreed. "Tracking, in a very short period of time, has really made some huge advances in our understanding." It also showed how glaring our deficiencies remain. Why were whimbrel numbers declining? Part of the answer came in 2011, when two of CCB's tagged whimbrels—nicknamed Machi and Goshen for landmarks close to Boxtree Creek—survived passage through a tropical storm in one case, and a hurricane in the other. Their storm-tossed crossings were the focus of breathless news coverage— as were their deaths when, exhausted and trying to make landfall on the island of Guadeloupe in the Lesser Antilles, they were shot by local sport hunters who know to hunt when there are tropical storms to the north, because the tired birds will be forced to land to rest. While North American biologists were aware that shorebird hunting remained popular in parts of the Caribbean, the whimbrels' deaths dramatically revealed the degree to which migrant shorebirds were still being gunned down in parts of the Caribbean and South America. Shorebird hunting (with the exception of woodcock and Wilson's snipe) has long been banned in North America, but it remains a significant problem on Caribbean islands like Guadeloupe, Martinique, Trinidad, and Tobago, and on the South American mainland in Suriname, French Guiana, and the north coast of Brazil. On Barbados alone, the annual kill had been estimated at up to 34,000 birds, about 19,000 of them lesser yellowlegs, the most commonly hunted species in the region. But American golden-plovers, short-billed dowitchers, pectoral sandpipers, and others, including much rarer species like red knots and whimbrels, are also taken, even in the few places where they are technically protected. In Suriname, on the northeastern coast of South America, the annual toll is likely in the tens of thousands of shorebirds, with semipalmated sandpipers a common target. Even where the birds are protected, they are not always safe; illegal shooting of whimbrels on commercial blueberry barrens in the Canadian Maritimes, by growers trying to protect their crop, is an increasing concern for Watts and other conservationists.

Brad Andres, the national coordinator for the US Shorebird Conservation Plan for the US Fish and Wildlife Service, said that by 2011 conservationists were already beginning to assess the impact of shorebird hunting. "But with Machi and Goshen getting shot, it pushed everything at lightning speed," he told me. Researchers scrambled to understand the magnitude of the problem; Bryan and his colleagues concluded that for species like whimbrels with low populations, the mortality just from the Caribbean hunts was likely enough to force them into decline. Since then, with heightened public attention, some jurisdictions have tightened rules; in Guadeloupe, where there are 3,000 licensed hunters, there are now seasons and bag limits where there were few before. In Barbados, where limits are set by the hunters themselves, Andres said weapons and ammunition are getting harder to obtain as the government cracks down on drug smuggling, and hunting is declining. Several of the artificial "shooting swamps" maintained by hunters—the only wetland habitat available for migrants—are now either sanctuaries (like the Woodbourne Shorebird Reserve) or no longer hunted. "There's actually another shooting swamp that just three months ago sent me a proposal to go to a no-shooting reserve," Andres told me. "The big problem now is coming up with money to maintain these sites year after year." Without intensive management of water levels and encroaching vegetation, the wetlands become far less valuable for birds, and if the hunting stops and no one steps up to maintain the swamps, the shorebirds may be in even worse trouble than they are now.

Without the knowledge of a migratory bird's full, annual cycle—the kind of information that improved tracking technology provides—it's easy to miss a problem as big and important as shorebird hunting proved to be. That's why the revolution in tracking technology comes just in time, especially for the vast majority of migratory birds that are substantially smaller than big, beefy whimbrels.

"Here they come!" Ned Brinkley yelled, interrupting my chat with

Bryan and Barry Truitt. The first flocks of whimbrels were rising out of the marshes to the south, loose lines and occasional chevrons of birds, mostly two dozen or fewer in each. Alex was calling out numbers and directions, tracking their movements while her seasonal technician, a young woman from Oklahoma wearing knee boots and a bright pink fleece against the cold, sat in a folding chair and recorded each flock on a data sheet.

A big flock of more than 100 whimbrels flew over, gaining altitude as they approached, the air ringing with their loud, rippling trills. In the old market-hunting days, shorebird gunners called whimbrels "seven whistlers" for their call, but here the notes formed a jumbled, musical chorus that cut through the wind, like someone enthusiastically ringing brass sleigh bells. "That's a rally cry!" Bryan said. "They're trying to excite other birds into joining them. Sometimes we'll see flocks come right up out of the marsh to meet the ones going over." The whimbrels looked sleek and muscular in flight, raking the wind with their long wings as they clawed for altitude. They would fly through the night, and at daybreak would be coming in across Lake Ontario near Toronto, where local birders would be waiting for them, marking the passage with their own annual ritual of observation. But the birds don't pause there; they push on north and west, a nonstop journey of five days.

"The next time they touch land, they'll be in the Arctic," Bryan said as still more flocks rose and headed north. Even after all these years, there was still a hint of awe in his voice.

As Bryan Watts said, the tracking age is upon us, revealing patterns we had no way of seeing before. One of the most interesting—and, for conservation, among the most important—is a fairly new concept known as migratory connectivity.

For a small bird, the blackpoll warbler has a huge breeding range, some 2.5 million square miles stretching from British Columbia and the edge of the Bering Sea to Labrador and Newfoundland. Because of the extraordinary distances it travels every year, the blackpoll (named for the male's black cap, or "poll") also has an outsized profile in migration science, one reason my colleagues and I have been studying its travels to and from central Alaska by placing geolocators on blackpolls that we catch in Denali National Park. We already knew that the park's blackpolls make a west-to-east flight of more than 3,000 miles to the East Coast, then arc out over the western Atlantic almost 2,000 miles to northern South America, where they winter somewhere in the Orinoco and Amazon basins. But where, exactly? When we pooled our tracking data with those of other researchers tagging blackpolls in Nome in extreme western Alaska, Whitehorse in the Yukon, and Cape Churchill on Hudson Bay, some interesting patterns emerged.

The Nome birds migrated along a northerly track, cutting across central Canada and leaving the East Coast in southern New England. Our Denali warblers, as well as those from the Yukon, followed a more southerly route, some of them angling as far south as Florida before flying out over the ocean, while the Churchill blackpolls migrated in a largely south-southeasterly direction, slicing diagonally across the other routes and cutting out over the Atlantic mostly in the Carolinas. (Because geolocator data aren't terribly precise, the exact locations are squishy.) Once in South America, each regional breeding population also segregated itself, with the Nome warblers the farthest east and south, wintering near the mouth of the Amazon in northeastern Brazil, Suriname, and the Guianas, while the Churchill warblers clustered about 1,000 miles away in the western Amazon, where Brazil, Peru, and Colombia meet. Our Denali blackpolls were between them, along with those from the Yukon, although the latter appeared to also use the Orinoco forests to the north in Venezuela.

That different populations would have their own migration routes and distinct wintering areas might seem like a common-sense arrangement, but for most of the centuries in which humans have been mulling over the intricacies of migration, no one really gave the idea much thought. It was as though the jungly forests of northern South America were a great bowl, into which North America poured an undifferentiated stream of blackpoll warblers that sloshed and swished and spread out evenly throughout the winter range. Our knowledge of migratory connectivity is still fairly rudimentary, but for some species, that does indeed seem to be the model—great reed warblers from Europe, for instance, whose migrations were studied using geolocators, show only modest connectivity on their African wintering range, with Spanish, Swedish, and Czech birds mingling along the Gulf of Guinea, while warblers from Bulgaria and Turkey migrated to eastern Africa. The latter two populations also have an unusual, counterclockwise loop migration, flying generally southwest into Africa in fall but detouring east to the Arabian Peninsula in spring—a pattern shown as well by smaller Eurasian reed warblers on their own migration from Africa, and probably tied to better food resources in the Middle East in spring.

Even a few hundred miles' distance between breeding sites can make a big difference in wintering locations. Take ovenbirds, for example, those thrushlike warblers common in eastern North America. A geolocator study showed that those from near the western extreme of the species' range, in Saskatchewan, migrated to southern Mexico, while a handful of band recoveries suggest that ovenbirds from the Great Lakes fly to Belize. Finer-grained information was lacking until 2013, though, when scientists at the Smithsonian Migratory Bird Center used a new tracking device, called archival GPS tags, to look more deeply into ovenbird migration. Unlike geolocators, which use a light sensor to record day length and sunrise/sunset times from which a rough latitude and longitude can be calculated, the GPS tag is programmed to turn on at a preset interval (in this case, once a month), get a highly precise fix from the orbiting GPS satellite system, archive that location in its memory, and go back to sleep until the next time, thus saving precious battery power. The Smithsonian researchers tagged ovenbirds in New Hampshire and Maryland, and recaptured about half the next year when they returned to breed. The GPS points showed that both populations flew generally south, but while the New England birds congregated on the island of Hispaniola and eastern Cuba, the Maryland ovenbirds took a more westerly track, wintering in western Cuba and peninsular Florida.

Other migratory birds show similarly close links between breeding and wintering grounds. Common nightingales from three parts of Europe—France, Italy, and Bulgaria—tagged with geolocators showed very tight connectivity with three separate wintering areas in western and central Africa. In North America, band recoveries from rose-breasted grosbeaks show that those nesting in the East winter in Panama and northern South America; those from the upper Midwest migrate to northern Central America; while those from the Great Plains go to central Mexico. But scientists have really gotten into the weeds with Swainson's thrushes, another North American species, which show fascinating patterns that split along both genetic and migratory divides.

Swainson's thrushes are smaller than robins, their breasts the color of soft buckskin beneath smudgy black spots, their faces with a wide-eyed look because of buffy spectacles around their eyes. Most of the continental population, those breeding from Newfoundland to Alaska and down through the Appalachians and Rockies, have dull olive-colored backs. Even those from the western edge of this vast region, like the ones we study in Denali, first migrate far to the east before turning south and spending the winter (depending on their local breeding population) from Panama to southern South America.* The other population, whose backs are a bright chestnut color,

---

* This oddly circuitous route is thought to be an echo from the end of the last ice age, when the olive-backed Swainson's thrushes were restricted to relic conifer forests in the Southeast, well below the continental glaciers. As the ice retreated and the spruce-fir forests moved north and west in their wake, the thrushes followed, but each autumn they retrace that route of colonization east and south. They are just one of many boreal forest species—fox sparrows, gray-cheeked thrushes, a wide variety of warblers, and others—among which at least some populations follow a similar route, presumably for similar reasons.

hugs the Pacific coast year-round, breeding from Southern California to the Alaskan panhandle and migrating to Mexico and Central America, taking a largely coastal route to get there. But even within these general divisions there are strong local bonds of connectivity. Swainson's thrushes from Marin County, California, migrate 1,600 miles to the Mexican state of Jalisco on the Pacific, while those from farther up the coast, near Vancouver, British Columbia, leapfrog over them and migrate clear to Central America.

Interestingly, in southern British Columbia there are a few valleys where both the rusty coastal form and the olive-backed inland thrushes occur together—and hybridize. Because migration is largely instinctive, written into a bird's DNA, it's not surprising that their hybrid offspring have warring genes that appear to tug them in contradictory directions. Thanks to geolocators, we know that some hybrids split the difference and follow an intermediate route—a poor choice and a challenging one, taking them over rugged mountains and deep deserts with few chances to rest and feed. Other hybrids may in the fall follow an inland-type route, inherited from one parent, swinging far to the east and south, but in spring use a coastal-type route inherited from the other parent, looping all the way to the Pacific before heading north. Or the reverse. Not surprisingly, given the extra trials and distance their bastardized navigation puts them through, the hybrids remain uncommon, maintaining the genetic divide between the two subspecies groups, and reinforcing each form's unique migration connectivity. (Similar challenges face European warblers known as blackcaps, where a migratory divide in Austria and the Czech Republic splits the population, one side migrating southwest to Iberia, the other around the eastern Mediterranean. Hybrids inherit a tough-luck map that sends them across the Mediterranean and into the Sahara.)

Where strong connectivity exists, it has serious implications for conservation, especially in parts of the world that are undergoing rapid changes. A joint Canadian-American study that tagged more than 700 wood thrushes, both on the breeding grounds in eastern

North America and the wintering grounds in southern Mexico and Central America—a herculean effort, requiring hundreds of hours of labor to try to relocate and trap the tagged birds—found extraordinarily tight connectivity among regional wood thrush populations. Those from New England south to Pennsylvania and North Carolina wintered almost exclusively in eastern Central America, an area that includes eastern Honduras, Nicaragua, and Costa Rica and that supports more than half of the entire species. Those wood thrushes that nested in the American Southeast and Midwest, by contrast, migrated primarily to the Yucatán Peninsula, El Salvador, and western Honduras, and only a relative few of the tagged thrushes—all from the Midwest—wintered in southern Mexico.

Especially given the rapid pace of deforestation there, the importance of the eastern Central American wintering area can't be overstated, the scientists concluded, and they urged greater conservation efforts in that region. But their analysis also showed that, regardless of what part of Mexico and Central America a wood thrush uses in winter, there's an excellent chance it will pass north through the central US coast of the Gulf of Mexico, a bottleneck through which nearly three-quarters of the species migrates in spring. Focusing land conservation work along the Gulf would pay huge dividends not only for wood thrushes across their breeding range but also hundreds of millions of other songbirds, of hundreds of species, that pass through that area every spring. Tracking is revealing many such stopover hotspots that no one suspected before, ideal candidates for targeted conservation. By pooling the results of several tracking studies, for instance, scientists realized that European rollers—striking pigeon-sized birds with turquoise-and-chestnut plumage, which have declined overall by as much as 30 percent in recent decades and which have vanished entirely from some parts of Europe—converge in autumn on the savannahs around the Lake Chad basin in Africa, where Nigeria, Chad, and Cameroon meet. These savannahs in the Sahel, the southern fringe of the Sahara, are also proving to be a critical stopover site for many other insectivorous birds migrating south from Europe, including shrikes, bee-eaters, and cuckoos. Because the Sahel is increasingly drought-prone, it may be no coincidence that many of these other species have suffered their own declines.

Even where, strictly speaking, migratory connectivity is weak, the insights from tracking studies can be shocking. Perhaps the best example of this involves prothonotary warblers—songbirds of swamps and lake edges in eastern (especially southeastern) North America, so riotously yellow that one ornithologist described them as looking like "butter dripping from the trees." Scientists had long assumed most prothonotary warblers wintered in coastal mangrove swamps in southern Central America and northern South America. But a study published in 2019, in which prothonotaries from across their wide range were tagged with geolocators, found that regardless of where they nested, from Louisiana to Virginia to Wisconsin, almost all of them gravitated to a small area along the Magdalena River in Colombia, far from the coast—and far from where conservationists were focused on land preservation. Technically, this pattern is almost the reverse of migratory connectivity, but the consequences for saving this species, which is already listed as endangered in Canada, are immense.

There's still a lot to learn about how migratory connectivity works, and how we can use that emerging knowledge to conserve migrant birds. Sometimes, though, science be damned; the results are just simply, breathtakingly cool.

In 2018, our crew caught a number of Swainson's thrushes in the spruce forests of Denali and fitted them with archival GPS loggers. A year later Emily Williams, the park's avian ecologist, and avian technician Tucker Grimsby, recaptured the first three of the tagged thrushes just after the birds had returned to the park. One of the three tags had failed; mechanical glitches are common when you're using highly miniaturized technology that will travel halfway around the world on a bird. But the other two tags told a remarkable story of connectivity. Both thrushes left their breeding territories, which lay less than a mile apart in the park, in August of the previous year, one bird about two weeks earlier than the other. They migrated south-east, crossing the Yukon, northern British Columbia, Alberta, and Saskatchewan. By September they had rounded the western edge of Lake Superior and were passing through the Midwest; one of them stopped for a week in the backyard of a suburban development north-east of Indianapolis. (The development, we noted with a chuckle, was called Feather Cove.) The thrushes crossed the Blue Ridge Mountains near Great Smoky Mountains National Park at the end of September, now traveling just a few days apart; they passed through the Florida panhandle and flew across the eastern Gulf of Mexico and western Caribbean to Central America; then hooked east, following the isth-mus to Colombia by late October. From there, the thrushes traveled another 2,300 miles along the eastern flanks of the Andes. When they finally stopped—after more than 8,000 miles of travel—they were within 20 miles of each other on the Bolivia-Argentina border.

A coincidence? In fact, the next four thrushes that Tucker and Emily caught followed basically the same path to almost precisely the same goal. When, for the first time, I watched the glowing green lines that represented their flights spread across the digital globe on my computer screen, all converging on the same slender band of foot-hills forest, all six coming to rest for the northern winter within 100 kilometers of one another, I actually gasped. It was yet another exam-ple of the tightly bound web of surprising connections that stretch across the globe and among the hemispheres, woven by the wings of migrating birds.

Four

BIG DATA, BIG TROUBLE

Dawn on the Alabama coast; the chuck-will's-widows are calling in the last of the twilight at Fort Morgan, a slender peninsula that juts out into the eastern side of Mobile Bay. Soft sand under my feet as I walk a well-worn path I could follow with my eyes closed, snaking through the live oak and pine, I stop every so often to slide open another set of mist nets. Except for the *chuck-will's-WIDow, chuck-will's-WIDow*, it was silent, but already I could sense a tension in the forest that was missing the previous few warm—and essentially birdless—April days. Some mornings it had been so still that, as I stepped out of the eaves of forest and up to the seawall that rims the edge of the bay, I could hear the subtle, pufflike exhalations of bottle-nosed dolphins surfacing in the shallows far beyond, a pod slicing through the calm water with their dorsal fins cutting the reflections of natural gas derricks out in the bay, misty and indistinct in the humid air.

This morning, though, was very different. The previous afternoon a cold front had swept down from the northwest with rain, blustery winds, and falling temperatures. While my friends and I ate dinner in a beach house a few miles from the banding station, listening to the rain hammering on the roof, we knew that the line of storms moving beyond us to the south would slow the invisible legions of migrants, which were, even then, crossing the Gulf of Mexico. Instead of a relatively easy passage north from the Yucatán Peninsula, maybe 18 or 20 hours of steady flight across the open ocean, these birds would have to claw their way through the storms and their windy aftermath for more than twice that length of time. Many would die; some mornings after similar storms we found their bodies washing up in the surf by the dozens or hundreds, little scraps of brilliantly colored feathers quickly scavenged by gulls or ghost crabs. (And, as one study showed not long ago, by tiger sharks cruising offshore.) As we slept, those that survived—exhausted and hungry—would have started to pile into the maritime forests of Fort Morgan: a fire escape, in the parlance of migration ecology, a pit stop where they could escape imminent death, and grab a quick bite and a little rest before continuing north.

As the light grew around me, the woods, I realized, were seething with birds, a classic Gulf Coast "fallout" that would make any birder's life list of red-letter days. They darted across the narrow trails I followed beneath the oaks, flickers and slashes of motion in my peripheral vision, and flushed before me like a bow wave as I walked—warblers and sparrows, buntings and orioles, catbirds and thrushes, flycatchers and grosbeaks. This was the kind of day for which my colleagues and I came here, when we might net and band 1,000 or more birds in a morning, documenting the migrant pulses that roll through this part of the world twice each year. As I turned back and retraced my steps, at each of the newly opened nets I found a dozen or more small birds already cradled in the mesh, which I quickly extracted and slipped into drawstring bags looped over my left forearm. By the time I completed half of the circuit, meeting my friend Fred Moore coming the other way, we each had dozens hanging from our wrists, which Fred took from me and hurried off to the banding table for processing as I turned back for yet another pass. It was going to be a very busy morning.

The Gulf of Mexico is one of the largest migratory choke points in the world, through which an average of 2,060,300,000 birds pass each spring. That may seem like a strangely precise number, but its precision reflects some of the astonishing changes that have swept the world of migration research, and science's ability to follow tiny birds across vast distances. Armed with greater and ever-faster computing power, ornithologists are crunching almost incomprehensible volumes of data—the output of the continental Doppler radar system, to take one example, which in addition to fine-tuning your nightly weather forecast also reflects nocturnal bird movements, creating an essentially moment-by-moment picture of bird migration so detailed that experts can calculate the number of birds per cubic meter of air space, distinguish between large, medium, and small species, and even tell a migrant's beak from its tail. I'm tempted to describe this as the Golden Age of Ornithology—but then, I suspect every age has fancied itself blessed. No doubt the museum ornithologists of the 1890s, shooting specimens for their collection cabinets, thought themselves living in a Golden Age thanks to the invention of smokeless gunpowder and cartridge shotguns.

But this does feel fundamentally different, as new ways to employ remote sensing and process immense data sets allow us to figure out just what species of birds are going where, and when, and in what numbers, in near-real time. It enables us to identify the exact places being used by the largest numbers of migrants, and thus target the most effective use of scarce land protection dollars. We're pooling the daily observations of millions of birders to reveal hitherto unknown migration routes and stopover sites; we're harnessing machine learning to teach computers to listen to the skies, identifying and tabulating the millions of travelers overhead by their nighttime calls. It's also showing where the greatest dangers to migrants lie; interestingly, in North America some of the riskiest places and the highest-value conservation lands both lie in and near some of the largest metropolitan areas.

The view, assembled from all these groundbreaking approaches, is both breathtaking and terrifying. The same technologies that are finally giving us a window into the true scale of bird migration also provide the glaring evidence that we are on the cusp of catastrophe. Indeed, some would argue we have already slipped over the edge of the abyss, as billions of migratory birds have disappeared in just the past few decades in North America alone, lost to habitat destruction, pesticides, building collisions, cats, and many other threats. But the realization that the hour is late and the situation increasingly dire has only further motivated conservationists, who are scrambling to use this new tech to find the answers that may allow them to reverse the downward plummet.

This new science has enormous potential for migratory bird conservation everywhere, but it's being put to the test first along the Gulf of Mexico, perhaps the single most critical link in the entire Western Hemispheric chain of habitat that makes migration possible. The Gulf Coast is hallowed ground for birders, the place which, when the weather is just right, can produce epic fallouts of thousands or hundreds of thousands of songbirds at places like High Island, Texas, or Fort Morgan and Dauphin Island, Alabama. It is one of the best places in the world to experience migration, and a region that holds a special meaning for me.

For more than 15 years I came to Fort Morgan every spring to net and band migrant birds, the happy consequence of a chance detour during my research for *Living on the Wing*. In 1997, I was doing prep work for a spring reporting trip through the South to write about trans-Gulf migrants, the flood of songbirds that leave Mexico and fly more than 500 miles nonstop across the Gulf to make landfall from east Texas to the Florida panhandle. A fellow named Bob Sargent had posted something on a bird-banding listserve to which I subscribed, talking about the years of work on the Alabama coast that he and his colleagues had done, and inviting any bander who might be passing through to stop in for a visit. A quick, cordial email exchange later, I'd added a stop to my growing itinerary, not realizing how much that little change would affect my life.

Bob, I soon discovered, was a force of nature—tall, bald, and powerful, a charismatic speaker who was a product of the rough-and-tumble mining camps of northern Alabama and who, after time in the air force, became a master electrician. Bob came to birds relatively late in life, first for pleasure and then for science. In the 1980s he and his wife Martha formed the Hummer/Bird Study Group, a cadre of like-minded bird-banders who, for almost 30 years, spent weeks each spring and autumn at Fort Morgan State Historical Park, banding tens of thousands of birds and educating countless thousands of visitors who crowded around the banding table to watch the process, or listen to them preach the gospel of bird conservation with, say, a stunning scarlet tanager or rose-breasted grosbeak in their hands. By the next season I was part of the team, and kept coming back every year until Bob's unexpected death in 2014 ended HBSG's long-running effort. (A different group of Alabama scientists continues banding there today under independent auspices.)

Every year we banded thousands of birds at Fort Morgan, and submitted the data—age, sex, weight, fat scores, a variety of measurements—to the federal Bird Banding Lab, our small fraction of the more than 1.2 million birds banded annually in the United States. Each reencounter, whether a day later in our own nets, or years later and hundreds of miles away, added even more to our understanding—of, to pick one example, how quickly weary migrants regain lost weight, and how that rate of recovery varies from habitat to habitat. (Along the Gulf Coast, banding has shown that the habitats most valuable to hungry migrants aren't the lovely beach dunes and graceful pine forests we've protected in national seashores and parks—they are the greenbrier tangles, oak thickets, and deep swamps that few people bother to visit, and which have largely been ignored for preservation.) All those banding records are an incredibly rich vein to mine, and ornithologists are continually finding new and creative ways to use banding data to answer fascinating questions, applying the techniques of so-called Big Data, analyzing huge data sets to tease out patterns and trends that aren't immediately obvious. With more than 120 million individual birds ringed (and 4 million encountered again) in North America over the years, banding certainly counts as Big Data—but it's slow and laborious Big Data, having required more than a century to build up that library of band returns and encounters. The amount of effort that has gone into producing it is mammoth and daunting.

The Gulf has long been a proving ground for pioneering migration studies, and has more recently become an important nexus demonstrating the power of emerging technologies to help us understand and conserve bird migration. In the 1940s, when skeptics derided the notion that small birds migrated nonstop across the Gulf, a young Louisiana State University ornithologist named George Lowery Jr. signed on to a freighter and made the voyage back and forth between the United States and Mexico, recording the multitudes of migrant birds he saw during each trip. In the 1960s a graduate student at LSU, Sidney A. Gauthreaux Jr., used images from the newly established line of National Weather Service radar stations along the Gulf Coast to track, for the first time, this trans-Gulf migration on a landscape scale. (In later years Gauthreaux, by then a professor at Clemson, used those old archived images to demonstrate that the number of migrants crossing the Gulf had fallen dramatically over the decades). Once the NEXRAD Doppler radar system came online in the 1990s, the amount of information available to ornithologists exploded. Doppler shows the speed and direction with which objects—be they raindrops or birds—are moving through the air, and a new generation of radar ornithologists trained by Gauthreaux and others used this technology to peel back the anonymity of the night sky and begin to understand the intricacies of nocturnal migration, when the vast majority of birds are on the wing.

One needn't be a specialist to watch the migratory pageant on radar; any National Weather Service radar website will do. It takes only a little practice to distinguish the more intensely colored bands of precipitation from what meteorologists call "bioscatter" or "bioclutter"—the almost ethereal clouds of pale blue and green formed by millions of birds, blossoming up and out on radar around the Doppler stations in the hours after dark in spring and autumn. But bioscatter comes in many forms. Swarms of insects, like the clouds of ladybird beetles that engulfed Southern California in 2019, or a 70-mile-wide swath of migrating painted lady butterflies in Colorado in 2017, can also show up on radar, as does the nightly emergence of millions of bats from caves and bridges in the Southwest.

As revolutionary as the original NEXRAD radars were for ornithology, the technology hasn't stood still, and advances in radar designed to aid weather forecasters, like new high-resolution upgrades a few years ago, have had unintended benefits for migration specialists, too. For example, the Doppler system now uses "dual polarization" radar, which instead of employing a single horizontal beam sends out a second, vertical signal, giving meteorologists the ability to see the size and shape of precipitation inside storms—distinguishing rain versus hail versus snow, for instance. But dual-pol radar can do the same for birds, allowing an ornithologist not only to calculate how many birds are flying per cubic meter of airspace, but to distinguish the head from the tail on each one—allowing, for instance, researchers to see how birds high in the air column, masked by darkness in the middle of the night, are subtly compensating for crosswinds and other forces.

The biggest obstacle in working with radar data is the raw computing power that it requires to work with the digital output of the 143 radar sites in the contiguous United States, each making a sweep every few minutes. Andrew Farnsworth, a research associate at the Cornell Lab of Ornithology who specializes in migration research, remembers how when the lab launched something called BirdCast in 2012—an attempt to create real-time, regional forecasts of bird migration—it took him four or five hours of arduous, often hands-on processing and manipulation of data from just 10 regional radar stations to create a single snapshot of the previous night's migration. "Whereas today, we do basically every [radar] scan that's produced every five or ten minutes, for 16 stations in the northeastern US— a whole night's data may take minutes, literally, to do the manipulation that took hours before," Farnsworth told me. Not only does that mean that the current version of BirdCast, which covers the whole of the lower Forty-Eight, is far more accurate and nimble than it was in the past, with live migration maps and expert commentary—more importantly, the algorithms that drive BirdCast are underpinning the dramatic new ways scientists are using radar ornithology to focus conservation efforts where they are needed the most.

Along the Gulf Coast, this means not only tracking the current migration season, seeing in extraordinary detail where and in what numbers the birds are moving—with a NEXRAD archive stretching back to the mid-1990s, it's also possible to travel back in time, to see whether the timetable of migratory movements has shifted, or how spring and autumn bird traffic responded to climatological changes like El Niño, the Atlantic multidecadal oscillation, or major storms. (By 2019, scientists from the Cornell Lab and the University of Massachusetts had created a machine-learning system they called MistNet, which automatically strips away the signal for precipitation in archived radar data, leaving only birds, bats, and insects, streamlining the process even further, whether looking at the air column over a small patch of land or the entire continent. Interestingly, MistNet uses the same kind of image-reading artificial intelligence as Merlin, the Cornell Lab's popular bird-identification software.) Most fundamentally, these advances allow scientists to calculate whether the number of migrants has changed through the years—and the news here is mixed, at best. Jeff Buler, director of the University of Delaware's aeroecology program and a leader in this field, has used Doppler archives to look at changes in migratory bird populations in a number of regions like North Carolina, where the number of autumn migrants dropped by 27 percent in 12 years, and the Northeast, where numbers fell 29 percent in just seven years. Yet along the Gulf of Mexico, where radar data allowed a team of scientists that included Buler, Farnsworth, and others (and led by Cornell postdoc Kyle Horton) to calculate very precisely that an average of 2,060,300,000 birds pass through the region each spring, they found that the intensity of the migration hadn't changed appreciably over a nine-year period ending in 2015.

That struck me as one of the most hopeful reports I'd read in a long time, especially given the gloomy statistics from other regions. But when I mentioned it to Farnsworth, he cautioned me that a lot of the birds picked up by radar along the Gulf Coast aren't migratory songbirds, the group in deepest trouble, but rather coastal species like ducks and wading birds whose populations are booming, and which are also physically large and thus show up very well on radar. "Some of these species are doing swimmingly well, like the explosion down there in the great egret population. And a single great egret is a huge radar reflector," he told me.

Radar data can also do more than simply track what's up in the air. Combining radar with other forms of remote sensing, like high-resolution satellite imagery, has allowed Jeff Buler and his team to look back and see how birds respond to huge disturbances like Hurricane Katrina in 2005. In that case, radar showed how migrant birds abandoned their preferred stopover sites in bottomland swamp forests, which had been stripped of their vegetation, shifting to less damaged, pine-dominated uplands—habitat that's normally a poor second choice for migrants, but the only alternative after the storm. Five or six weeks later, as the bottomlands began to green up, the birds shifted back into those recovering forests. But perhaps the most consequential new application is using radar to figure out exactly where the most important stopover habitat exists on the ground, so it can be protected and restored. Buler realized that by focusing on the very lowest beam of energy emitted by each Doppler unit—the beam that scrapes along just above ground level—he could detect birds moments after they rose into the night sky to begin their migration. Logically, the places from which the largest numbers of birds were emerging would likely be the best stopover sites—and Buler and his colleagues have been able to zero in on patches of habitat as small as 25 acres. In a region like the Gulf of Mexico, where every acre of coastal habitat is at once precious and increasingly expensive, knowing what land the birds are actually using most heavily means spending scarce conservation dollars far more wisely. This technique is now being scaled up dramatically. In 2018, Buler and the US Fish and Wildlife Service published a massive study looking at autumn stopover in the entire mid-Atlantic region and New England. Certain areas, where bird density was reliably high throughout the season, lit up like bulbs on a Christmas tree—northern New England, the

Adirondacks and Catskills, southern New Jersey, parts of the Delmarva Peninsula, and the western shore of the Chesapeake Bay. For agencies and nonprofits looking to get the most bang for their land preservation buck, this provides a road map for effective migratory bird conservation.

The obvious drawback about radar is that, as Andrew Farnsworth likes to say, it is taxonomically agnostic. You may be able to count birds, calculate biomass, and tell beak from butt with dual-pol radar, but you can't tell if that beak belongs to a red-eyed vireo or a hermit thrush, a hyperabundant white-throated sparrow or a federally endangered golden-cheeked warbler. If radar is a black-and-white snapshot, you need other forms of Big Data to fill in the colors, to map out the variety and complexity of the migration at the species-by-species level—cerulean warblers or yellow-billed cuckoos, indigo buntings or scarlet tanagers, or hundreds of other migrants. At the most basic level, you can supply those details either by looking or by listening.

A migrating songbird, flying through the night, gives short, simple call notes—essentially anticollision alarms, a way to avoid hitting any of the thousands or tens of thousands of other birds using the same airspace. (Although huge numbers may be aloft together, they do not fly in cohesive, coordinated flocks; each is migrating on its own.) Such vocalizations become especially frequent in the hours just before dawn, when the migrants drift lower and lower, looking for a place to set down for the day. It wasn't until the 1980s and '90s that birders and scientists—led by ornithologist Bill Evans of New York, the pioneer in this field—really began to pay attention to nocturnal flight calls. At the peak of migration, it's possible to hear hundreds or even thousands of flight calls in a few hours—Swainson's thrushes piping like choruses of spring peepers high in the sky, for instance, giving you a vivid sense of the number of birds hidden above in the darkness.

For less than $100, you can even build a rooftop microphone of Evans's design to record the calls of birds going over your house at night. The design is intentionally cheap and delightfully Rube Goldberg–esque. You mount a microphone and some inexpensive electronics onto a plastic dinner plate, glue it all into the bottom of a plastic bucket (a large flowerpot works as well), and stretch plastic wrap over it for waterproofing. Mounted on a roof and connected to a computer, it will record the calls of songbirds, waterfowl, rails, shorebirds, owls, and other nighttime migrants, many of which are species you would never see by day.* Multiply that one microphone by thousands—and create algorithms that automatically identify and tally up the calls they record—and you'd have a network that provides identity to those anonymous blips on Doppler radar.

That kind of continental audio network, as a complement to radar, has been a dream of ornithologists for decades, but the technical challenges have been daunting. First, it required some serious detective work just to decipher which calls were being made by what species, since the identities of the callers were masked in darkness. (There are still a few lingering mysteries, like the identity of an unknown bird that's been recorded a number of times in migration over eastern Mexico, but nowhere else in North America.) Finding ways to filter out the noise and teach machines to pick out and identify flight calls—something at which the human ear and brain excel—has been an enormous hurdle. But the same advances in computing power that made radar analysis quick and accurate are being brought to bear on audio analysis, and Andrew Farnsworth at Cornell believes they are getting close. I've spoken with Farnsworth several times over

---

* Some years ago, I was traveling with my friend Jeff Wells, an expert in nocturnal flight calls then working for the Boreal Songbird Initiative. We were flying to the Northwest Territories, where Jeff was going to set up several flight call recording stations in a remote Native village on Great Bear Lake. As we went through Canadian customs in Toronto, a border officer unzipped Jeff's duffel bag and pulled out a couple of big plastic flowerpots. He looked up with a long-suffering expression that spoke volumes about the ways Americans misperceive his country and said, "You know, sir, we *do* sell things like this in Canada." I'm not sure Jeff's explanation carried much weight with him.

the years about advances in audio monitoring, and in 2016, when I interviewed him for a magazine piece about emerging migration science, he had the zeal of an evangelist, with a multimillion-dollar National Science Foundation grant to back up his enthusiasm.

"Put a microphone out your window, run an algorithm [on your computer], and in the morning, you have a histogram with the number of calls and the species of callers. And you'll do that not just for a single site, but an entire region or across the country," he told me. "That's gonna happen in five years. It's an unimaginable leap. It opens a whole new realm of information at a huge scale. There's no other technology that can do what acoustics can do, on the scale of a continent, to tell us what species are migrating."

Three years later, I checked in with Farnsworth to see how things were progressing on what was now known as the BirdVox project, and found that his excitement, while tempered by some of the technical challenges, was largely undiminished. BirdVox is a collaboration between the Cornell Lab and NYU's Music and Audio Research Laboratory, and Farnsworth is the only birder in the bunch. "They're all music people, people who are interested in, 'How do I classify the chords in the Beatles' "Let it Be," and how do I properly identify it when a band covers it versus the original, versus the version that was alternate take 12 on the unmastered, unreleased version?' They're interested in signal processing, how do you make sense of information. Whether it's a bird, or the Beatles, or the urban sound of a gunshot, or a horn or alarm or whatever, they just want to understand how to extract the information." The BirdVox team has also partnered with experts at Google, because getting a machine to recognize someone talking to a smart speaker, and getting a machine to recognize a Townsend's warbler's flight call, turn out to have many similar technical challenges.

He was still bullish about his original five-year time frame for an accurate, automated analytical foundation for a continental acoustic migration monitoring system. By the middle of 2020, the BirdVox team had its first prototype for automated workflow, a way to suck in nocturnal recordings, filter out the background noise, and identify the birds, with little or no human interaction. "Once we prototype this, we can start to deploy it to people and say, 'Hey, beat this up, tell us what you think.' So things have come a long way, but it's by no means as advanced as you've seen on the radar side of the BirdCast project in terms of the level of confidence, and being able to run it for giant amounts of data—yet."

Still, the potential for fundamentally changing the way scientists monitor and track migration by listening to the night sky is tantalizingly close. And we have an example of how Big Data can radically transform our understanding of bird movements, using an even more basic tool: our eyes. Birders have been watching birds for a long time, but only recently have all those eyeballs, and all those observations, been harnessed in a way that revolutionized our ability to grasp migration as a whole. That's because there's never been anything quite like eBird.

No one planned for eBird to take over the world. It just sort of happened.

The birding world, at least. When eBird was launched in 2002, a joint project of the Cornell Lab of Ornithology and the National Audubon Society, it was seen by its creators as a citizen-science portal, where interested birders could provide sighting data that would help researchers better understand bird populations and movements. Birders have, as a group, always been a little fanatical about keeping records—daily checklists, life lists, state and county lists, records of first arrivals, seasonal late dates, out-of-range rarities, breeding confirmations, and much more. Almost every birder had a couple of shoeboxes stuffed with decades' worth of old checklists and notebooks or (more recently) computer files on their home PC containing much of the same. In aggregate, across millions of birders, those records represented a staggering amount of potentially valuable information, but it was all essentially useless because almost none of it was accessible. Worse, those records are usually doomed to be discarded or destroyed when the original owner dies or quits the hobby.

The goal of eBird was to give birders an easy way to submit their records into a single vast database where, together, they could paint an incredibly detailed picture of bird populations and movements. It was a brilliant idea, but initially, it was also a little bit of a dud. Helping science is all well and good, but eBird didn't really start to take off until Cornell and Audubon added features that allowed birders to capitalize on the sportier or more competitive aspects of their hobby—tracking their time in the field and easily managing their various lists; creating maps of their sightings or bar charts showing seasonal abundance for locations around the world; finding where particular species are being seen in close to real time.

That did it. Soon, hundreds of thousands of birders—first in North America, and then worldwide—were submitting electronic checklists that included what scientists call metadata—where, when, and for how long each outing lasted, and how many of all species (not just the rare goodies) they had seen. Birders could submit their data from the field, on their smart phones. Automated filters flag any unusual or suspicious sightings; regional human reviewers later perform a quality check on anything out of the ordinary. You can upload photos, video, and audio recordings to back up your identification (or just because you got some cool photos that day). Each checklist provides a detailed snapshot of the birdlife in a single place at a single time— multiplied by millions. eBird has been on an exponential growth curve ever since. In 2012, after a decade, it marked its 100 millionth observation; the next 100 million took just two more years. By 2018 the total had grown to 590 million observations, 17 million of them submitted in the month of May that year alone. eBird's database continues to grow by 30 or 40 percent a year, and now includes data on all but a handful of the world's roughly 10,300 species of birds. (Full disclosure: I am a pathetically bad eBirder, as my laughably poor personal statistics remind me every time I log on to the site.)

As eBird's creators hoped, all that information has proven to be a mother lode for scientists and conservationists to mine. One of the first, and most eye-catching, developments using eBird data was animated "heat maps" that the Cornell Lab produced, showing the seasonal spread of migratory birds across North America—a northward flood of yellow and orange across the map, showing the week-by-week movements of species like wood thrushes or lazuli buntings, averaged from millions of checklists, and using computer modeling based on habitat and environmental conditions to fill in the gaps. Areas where the species are especially abundant, based on eBird records, glow brilliantly, but wisps of color shade even regions where a species is rare. Such data visualizations have revealed previously unrecognized stopover areas and migratory corridors, or shown that some species actually have distinct, disjunct regional populations that require targeted conservation action. They've allowed researchers to map how much of a given species' range is protected as public land, and how much is held by private landowners.*

The first big, real-world test of how eBird could drive on-the-ground conservation, though, came in the Central Valley of California. Known today as an agricultural powerhouse, the valley was once an avian paradise, with more than 4 million acres of wetlands through which passed incomprehensibly large numbers of waterfowl, shorebirds, and other migratory species—up to 80 million birds, by some estimates. Now just 200,000 acres, less than 5 percent of the

---

* Perhaps inevitably, eBird has also given rise to Fantasy Birding, the brainchild of birder and web developer Matt Smith in Virginia—like fantasy football or baseball, except that instead of building a team, the player chooses a different real-world location each day, like Cape May, New Jersey, or Attu Island in the Aleutians, and the fantasy results are the product of actual checklists submitted by birders in those locations on those days. Like a Big Year, in which birders compete to find out who can see the most species within a geographic region like a state or continent, Fantasy Birding is all about racking up the biggest list, driven by eBird data. And if that seems esoteric, consider that some birders have compiled a list, now totaling more than a thousand species, of birds inadvertently photographed by the Google Earth Street View camera cars and later spotted online by exceedingly patient enthusiasts.

original wetlands, still remain. Into them are crammed 3 million ducks, 1 million geese, and half a million shorebirds that overwinter in the valley, plus many more that pass through during migration. As a result, the Central Valley is ranked as one of the most important—and threatened—bird habitats in the country. Converting farmland back to permanent wetlands for birds would be nice, but given the value of Central Valley real estate for farming, it's prohibitively expensive to purchase land there for conservation, even for deep-pocketed organizations like the Nature Conservancy.

But why buy when you can rent? Land managers at TNC had a brainstorm—they knew that some ag lands, like rice fields, can provide good habitat for waterbirds, provided they're flooded at the right time and to the proper depth. Looking at eBird data, they also realized that many migrants, especially shorebirds, were only using the Central Valley for a few weeks at a time, and they had high-resolution data from NASA on surface water conditions, which allowed them to predict where and in what quantities water would be available exactly when the migrants needed it. Combining the data sets, in 2014 the conservancy for the first time held a sort of reverse auction, where farmers bid on the chance to flood their rice fields—something they normally do anyway to help clear the stubble from the previous year, although often at depths too deep for shorebirds to use effectively. Farmers would be paid to keep a few inches of water in their fields in late summer and early autumn, when shorebirds are migrating south through the region, and again from about February through April, when NASA data showed that surface water was at its scarcest.

From a small start that first year, the so-called pop-up wetlands program (more formally known as BirdReturns) has blossomed across more than 50,000 acres of Central Valley farmland, allowing conservation managers to create shallow wetlands from autumn through spring, as and where needed. Financially, it's been a huge savings. That first year, BirdReturns was able to lease more than 15 square miles of habitat for $1.4 million, keeping it in water during the eight weeks or so that the birds needed it. Buying and restoring that acreage, by TNC's calculations, would have cost $175 million, or 125 times as much. The birds needed no convincing. Observers found three times as many shorebird species, and five times as many individual birds, on the pop-up wetlands compared with fields that weren't enrolled in the program, and numbers were an order of magnitude greater on the BirdReturns tracts compared with rice fields that were flooded as part of the normal farming rotation. The fact that BirdReturns started in the midst of California's historic drought only amplified the value of the initiative—as well as its flexibility, because when the state went from record drought to record winter moisture a couple of years later, TNC simply scaled back the program accordingly.

The folks managing BirdReturns were using radar to track where and in what numbers the migrants were moving in the Central Valley, but eBird fleshed out the picture, providing the details of what species were using which areas at what times, all thanks to thousands of birders who were simply passing on the results of their recreational visits to the region. It was the first indication that data from eBird could make a demonstrable difference for bird conservation, but it wouldn't be the last. Along the Gulf of Mexico, the radar analyses from scientists like Jeff Buler and others were identifying important stopover sites—but what species were in those clouds of birds being caught on radar as they climbed into the sky each night to begin their migrations? Here again, eBird held many of the answers—although a lot of the places radar pinpointed as critical sites for birds were well off the beaten path, and even the most dedicated birders have a hard time getting into the remote, swampy bottomlands that often comprise the best bird habitat in that part of the world. But by combining eBird data from places that *are* birded regularly, along with high-resolution, fine-scale satellite imagery showing land cover and habitat types all along the Gulf, it's possible to extrapolate from heavily birded places to rarely birded sites. If you know from eBird records that a particular combination of topography, forest cover, and habitat types usually holds hooded, Kentucky, and yellow warblers, but not a lot of gray-cheeked thrushes or white-eyed vireos, you can train a computer model to find similar patches of habitat and fill in the blanks, explained Daniel Fink, a computer statistician at the Cornell Lab who is another of the many specialists working on the Gulf of Mexico project. "You can train the model to learn the associations between an observed pattern of occurrence of a species, and the types of land covers and habitats where it tends to occur and tends not to occur. Then we can make predictions about a set of locations, most of which eBirders have never been to, about the expected occurrence or abundance of that species at that location."

Fink, I was surprised to learn, is not a birder—he's a numbers guy, and in this case, the interesting numbers around which he's modeling just happen to be birds, and the fascinating patterns their movements create. When I first spoke with him in late 2016, he mentioned one intriguing pattern that didn't make a lot of sense. Modeling habitat preferences for autumn migrants in the East based on eBird data, a team that included Fink found that forest-nesting songbirds seemed to be found most often in human-modified habitats, including urban areas. Look at the models for wood thrush, he said, "and what you see as fall comes in is that deciduous forests decline [in importance] and the cities light up" with birds. "You see this urban footprint emerging. Something's happening in migration, but what?"

As has become plain since that conversation, the reason is as simple as it is pervasive—urban lights. In the past few years, it's become clear from several lines of Big Data evidence that city lights are dramatically reshaping migration, especially in autumn when there are millions of naive, young birds on their first journey south. Early indications of this issue came, as Fink said, from eBird, which reported some of the highest densities of migrants in urban parks. Initially, that was chalked up to a bias caused by lots of urban eBirders, but radar data later showed that huge numbers of migrants were, in fact, being drawn toward cities.

Remember, birds evolved to navigate and orient with the faint light of stars, not the pall of wasted illumination that bathes the night sky near even the most modest urban center—light that is visible to a flying bird from as far as 190 miles away. Nor can a bird really escape it; skies over at least 70 percent of the United States, and 40 percent of global land area, are so badly polluted by light that the Milky Way is no longer visible. Ornithologists have known for generations that artificial light disorients birds; as far back as the 1800s, lighthouse keepers described huge bird kills on foggy nights, when migrant songbirds battered themselves against the glass. Lighted skyscrapers remain a major cause of mortality during migration, so much so that campaigns have been launched in many cities to convince building managers to turn off the lights during the peak of migration, while volunteers each morning pick up the hundreds or thousands of dead and dying songbirds that have tumbled to the sidewalks below. Some 90,000 birds a year die from building collisions in New York City alone, where even the Tribute in Light—the 9/11 memorial each September that features two dazzlingly bright beams of light shining up from lower Manhattan—has proven to be a hazard to migratory birds. New York City Audubon raised the alarm when the tribute was first proposed in 2002, because September 11 coincides with the peak of passerine migration in the Northeast. Some years the heaviest flights have occurred on the same night as the tribute—as happened in 2010, when after days of rain dammed up migrants to the north, the skies cleared and birds flooded over New York the night of the Tribute. Radar studies have shown that the twin beams concentrate migrants at 150 times the normal rate. Since 2005, New York City Audubon has had an agreement with the Tribute organizers that when monitors detect at least 1,000 songbirds, bewildered and trapped whirling within the beams, the spotlights are turned off for a period of time to allow the weary fliers to disperse.

It now appears the same thing is happening at a continental scale, where Big Data is showing both the scope of the problem, and some possible solutions. One radar study involving Jeff Buler found that the density of migrant birds in autumn increased with proximity to urban light sources, even though the best habitat was farther away in darker regions; birds were being pulled toward cities like moths to a flame, to areas where there was less quality stopover habitat for them to use, and where the danger from collisions with buildings, communications towers, and other obstacles was much greater. A Cornell study led by Kyle Horton, also using archived radar data, found that migrants' exposure to artificial light was greatest in autumn in the East, but in spring in the West, especially along the Pacific Coast. (Horton's team also found that just 5 percent of US land area produces almost 70 percent of the artificial light, with Chicago, Houston, and Dallas the worst offenders among the 125 largest cities, given their light pollution levels and their locations in the middle of major migratory pathways.)

There may be a lesson, and a silver lining, in this news. The lesson is that urban land conservation may be far more important for migratory birds than anyone has ever realized—not just protecting remaining land from development, but improving and restoring urban parks (many of which are overrun with exotic invasive plants of limited value to birds, and which are managed more for esthetics and human recreation instead of wildlife). In terms of producing the maximum value to birds in the greatest need, restoring habitat in a fairly small urban park may be more important than setting aside a significantly larger tract of land in some more distant area.

As for the silver lining, Horton's team found that the great bulk of the migration, spring and fall, occurs in a very short period—each season, about half of all migrants move through any particular urban area in just six or seven days. Andrew Farnsworth, who was part of Horton's study, believes that fact opens the door to very specific, highly targeted automated alerts—a way to get cities and landowners to turn off their lights, for just a few nights every season when the great majority of migrants are passing through. It's a way of turning those BirdCast migration forecasts into a force for conservation, at a scale greater than the local lights-out alerts already given in cities like Houston. "Being able to use Big Data to generate baselines [for migration], we'll be able to look at what's forecast and say, 'This is going to be a night of migration that's off the charts, so we need to make a concerted effort to shut off lights,' then push through an automated message to the Midwest to turn out the lights in Chicago, or Cincinnati, or wherever."

But the fact is that sometimes conservationists feel as though they're playing whack-a-mole, as new challenges arise just as they begin to grapple with old ones. During the course of our conversation, Farnsworth mentioned the launch, a few weeks earlier, of the first of what is expected to be *12,000* small, internet-servicing satellites from Elon Musk's company SpaceX, a project called Starlink with the promise—or threat, depending on your perspective—of creating an artificial galaxy blanketing the sky. Astronomers went ballistic when the first of these small, brilliant objects were lofted into low-earth orbit in 2019, concerned that the eventual "mega-constellation" (as it's been described) would interfere with their ability to study the stars, and alter the character of the natural sky to everyone, everywhere on the planet's surface—and that was before Musk said he was seeking permission to add a further 30,000 satellites to the total. The Federal Communications Commission, which approved Starlink, assured the public that SpaceX would "take all practical steps" to protect astronomy. But no one, so far as Farnsworth and I could tell, stopped to wonder what effect a "mega-constellation" would have on billions of migratory birds, already trying—with ever-diminishing success— to find their way through a night sky bleached of its darkness.

Big Data has opened our eyes to the magnitude of migration; it has pinpointed problems and suggested avenues to rectify them. And it may finally be giving us a sense of how enormous the stakes are for migratory birds and those of us who cherish them. In the final months that I was working on this book, I heard rumors of a blockbuster analysis working its way through the peer-review process toward publication in one of the world's premier journals, a paper that would finally try to put into numbers how North America's bird populations have changed in the past half-century. When the paper was published in *Science* in September 2019, it proved to be as explosive and sobering as the rumors suggested, and confirmed the experience of every long-time birder who ever lamented that there just aren't as many birds as there used to be.

The analysis was a joint effort of some of the leading migratory bird researchers, a who's who from the Cornell Lab of Ornithology, the Smithsonian Migratory Bird Center, the US Geological Survey's Patuxent Wildlife Research Center, the Canadian Wildlife Service, and the Canadian National Wildlife Research Centre, among others. They drew from dozens of lines of Big Data evidence—decades of Doppler radar archives from the entire lower Forty-Eight; long-running censuses like the Christmas Bird Count dating back to 1900; monitoring schemes spanning half a century like the Breeding Bird Survey, which methodically counts singing birds every year at thousands of locations; and more esoteric and specific efforts like annual tallies of breeding goose colonies in the Arctic, migrant shorebirds, coastal sea ducks, swans, and many other groups of birds. Their conclusion: Since 1970, some 3.2 billion birds—almost 30 percent of the continent's total breeding population—have disappeared from North America, a reduction that was almost perfectly reflected in a similar decline in nocturnal bird migration captured in the radar archive. The losses were heaviest in the East, where the birdlife is dominated by migrants traveling between the tropics and temperate or boreal forests.

The heaviest losses were in 12 families of birds, including sparrows, warblers, blackbirds, and finches, all down 37 to 44 percent; in all, 38 families of birds each lost at least 50 million individuals. Shorebirds, the authors warned, are experiencing a "consistent, steep population loss" of 37 percent, but the group in the greatest peril are grassland birds like meadowlarks, grasshopper sparrows, and longspurs, which lost more than 700 million breeding individuals since 1970—more than half their original population. Three-quarters of all grassland bird species are declining. Perhaps most surprisingly (and worryingly), the analysis showed that even introduced species— birds that conservationists considered little more than feathered rats

because they were so adaptable—are losing ground; European star-
ling populations have fallen by half, and house sparrow numbers
have fallen more than 80 percent since 1970. So are species like red-
winged blackbirds that are habitat generalists. As the authors pointed
out, passenger pigeons once numbered in the billions but they vanished in a matter of decades—"a poignant reminder that even abundant species can go extinct rapidly."

The news, however grim, was not all bad. Some groups, like vireos, woodpeckers, nuthatches, and gnatcatchers, bucked the trend
with often respectable increases. Raptors, which had suffered from
both direct persecution and pesticide contamination in the early and
middle twentieth century, have rebounded in recent decades. Wetland birds, especially waterfowl, showed the biggest gains, and why
isn't a mystery—waterfowl have large, well-funded, and politically
powerful constituencies like duck hunters behind them, and billions
of dollars (both from private groups like Ducks Unlimited, and from
state, provincial, and federal resource agencies) have been directed
into wetlands protection and restoration.

Globally, things aren't much better. Similar studies have outlined
massive declines in European birds—a loss of more than 400 million
individuals between 1980 and 2009, according to one analysis, especially among more common species. Again, it's not all gloom—in
Europe as in North America, targeted conservation action for the
benefit of rare species has enjoyed great success, but the cost of waiting until a species is nearly gone can be enormous, and sucks up much
of the available funding for conservation work in general. Farmland
birds have been hit especially hard in Europe, where agricultural
intensification has crowded out nature almost entirely in favor of
chemically soaked crop monocultures, a trend that has been especially evident in eastern European countries like the Czech Republic,
which saw steep declines in farmland birds after they joined the European Union. In 2018, alarm bells rang all across the continent when
French scientists announced that bird populations there had dropped

by one-third in barely 15 years, with some species, like meadow pipits, down by 70 percent. "The situation is catastrophic," one of the
scientists said. "Our countryside is in the process of becoming a
veritable desert." The French scientists blamed the use of ever more
powerful insecticides, just as German entomologists were saying they
had documented a more than 80 percent fall in summer insect numbers in that country—the so-called insect apocalypse, evidence for
which is pouring in from around the world, including places like
seemingly pristine tropical forests. You can't have birds if they don't
have enough to eat, and evidence is building around the world that
the very foundations of the planet's ecosystems are crumbling.

Shortly before *Science* published the North American analysis,
I spoke with Dr. Peter Marra, one of the coauthors, who had just
stepped down as the director of the Smithsonian Migratory Bird
Center to take a teaching position at Georgetown University. He'd
shared a draft of it with me, and agreed that the paper made pretty
somber reading. "But now we know the score," Marra told me. "The
fact that introduced species and generalists are in trouble ought to
really scare us. But we also know that it's possible to turn things
around—look at how waterfowl responded to wetlands protection."
There is a lesson for us, he said, even at this late date, that targeted
conservation can work if we put the money and muscle into making
it happen. For the groups of birds in the most desperate shape, like
grassland birds, that model of intense focus on habitat restoration
actually has the most immediate chance for success.

Still, it's hard to look at the picture Big Data has painted, to know
that in my lifetime a third of the birds on this continent have disappeared, and not feel more than a little despair. So I was grateful
for another conversation I had, just a day or two after I talked with
Pete Marra—one that was a reminder that every once in a while, the
world offers up a vision of what once was, a glimpse of what the raw,
undiminished force of migration must have been like in an earlier
and more complete age. Every once in a while, the world shows us

what we're fighting for, what—with a little luck, and the will to make some difficult changes—could possibly be again.

The Côte-Nord, the north shore of the St. Lawrence River in Quebec, isn't a world-famous birding spot, but it should be. The town of Tadoussac, where the Saguenay River empties into the St. Lawrence, has a reputation (mostly among Canadian birders) as a notable autumn migration choke point for hawks, owls, waterbirds, and songbirds—a place where one might see 6,000 pine grosbeaks in a day, or 8,000 Bohemian waxwings. A local bird observatory, Observatoire d'Oiseaux du Tadoussac, has been doing counts and banding there since the mid-1990s. Few birders outside Canada, though, had ever heard of Tadoussac until recently; few knew that, when geography and weather align just right, it can produce a unique springtime migration spectacle. Ian Davies knew, and that's what he was hoping to see for himself.

Davies, a project leader at eBird, had seen the checklists coming out of the Tadoussac Bird Observatory's spring counts, describing occasional days, maybe once or twice every few years, when tens of thousands of migrating songbirds—mostly colorful warblers—concentrate along the glacial sand dunes that rise as sheer bluffs more than 200 feet above the St. Lawrence. That's a sight increasingly rare in these days of globally diminished migration, so Davies and four other birders—all of them, like him, employees at the Cornell Lab of Ornithology—headed north from Ithaca, New York, at the end of May 2018 in hopes of seeing it for themselves.

At Tadoussac the St. Lawrence is more than 15 miles wide, its frigid waters populated by blue whales, cod, and belugas. The northern edge of the river valley represents an abrupt transition, as the land climbs away from the estuary to the plateau known as the Canadian Shield, its cold boreal forest stretching north all the way to the subarctic tree line. Even in late May, spring is a reluctant season here, the scrubby birches that cling to the escarpment clothed in small,

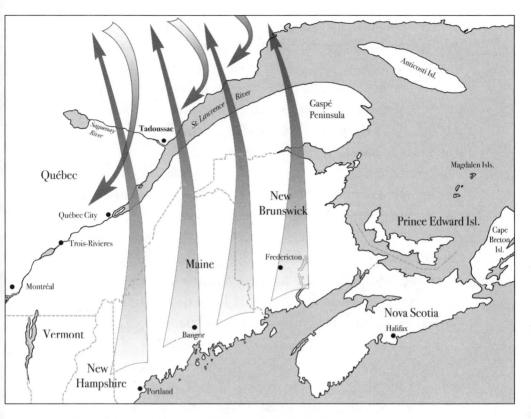

On occasion in spring, tens of millions of songbirds migrating north at night cross the St. Lawrence River and, at daybreak, encounter wintry conditions in the boreal forest beyond. They must reverse course, piling up along the north shore of the river in staggering numbers.

freshly opened leaves. Go a few dozen miles farther north, and you find a land still deep in winter—which, for migrating birds, can be a calamity. Songbirds riding warm southerly winds through the night often cross that boundary without realizing it, and at dawn discover they are among spruce woods still buried in snow. For seedeaters like sparrows, that's not too big a deal; they can scratch through the crust and find food. But insectivores like vireos, flycatchers, and warblers have no choice. They must turn around and fly back to the warmer St. Lawrence lowlands where the ground is bare, the trees have already leafed out, and insect food is available. This is known as a reverse migration, and because they would rather not cross the wide river, the birds turn when they hit it and move southwest along the shore in great waves. That's what the Cornell team was hoping to see.

"So we went up there for eight days, basically hoping we would hit the right combination of factors and get maybe 50,000 birds," Davies told me. "Fifty thousand in our wildest dreams. The first three days had just been totally blah—foggy, winds from the wrong direction, only tens or hundreds of birds moving. After a couple of hours we'd give up and go birding somewhere else." Then the winds changed, coming from the south; it looked promising, but the Cornell team couldn't be sure, because one of their best tools, Doppler radar, was unavailable—Canadian radar sites automatically filter out any bio-clutter like birds or insects. "In Canada there's no public, unfiltered radar, so you really don't know what happened the night before," Davies said. "There's this unknown, this mystique."

The next morning—May 28, 2018—it seemed at first that the string of blah days was continuing. The team arrived on the dunes at 5:45 in the morning to find more rain, and two Quebecois birders who happened to be there as well, François-Xavier Grandmont and Thierry Grandmont. "The first couple of hours were just basically nothing, a little bit of drizzle, and we were like, 'Okay, this is it. It looked good, but I guess not.'" Then small bunches of birds began to appear in groups of five or ten, raising everyone's spirits. At about 6:30 a.m., Davies said, "there was this point where we were looking out from this dune over the peninsula, across the St. Lawrence, and there was just this wall of dots. Just, the sky was covered with birds, basically. And it was like that for the next nine hours."

What followed was, as Ian Davies began the now-famous eBird checklist he posted that night, "the greatest birding day of my life." He and his friends—using some simple math to record the rate at which birds were passing, and estimating what proportion of the flight each species comprised—kept up a systematic count through the long, exhausting, exhilarating day. The songbirds, almost entirely warblers, were often flying by at a rate of 20 per second, more than

1,000 a minute. A single binocular view might encompass several hundred to several thousand songbirds. The video clips that Davies shot and posted to eBird as part of his checklist give only a sense of the scale of the event. It wasn't as though, he told me, there was a solid phalanx of birds moving in huge, coherent flocks; rather, there was a constant sheet of birds flowing over and through the landscape, following the edge of the river to the southwest, a universal flickering of movement at every level, from flying high in the sky to scraping the ground at the observers' feet—and legs, between which birds constantly were passing. Warblers flew into Davies's head, they bumped into his camera and arms. Hundreds more foraged on the ground or in the shrubbery all around them, looking hungrily for food.

The birds were everywhere, momentary flashes of color and pattern, each one gone in a second, leaving the briefest of impressions. Vireos, thrushes, and flycatchers stayed low in the brush, along with warblers like black-throated blues, magnolias, and Canadas, sheltering from strong winds. More powerful fliers like Cape May, baybreasted, yellow-rumped, and even tiny little Tennessee warblers were high up in the sky. The air was filled with what Davies called "just a nonstop sonic ocean of flight calls" that the birders had to consciously try to tune out. Trying to identify the darting shapes, the team used field marks that they'd never relied on before when watching warblers perched or at rest. Binoculars were almost superfluous. "It was crazy. It was numbing. It was one of those things that feels like a dream," he told me. He and his friends had no plan in place for trying to count such overwhelming numbers, since they never expected to encounter anything like this. "We were basically just coming up with this in the moment," he told me later. Every time the flow rate changed—more birds in the stream, fewer migrants passing their spot—the counters would start a fresh estimate, noting the time. When the winds shifted strongly from the southwest in early afternoon, the intensity reached its peak, with warblers passing at a rate of 50 per second—some 72,000 in half an hour. It was so massive a flight that even the algorithms on Canadian radar sites, unable to filter out so much bioscatter, showed massive blobs of migrating birds all along the north shore of the St. Lawrence.

"There was this war between trying to document and understand what was happening, while also trying to just enjoy the most incredible natural spectacle any of us had ever seen," Davies said. Except for sending someone to make a few lightning food runs into town as the day wore on, no one stopped counting and estimating and tallying until the flight slowly diminished by about 3:30 that afternoon. It wasn't until then, when the team began extrapolating their minute-to-minute flow rates and assigning percentages for each species, that the true enormity of what they'd seen sunk in. In all, the team estimated that more than 721,000 warblers had passed their observation point on the dunes, plus or minus about 100,000 birds. The total included an estimated 72,200 Tennessee warblers (10 percent of the flight); 50,500 American redstarts (7 percent of the flight); and 108,200 each of magnolia and Cape May warblers (15 percent each). The most common species that day was bay-breasted warblers, of which some 144,300 (20 percent of the total) were counted. (If that total is accurate, it represents 2 percent of the species' global population. Tom Auer, an eBird researcher who was on the team, put it more simply on Twitter: "Have seen more Bay-breasted Warblers today than my entire life X 100.") There were some 28,900 Blackburnian warblers, 72,000 yellow-rumps (which dominated the earliest and latest hours of the flight), and more than 14,000 Canada warblers—and adding to the sheer eye-bulging spectacle of it all, a lot of the warblers were adult males in breeding plumage. Almost 110,000 were simply recorded as "warbler spp." because they couldn't be identified. Even the staff at the bird observatory, which has been monitoring the spring migration for years, said they'd never seen the equal of that day.

"No one's going to believe us," Ian Davies remembers the team saying as they added up their totals. Actually, no one seriously questioned the totals; this was the perfect conjunction of a team with the credentials and sophisticated skills necessary to identify dozens of species of small birds, on the wing and passing in overwhelming numbers, while also making constantly changing estimates of both flow rates and species proportions. Still, I may be one of the few people who knows exactly how he and his friends felt in the immediate aftermath, how surreal the situation becomes when the magnitude of what you've witnessed begins to sink in, and when you're unsure whether you can convince anyone who didn't see it that you're not cracking up. In 1992, I was part of a team in eastern Mexico tasked with documenting the only recently recognized autumn raptor migration through the coastal state of Veracruz. At the time, no one knew that some 4.5 million hawks, eagles, kites, falcons, and vultures poured each autumn through the narrow choke point between the inland mountains and the Gulf of Mexico, and as our daily counts kept climbing—40,000 raptors, 60,000, 88,000, each one far surpassing the highest one-day totals for raptors at any count site in the world—we were already getting slightly snarky, slightly disbelieving comments from colleagues back home. Then, after several days of poor weather that clogged the migratory pipeline, much as happened at Tadoussac, the skies opened up, and we counted nearly half a million raptors in a single astounding, grueling day. We now know that such half-million-to-1-million-bird days are typical in Veracruz when the weather is right. But at the time, sitting in shock in the humid twilight at the end of the day in Mexico, staring at the numbers we'd tallied, my friends and I, too, simply thought: no one will ever believe us.

If there was anything truly unbelievable about that May day along the St. Lawrence, it was the feeling many of us had that such spectacles were long gone; it felt almost naive to accept that such numbers were still possible. Ian Davies's checklist was like a gift from a lusher, richer past, reminding us of what might be possible if we can check the slide toward oblivion. News of the mega-migration at Tadoussac spread rapidly through birding circles and beyond, including the pages of the *New York Times*. Perhaps not surprisingly, it was held up by at least one climate change denier, who also disputes claims about declining bird populations, as proof that the bunny-hugger doomsayers of all sorts are wrong. And did he have a point? As Davies explained to me, there was nothing especially unique about their location on the Tadoussac dunes that day—they would have seen much the same thing anywhere along the 186-mile (300-km) shore of the St. Lawrence, where small birds were concentrating in such numbers that nonbirders posted photos of hordes of exhausted warblers crowding highway median strips and backyards. There were millions and millions of songbirds passing through southern Quebec that day, enough to make anyone doubt the predictions of doom. The difference, of course, is that what today is astounding would, as we know from radar archives, have been fairly commonplace even just a few decades ago, had anyone known to go to Tadoussac and watch for it.

Bird migration is a shadow of what it once was, but that shadow is still mighty enough to leave us slack-jawed and awestruck, at the right time and in the right place. There are still billions of migratory birds. Although the hour is late at least, as Pete Marra said, we know the score. And that includes the realization that each small bird flying north through the Canadian woods carries with it the echoes of the previous winter, where conditions in tropical lands thousands of miles away and many months earlier may predestine it for success or failure—an aspect of migration that is only now coming into focus, and providing yet another critical element in our understanding of how to keep the billions aloft, and safe.

Five

HANGOVER

You are a Kirtland's warbler, the newly risen sun shining off your pale yellow breast and illuminating hundreds of acres of low, scrubby jack pine forest in every direction. Juncos, hermit thrushes, and savannah sparrows sing from the thickets, an upland sandpiper wolf whistles from the top of a dead snag, but you spare them not a thought. Having arrived overnight after your long migration north, your only urge is to sing your own song to the dawn—a driving, six-syllable phrase that is both a warning to potential rivals and an advertisement to potential mates, and which declares this patch of northern Michigan yours.

The weather is gentle, the food abundant, nest sites ample; conditions could not be more promising. Yet none of that may matter, because months earlier and 1,500 miles away, the winter rains failed to fall in the Bahamas. As a result, food there was scarce, and you struggled to put on the fat you needed as a migratory fuel. You were late leaving on migration, and late arriving in Michigan. Even before you sing the first note of your first spring song, you are behind the eight ball.

Scientists once thought of winter as a respite for a migratory bird, an easy-living, tropical hiatus from the serious work of migration and reproduction. But they're learning that a bad winter casts a very long shadow indeed, an ecological hangover that can linger for many months and across thousands of miles. Sparse rain and limited food create a caloric deficit that delays the start of a bird's migration and may even force the migrant to actually cannibalize its own muscle and organs to make the trip. It raises the already substantial odds of dying on the journey, and even if the bird arrives on the breeding grounds to find ideal conditions, those shortages may sabotage breeding success.

Finally, given that tropical wintering areas on which hundreds of millions of migratory birds depend are warming and drying— a trend that is expected to accelerate in the decades ahead—this discovery has ominous implications at a time when migrant populations are already in steep decline.

It's ironic, therefore, that the species likely to shine the most illuminating light on what scientists call "carry-over effects" is Kirtland's warbler, a bird that was nearly extinct just a generation ago, and which is routinely hailed as an unparalleled conservation success. The very aspects of its ecology that brought it to within a hairsbreadth of extinction—its highly specialized habitat requirements, the incredibly restricted size of its breeding and winter ranges—make it the ideal lens through which to understand the causes and consequences of carry-over effects. Those same elements, though, may also make its future as uncertain as its past.

So to understand what is happening in the jack pines of northern Michigan on this early May morning, it's necessary to turn back the clock a couple of months, to the end of winter on a small island in the central Bahamas.

Nathan Cooper is driving as fast as he dares, through murky twilight along a twisting road with an unsettling number of pedestrians, free-range chickens, loose dogs, and feral cats. The grandly named Queen's Highway is a narrow, unmarked strip of potholed macadam that runs the 48-mile length of Cat Island. We need to be at the far southern end by sunrise, and we're late.

Cat Island is well off the main tourist drag in the Bahamas. Besides a fishing marina and a few small resorts, it's best known (to the extent it's known at all) as the boyhood home of Sidney Poitier. At one time scholars thought it might have been Christopher Columbus's first landfall in 1492, but historians generally abandoned that notion a century or more ago, leaving Cat in obscurity. Shaped like a long, narrow fishhook, the island covers just 140 square miles, and is so slender that for much of its length it's only about half a mile wide. From the window of a cramped puddle jumper from Nassau the previous afternoon, the island looked flat and largely featureless, a lot of dry scrub forest bisected by a few, mostly sand roads, and rimmed by beaches, white breakers, and blue water. There are barely 1,500 residents, and I had to look hard to spot the relatively few houses, which mostly hugged the coast or the Queen's Highway. Now, as we drive through the dawn, I'm surprised that so many of the buildings I do see along the road are abandoned.

"Yeah, sometimes I think there are more empty houses on this island than occupied ones," Cooper says. We pass dozens of ruins as the miles click by, their roofless, gray limestone walls open to the sky. The population here is half what it was in the 1950s, as the bigger resort islands with better jobs (or the American mainland, just 300 miles away) have drawn off young people who see few prospects on Cat, where slash-and-burn farming, raising goats, or fishing for conch are among the only options for subsistence.

But what makes Cat Island a tough place for people—its hot, dry climate and hardscrabble soil, the scrubby forest full of highly toxic poisonwood trees, even the herds of ravenous goats—make it arguably the best wintering spot in the world for the exceedingly rare Kirtland's warbler, a handsome bird with a lemony breast, blue-gray back, and broken white eye-rings. Perhaps 1,000 of these half-ounce birds, a quarter of the global population, migrate to this relative speck of land, whose arid scrub forest provides exactly the habitat the warblers prefer. And it's why Dr. Nathan Cooper and his crew are back on Cat for their third winter. A postdoctoral fellow at the Smithsonian Migratory Bird Center in Washington, DC, Cooper is capitalizing on the warbler's unique biology to learn more about how carry-over effects dictate the lives of migratory birds to an extent no one would have believed even a few years ago. For the first time anywhere, new technology will let his team follow individual warblers from the wintering grounds to their breeding territories, and directly measure how their condition here in the Bahamas impacts their later nesting success in Michigan.

Cooper had previously tracked Kirtland's warblers with light-sensitive geolocators of the sort my colleagues and I used in Alaska, a study that made a splash in the popular press because his study subject is so rare. That project revealed hitherto unknown details about the Kirtland's broad migratory route, but geolocators have their limits; they provide only a very rough estimate (accurate to within about 150 kilometers or so) of the bird's location, and only after one recaptures the bird and downloads the stored data the following year—assuming you can recapture the bird. This season, Cooper is using nanotags, those tiny radio transmitters that weigh as little as two-tenths of a gram, which are tracked by the Motus system of automated receiver stations that my friends and I have been installing all over the Northeast in recent years, part of what has become an international network stretching from the Arctic to South America. It is the middle of April, and if all goes well, when the warblers leave Cat Island in a few weeks those receiver sites will allow Cooper to follow the birds as they migrate north, making landfall in Florida and Georgia and pushing on to the Great Lakes. Once they reach Michigan, directional antennas atop 11 carefully sited Motus towers, covering basically the entire core breeding range of the species, will allow him and his crew to quickly relocate the tagged birds and begin to monitor their nesting success.

This plan wouldn't work with any other songbird, because no other North American passerine has such a circumscribed range, thus making it plausible to find the same individual at both ends of its migration. Kirtland's warblers seem always to have been hyper-local, and were long an enigma. The first specimen was collected in 1851 in Ohio, named for renowned naturalist Jared Kirtland, on whose farm it was found, but only four more surfaced over the next quarter-century. Twenty-eight years later scientists found their wintering grounds in the Bahamas, but it wasn't until 1903, more than half a century after its discovery, that the first Kirtland's nest was found in northern Michigan and the mystery of their breeding grounds resolved. The bird was, scientists realized, the ultimate habitat specialist. It nests almost exclusively in dense stands of young jack pine, a scrubby, short-lived conifer that thrives in sterile, sandy soils, and which reaches the southern limit of its boreal range in Michigan's northern Lower Peninsula. The jack pine, in turn, is a fire specialist; its curved, pecan-sized cones cling to the trees season after season, in ever-growing numbers, opening only after they have been scorched by flames, and then releasing seeds at a staggering rate of 2 million per acre. A few years after a fire sweeps away a mature forest, millions of fast-growing jack pines form a head-high ocean of brambly, green-gray boughs, beneath which the warblers build their well-camouflaged ground nests among the lichens and blueberries.

It was a system that worked well in the Pleistocene, when Ice Age glaciers pushed the jack pine ecosystem down to the sandy coastal plain of the Southeast. From there it was an easy flight to the nearby Bahamas, which—because sea levels were hundreds of feet lower than today—were 10 times their modern land area. As the glaciers retreated, the jack pines emigrated north, century by century, their dependent warblers in tow, each year stretching the birds' migration a bit farther to the ever-shrinking Bahamas. But the system still worked, as natural and Native-set fires maintained the jack pines. Later, the almost complete clear-cutting of the Michigan forests in the late nineteenth century, and devastating wildfires that swept millions of acres and took hundreds of human lives, may have unintentionally created still more nesting habitat.

The twentieth century was a very different story, though, with Smokey Bear's finger-wagging message about fire suppression. The

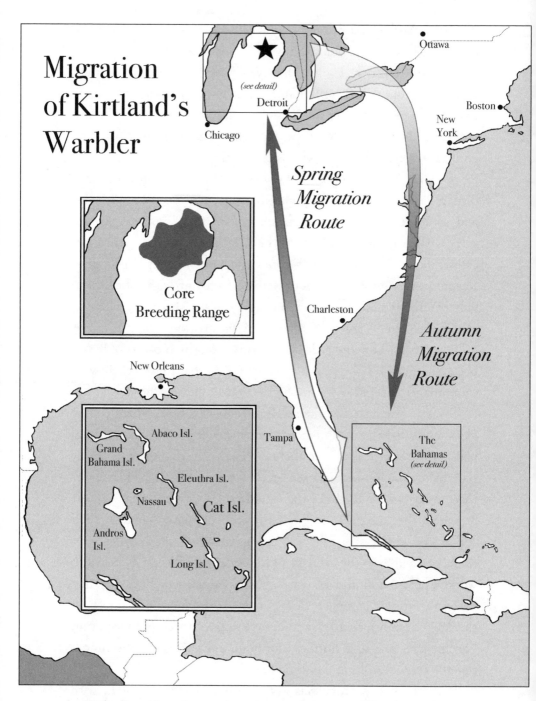

Migration of Kirtland's Warbler

Core Breeding Range

Spring Migration Route

Autumn Migration Route

Ottawa

Boston

New York

Detroit *(see detail)*

Chicago

Charleston

New Orleans

Tampa

The Bahamas *(see detail)*

Abaco Isl.

Grand Bahama Isl.

Eleuthra Isl.

Nassau

Cat Isl.

Andros Isl.

Long Isl.

Kirtland's warblers migrate each year between their very small breeding range, primarily in northern Michigan, and their winter home on a few tiny islands in the Bahamas.

forests grew and thickened, without cleansing fires to reset the clock. Suitable habitat disappeared, and in the few places where fire did create fresh jack pine stands, nest parasitism by brown-headed cowbirds (which were newly arrived in the region) further pared productivity. Probably never very high, the warblers' numbers fell, and in 1967, when the first federal endangered species list was made, Kirtland's was on it. Four years later, when a census was conducted, scientists were shocked to find that the bird's numbers had dropped 60 percent to just 201 singing males. The low point came in 1974, when only 167 could be found.

Wildlife managers scrambled. They tried generating new habitat using prescribed fire, but in 1980 a fire that was supposed to be restricted to about 200 acres escaped, consuming 24,000 acres, destroying dozens of homes and killing a 29-year-old biologist named Jim Swiderski who was fighting the blaze. The Mack Lake fire, as it was known, did two things. It caused agencies to shy away from using fire to create Kirtland's warbler habitat, but when, 10 years after the conflagration, the warblers moved into the regenerating young forest in droves, it also demonstrated that managers had been thinking too small, that the little patches of young jack pine that they'd been creating weren't big enough. The Mack Lake fire, tragic as it was, likely saved the Kirtland's warbler from extinction. Although wildfires from lightning and careless humans still occur in the jack pine region, managers today use clear-cutting and replanting to create a mosaic of young pine forest across 23 huge blocks of federal and state land. The approach has succeeded beyond almost anyone's dreams. Today there are roughly 4,000 Kirtland's warblers, and in 2019, the birds were removed from the list of federally endangered species.

Almost all the scientific and conservation attention, however, has been directed to Michigan. Aside from shooting more than 70 of the birds for museum collections in the late 1800s in the Bahamas, all but a few scientists have ignored the more than seven months each year that the warblers spend in the islands. If Cooper and his colleagues are right, though, conditions on islands like Cat deserve an outsized piece of the credit for the Kirtland's warbler's recovery. And, the biologists fear, those conditions may also prove to be the species' Achilles' heel in a fast-changing world.

The sun is up by the time we reach the southern end of the island. Cooper makes a few turns onto progressively smaller, unpaved tracks, then parks. The ocean is just 100 yards away, the suck and boom of the wind-riled surf loud in our ears, the tops of tossing coconut palms visible, but we're surrounded by low, thick woodland. Hardly the stuff of calendar photos, it's scruffy and jumbled, the tallest trees maybe 15 feet high, the understory impenetrable. A parasitic plant called dodder is abundant in these woods, draping shrubs and even entire trees in dense blankets of fine, orange tendrils, so tightly grown that you can barely see the vegetation under the thick mats.

"We work on the roads, mostly," Cooper says as he shrugs on a bulky pack, sliding an unsheathed machete, edge up, through the straps behind his back. "We tried to bushwhack through this stuff the other day, and it took us two hours to go 600 meters."

Cooper is 37, with unruly brown curls, chin whiskers, and the compactly muscular torso of an avid rock climber. On the inside of each bicep he sports elaborate tattoos—one of a vintage fire truck commemorating his grandfather's career as a Michigan firefighter, the other with a sort of steampunk theme to mark his late father's passion for sci-fi. The scratches and bug bites scabs that overlay them are testament to the difficulties of working in the Bahamian scrub.

The other half of the team today is Chris Fox, quiet and darkly bearded, taking a break from his desk job with an Indiana conservation district to do some field work. Fox slings a heavy radio receiver over his shoulder and grabs aluminum mist net poles, while Cooper turns on a handheld predator caller and strides down the sand road at a brisk pace, blasting the song of a male Kirtland's warbler into the brush. (The always reliable David Allen Sibley transliterates the Kirtland's song as a "rich, emphatic *flip lip lip-lip-tiptip-CHIDIP* rising in pitch and intensity." It is a sound I will come to hear in my sleep in the days ahead.)

We trudge up one road, over a small hill past a new vacation home with hurricane shutters in place, and down another. A fusillade of angry *chips!* comes from the underbrush, the sign of a territorial warbler, and minutes after the two biologists set up their net along the edge of the woods, the Kirtland's launches itself in righteous rage at the sound of what it thinks is an intruder, and into the net.

The warbler, a male, weighs 16.5 grams, about six-tenths of an ounce—a couple of grams heavier than normal, and a sign that this bird has been finding lots of fruit and bugs. "He may be ready to go in the next couple of days—I think we've caught him as he's bulking up" for migration, Cooper says. In addition to routine measurements, the researcher takes a few drops of blood from a vein in the bird's wing, and when it splatters a wet dropping on his pants leg, Fox is ready, scooping it up with a small plastic spatula, bottling some in a vial of preservative and smearing the rest on a square of blotter paper—a colleague at the Field Museum in Chicago is looking at how the microbiome of the warblers changes from winter to summer, and how it may affect their survival and productivity, the first time anyone's been able to do that. Finally, the warbler gets a unique set of colored leg bands so it can be visually identified at a distance, and a nanotag, which sits low on its back, held in place by elastic loops around the tops of its legs. This radio transmitter weighs just a third of a gram and broadcasts a coded identification pulse, like Morse code, which is unique to this warbler, allowing any receiver station in the expansive Motus network to pick it up and identify it in passage.

As Cooper adjusts the transmitter and Fox fiddles with the receiver to make sure the tiny tracker is working, photographer Karine Aigner, who is photographing everything for a magazine article, glances up from her viewfinder. "There's another one," she says, and sure enough, a second Kirtland's is quietly investigating the source of the now-silenced song, bobbing its tail rhythmically as it worms through the thick brush near the net. Shortly after Fox starts the audiolure, this new male is in the hand—and before they finish processing and tagging that one, a third warbler, a female, is caught.

"I think this is our sixth female," Cooper says as he extracts the small bird from the mesh. "They just don't respond as well to the playback as the males, so this one is, in some ways, unusually valuable." Although she is (like most female songbirds) overall a bit smaller than the males of the species, Cooper finds she actually weighs a bit more than did the previous male we caught, a sign of excellent physical condition. Overall, this is proving to be one of the best days Cooper's crew has had all month, so we don't feel too bad when, as we eat a late-morning lunch of smooshed and gooey peanut butter and jelly sandwiches and chug water under the broiling sun, a fourth bird circles the net but refuses to commit itself.

The next day, though, I get a taste of how it usually goes. There was heavy rain overnight, and the air is muggy and warm even before sunrise. Biting flies and mosquitoes swarm, but the bugs are nothing compared with what the scientists had been dealing with a week earlier, when the weather was much hotter and tiny sand flies came right through the screens of the crew house. "We had to keep the windows closed and wear long sleeves and our pants tucked in our socks, even inside. And I was still all chewed up," Cooper said. "It was awful."

We fishtail back another sand road, through vegetation that's barely waist-high—perfect for Kirtland's, a bird that often forages on or near the ground. "This is what they're eating," Cooper says, pointing to a low shrub, and I do a double take trying to see what, exactly, he's talking about. "Right here—this is black torch, one of their favorite food plants," he explains, showing me minute, desiccated fruits, each barely one-sixteenth of an inch in diameter. The bird's other staple, a lantana sometimes known as wild sage or button sage, has similarly diminutive, purple fruit that are about the size of pinheads. The best habitat for these shrubs—and thus for the birds—is abandoned fields and old goat pastures, disturbed habitat that is maintained, however accidentally, by the rudimentary farming practiced on Cat.

The hot, hot sun is now overhead, reflecting off the white coral sand. We've walked a couple of miles, slip-sliding with every step. Sweat is oozing down into the corners of my eyes as the audiolure belts out the Kirtland's song and my thigh muscles ache. *Flip lip lip-lip-tiptip-CHIDIP.* Pause. *Flip lip lip-lip-tiptip-CHIDIP.* Pause. *Flip lip lip-lip-tiptip-CHIDIP.*

"Doesn't that playback make you crazy?" I ask.

"I don't even hear it anymore," Cooper says. "But yeah, on Abaco it drove me nuts—this is all we did from six in the morning until two or three in the afternoon. We were so fried that eventually we didn't even look up." He was referring to 2015, his first field season in the Bahamas, when they focused much of their survey efforts on the large, northern Bahamas island of Abaco, thought to be a stronghold for Kirtland's. It was a bust; in more than three weeks of nearly continuous searching, walking countless miles every day under a beating sun, Cooper and his crew found exactly four warblers. Abaco, they realized, has too much pine habitat and not enough of the low deciduous scrub Kirtland's prefers, and most of the old records were probably migrant warblers, not winter residents. (There might also have been a fair bit of misidentification at work, since the Bahamas race of the yellow-throated warbler—which loves pine forests like those on Abaco—is easily mistaken for its rare cousin.)

Given their poor track record on Abaco, Cooper all but fell to his knees in thanks when, at the end of that wearying season he and his crew came to Cat and immediately found dozens and dozens of the birds. They'd stumbled onto what is probably the densest wintering population of these rare warblers in the world.

Finally, we hear the distinctive chip notes of a Kirtland's, get him trapped and tagged, his blood archived, and his poop scooped. "One of the questions we're looking at is just how does the microbiome of an individual change as it moves 2,000 miles?" Cooper says as he smears the dropping across the sample paper. "How does the blood parasite load change within individuals in two different periods in the migration cycle? We don't know the consequences of having different organisms in your microbiome" at different times of the year.

This is the fifty-ninth warbler for the season, and with only a few days left, it's clear Cooper won't hit his goal of putting nanotag transmitters on 100 Kirtland's. (In the end, his team will tag 63.) It's satisfactory but not great, a result that shows in the sag of Cooper's shoulders, even as he looks ahead to months more of hard work up in Michigan finding, retrapping, and monitoring these same birds, then trying to tease out how their winter condition may have affected their later migratory speed, survival, and breeding success.

That afternoon, we sit on the covered deck of the beach house they've rented for the season, intensely blue water down a short bluff and curly-tailed lizards skittering around our feet. Fox is snorkeling in the chilly winter water, as he does almost every day, while intern Steve Caird—tall and lanky, with a scraggly beard and a bit of a Johnny Appleseed vibe—sorts through dozens of fecal sample cards which, like the vials of blood they have collected, will require rafts of import-export paperwork to bring back to the United States.

Cumulus clouds gather as Cooper outlines the history of research into carry-over effects, putting this latest effort into a broader scientific context. Soon rain is drumming on the corrugated plastic roof—fitting, since a lot of what he tells me has to do with how precipitation during the traditional winter dry season may ultimately determine a warbler's chances.

As far back as the 1970s a few waterfowl scientists—some studying swans in Sweden, others snow geese in Canada—had an inkling that wintering ground conditions might carry over into the nesting season. But most experts assumed that the real driving force in a bird's reproductive success was the quality of its nesting habitat, not what had happened to it months earlier. The breakthrough came in 1998, when Cooper's boss at the Smithsonian Migratory Bird Center, Pete Marra—then a PhD student at Dartmouth—published his research on American redstarts and sparked an intellectual gold rush into carry-over effects.

Redstarts are hyperactive little sprites that flash their brightly marked wings and tails—orange against black for males, yellow against a base color of gray for females—to flush the insects they eat from the forest canopy. Redstarts are one of the most abundant and widespread songbirds in the hemisphere; they are easy to find, highly vocal, and nest fairly low where humans can observe them, and are thus in many ways an ideal study subject—a "model species," as researchers say. Scientists have used redstarts to explore nesting ecology, sexual selection, population declines, and a host of other questions whose results may apply to migratory songbirds in general. Marra's PhD advisor at Dartmouth College, Richard Holmes, was himself a leader in redstart research, so it's no surprise that Marra picked up one of those threads for his own doctoral work.

In 1998 Marra and Holmes, with Keith Hobson from the Canadian Wildlife Service, published a groundbreaking paper in *Science* based on their redstart study. On the Jamaican wintering grounds, they found that older male redstarts dominate the better-quality, food-rich wet mangrove forest habitat, largely forcing females and younger males into drier, second-growth scrub—the kind of sex- and age-based habitat segregation that has proven common among many migratory songbirds. Regardless of sex, though, redstarts in wet forest habitat fared well, maintaining or gaining weight, while those in dry scrub lost weight and showed other signs of deteriorating physical condition.

When the redstarts in the study areas left Jamaica, though, there was no way to track them north. Instead, Marra and his colleagues captured different redstarts in northern forests, drew a bit of blood, and looked at the ratios of stable carbon isotopes in the samples. Such ratios, they knew from their work in Jamaica, would reflect whether the bird had spent the winter in good forested habitat or poor scrub habitat. Marra found that those birds arriving earliest—therefore the ones with the best selection of territories and mates—were those that had wintered in wet forest, while birds with the scrub signature showed up late to the game, after the best spots were claimed; they also weighed less than early arrivals and were in overall poorer condition. Females from dry habitat had fewer chicks, and those chicks fledged later than those of mothers from good winter habitat.

Marra's students in Jamaica have since shown that, in dry winters, female redstarts in wet habitat maintain their body condition far better than those in dry scrub, while in wet winters both populations fare rather well. They also found that if redstarts from dry habitat are allowed to move into wet mangrove forest, their condition improves and they depart on migration earlier than those that remain in scrub—what's known as an "upgrade experiment." Cooper, who spent six winters in Jamaica working on his own PhD, flipped that study and did a downgrade, using a mild insecticide to reduce insect populations on some study plots. That caused redstarts there to sacrifice muscle mass and significantly delayed their departure on migration by a week. ("I was the Grinch," he says with a pained smile, recalling that necessary but unpleasant work.)

Still, there was no way to follow the same individual redstarts back and forth, and directly measure how changes in winter helped or hindered the birds on the breeding grounds. While redstarts have been good study subjects because they're so common, in some ways they're *too* abundant and widespread. There are an estimated 39 million redstarts in North America, nesting as far south as Georgia and Texas and as far north as Labrador and the Yukon. Their winter range extends from coastal Mexico through Central America and northern South America, and most of the Caribbean. If you're trying to relocate a single known individual at both ends of its migration route to directly measure the carry-over effects of winter, that's one helluva big haystack. But by happy accident, the very characteristics that have made Kirtland's warbler such a conservation headache for half a century—its small population size and exceedingly limited range—also make it the perfect species through which to understand carry-over effects, a model of a different sort. There are only a few thousand of them, not tens of millions, and their range is a global pinprick. In winter, almost every Kirtland's can be found on a handful of small islands—Cat, Eleuthra, San Salvador, and a few more—totaling less than 1,000 square miles. In summer almost all of them migrate to a few counties in Michigan, concentrating still further in carefully managed plots of young jack pine. You couldn't design a better study subject to learn how the impacts of winter drag into the nesting season.

Watching the migrants on Cat Island was a constant reminder that for a small bird, every moment matters. From well before sunrise until the waning minutes of the day, they are engaged in a relentless, frenetic search for food. One morning while waiting for a Kirtland's to respond to the call, I watch a palm warbler, a slim, yellowish migrant that nests across the central boreal forests of Canada, a rusty cap on its head and a perpetual wag to its tail. The bird was working its way through the dense scrub along the road, flitting from perch to perch, peering beneath leaves, poking into thick tangles, fluttering near the ends of branches to reach the uttermost tip. By my count, it pecked up something to eat every three seconds or so, for as long as I watched it.

But what if food is a little scarcer, the habitat a bit more marginal, the climate a shade drier, forcing the warbler to hunt just a little bit harder? What if, instead of finding a morsel every three seconds, it catches something on average every four seconds? Sounds like a minor difference, but that's a 25 percent decrease in what the warbler can consume over the course of the full day—an enormous deficit, and one that may not allow the bird to claw back the reserves it spent flying down here from Manitoba or western Ontario, much less lay on the extra fat it needs to migrate back north again in March or April. On such razor-thin energetic margins does success or failure hinge for a migratory bird.

The work by Pete Marra and his associates set off a rush of research into seasonal carry-over effects in a range of migratory birds. British researchers netting black-throated blue warblers in the Bahamas, and assessing their isotopes, found that those that had wintered in moister habitat were in better condition during spring migration. Scientists working with black brant in Alaska found that those geese wintering the farthest south in Mexico were the latest returning to the sub-Arctic, and produced smaller clutches. Among Cassin's auklets in British Columbia, isotopic analysis has shown that females that spend the winter feeding on high-quality copepods nest earlier and lay larger eggs than those that feed on lower-quality rockfish.

Carry-over effects, it should be noted, aren't strictly avian, or unique to migratory organisms; they've been identified or suspected in gray whales, elk, red squirrels, several species of fish, and sea turtles, among others. And as we'll see, even among migratory birds scientists have found a few intriguing exceptions. Still, the weight of evidence suggests that for most migrants, bad luck in one season can carry through to the next, or even beyond. Scientists studying Old World songbirds have shown especially stark correlations between breeding success in Europe and the amount of winter rainfall in the Sahel, the arid, southern fringe of the Sahara where millions of the birds winter. What happens on the wintering grounds does not stay on the wintering grounds. (Such seasonal interactions, as Marra calls them, can also work at the population level, and in reverse; a highly successful breeding season up north may mean more competition from more individuals once they reach the wintering grounds.)

Rainfall—and by now, as Cooper talks, it's coming down in torrents outside, cooling the air and turning the usually turquoise sea a dark, chilly Prussian blue—appears to be the critical factor for Kirtland's warblers, too. As far back as 1981, Cooper tells me, scientists noticed that more male warblers showed up on the Michigan breeding grounds after wet winters in the Bahamas than after dry winters. More recently another scientist in Pete Marra's lab, Sarah Rockwell, was able to show specifically that if the rains fail to come in March, already the driest part of the year, the warblers' migration mortality rate jumps significantly. In such situations the birds arrive later on the breeding grounds, they are late in starting nests, and the number of chicks they fledge successfully tumbles. Young, less experienced males are especially hard hit by poor winter rains, she found.

Most alarmingly, Rockwell concluded that as little as a 12 percent reduction in average winter precipitation in the islands could reverse the Kirtland's decades-long and much-celebrated climb from near-extinction, and tip it back into decline. Nor is that loss of precipitation merely theoretical; she noted that since the 1950s, rainfall in the Bahamas had already decreased up to 14 percent. Rockwell's findings have profound implications not just for Kirtland's warbler, but for hundreds of millions of songbirds, of hundreds of different species, which also migrate into the Bahamas, Antilles, and Central America. Climate models suggest the Caribbean region, one of the most important wintering areas on the planet, will continue to dry out as the planet warms. (A few weeks before my Bahamas trip, Marra told me that the winter in Jamaica had been "ridiculously dry" even by recent parched standards.) The future is under a worrisome cloud. Cornell researchers, layering climate modeling over eBird checklist data on songbird distribution, predict that less rain on the wintering grounds, combined with warmer and rainier conditions in the north, will challenge many species that winter in the Neotropics. In the end, the carry-over effects from such changes may be the make-or-break factor in the long-term survival of hundreds of species of migrants that winter in the region.

Until now, though, everything about carry-over research has been indirect and at one remove—studying isotopic signatures, inferring past history from current conditions. "Pete and Sarah found strong evidence that carry-over effects exist at the population level," Cooper continues. "But at the individual level, Sarah didn't have the ability to look at them. For the first time, we will. We can look at individuals instead of just at the population level—we don't have to use indirect techniques like isotopes, but can directly study individuals, which removes a lot of the variance and assumptions you have to make."

Cooper and Marra are also shifting the focus on carry-over from a simple winter/summer question to one that includes migration, easily the most dangerous part of a bird's annual cycle. In songbirds like Kirtland's warblers, biologists estimate that 50–60 percent of the annual mortality occurs during spring and fall migration. But migration in small birds, because they've been hard to track, is still pretty much a black box. "What we'd like to know is how much of that [mortality] is driven purely by things happening during migration, and how much is set up by how good a winter you had, by what kind of condition you leave in, and what time you leave," Cooper says.

This work in the Bahamas, and what will follow over the summer in Michigan, is a proof-of-concept test. If it works, Cooper says, he and Marra want to launch a significantly more ambitious, four- or five-year study across multiple islands, one that closely tracks insect and fruit availability, winter home range size among the birds, and how those that hold a small, stable home range compare with warblers that roam over wide areas, as (for unknown reasons) some do. Most of all, they want to compare how different rainfall patterns on different islands impact the warblers' condition and productivity later in the year in Michigan.

The rain has passed, almost as if it never occurred. The sun is out, once again turning the surrounding ocean a radiant morning-glory blue. The heavily eroded limestone and gravelly soil around the house sucked up every drop, and looks as dry as they had before the storm. Cooper runs his hands through his hair, and turns his attention to the export paperwork needed to get those blood and fecal samples back home. Home: the subject brings a tired sigh. "I don't know which I miss most, my dog or a real beer," he says.

Two days later, the last warblers tagged, Cooper and I are on the puddle jumper back to Nassau. Fox and Caird are staying behind for a few weeks, using handheld radio receivers to track the tagged birds and determine exactly when they leave. Some have already gone, and as our flight crosses the hundreds of miles of water toward Florida, I suspect we're not the only ones in the air.

Carry-over research is by no means restricted to the New World. Leo Zwarts is a Dutch shorebird expert, an associate of my Yellow Sea friend Theunis Piersma, whom I had the pleasure of getting to know a few years ago during a conference and long birding expedition in Israel. Leo was involved in some of the most important carry-over studies in the world in sub-Saharan Africa. It also entailed what must

count as some of the most brutally difficult field work I've ever heard
described, judging from some of the hair-raising stories Leo told. He
and his colleagues, dressed in local garb to avoid the notice of Isla-
mist militants, Tuareg separatists, corrupt officials, and other dicey
elements, would crisscross the arid Sahel region along the southern
fringe of the Sahara, through frequently unstable countries like Mali,
Niger, and Mauritania. They were in the field for many months at a
stretch in the blistering dry season, running extraordinarily long survey lines across this sere landscape of scrub, grassland, and incipient
desert, visiting thousands of roughly four-acre study plots where each
individual tree and shrub—a third of a million of them, by the time
they were done—was measured for height, crown width, canopy volume, and other parameters; and in each tree, every bird of all species
was carefully logged. For nine years beginning in 2007, their exhaustive (and exhausting) research in the Sahel illuminated the changes
they and other scientists saw each summer back in Europe, where
species like common redstarts* have suffered catastrophic declines
during major Sahelian droughts, especially in the 1970s and '80s.
The result of this epic research was a raft of scientific papers showing the varying degrees of correlation between African rainfall and
breeding success of songbirds, swallows, shorebirds, and other species
in Europe. (They also laid out the early results of their work in *Living
on the Edge*, a magnificent, much-lauded book that Leo coauthored
in 2009.)

I had made plans to join Leo in the field—though I wasn't sure

---

* Despite their names, the common redstart of Europe, and the American redstart Pete Marra has been studying for decades, are not close relatives. Early European colonists often applied the names of familiar Old World birds to unrelated
but vaguely similar North American species. Common redstarts are Old World
flycatchers (which are not, in turn, closely related to the tyrant flycatchers of the
Western Hemisphere), while American redstarts are wood-warblers in the family
Parulidae . . . which is not the same as the Eurasian wood warbler of Europe, a
wholly different bird that belongs to an unrelated family, the *Phylloscopidae*, or leaf
warblers. Yes, it's a bit of a linguistic mess.

how a tall, blonde, very pale American was going to blend in enough for safety, even with the help of a turban. As it turned out, Leo and his colleagues decided to wrap up their long-running African studies after their 2015–16 field season, having traveled for three months through Mali, Burkina Faso, and several other countries. "It is getting more dangerous," Leo emailed me after returning home. "There are now hardly [any] western people left due to the risk of kidnapping and murder." Even the policemen, he said, were getting nervous. "As a tourist you can hide yourself with local clothes (e.g., turban half hiding your face if necessary), but this does not help us because we are so conspicuous measuring trees with a Nikon laser and searching for birds with binoculars." In the end, their primary funding source withdrew its support, telling the scientists it no longer wanted to be responsible for their safety.

Not everyone is convinced that carry-over effects are as pervasive and paramount as Pete Marra, Leo Zwarts, and many others in the field believe. There have been dueling studies in Europe, trying to decipher the degree to which conditions on the African wintering grounds, during migrant passage through the Mediterranean, and on the northern nesting range play a role in breeding success. One team, reviewing nearly 50 years' worth of nest records from the UK and winter precipitation data from Africa, concluded that breeding ground conditions were more than three times as important as winter rainfall in predicting the timing of egg-laying and the size of a migrant's egg clutch. Another study, though, zeroing in much more closely on three species of conservation concern, found mixed signals. For common redstarts, wet winters in the African Sahel meant an earlier nesting season and larger clutch sizes in Britain. Warmer British springs meant earlier nesting for redstarts and Eurasian wood warblers, but not for spotted flycatchers. All three species, though, benefited from warm springs when migrating through the Mediterranean, including larger brood sizes for the flycatchers and the warblers.

Regardless of the degree to which particular climate variables affect particular birds, scientists have found some intriguing exceptions to carry-over effects in general, notably among godwits. These are, you'll recall, those large shorebirds that perform some of the most astounding long-distance migrations on earth—including the Hudsonian godwit, a species with a 30-inch wingspan that nests in a few widely scattered locations in Canada and Alaska.

Nathan Senner, at the time a PhD student at the Cornell Lab of Ornithology (and the son of my good friend Stan Senner, himself a respected shorebird expert), studied the godwits that breed in southern Alaska. Using geolocators, Nathan found that these birds perform an extraordinary loop migration every year. Leaving Alaska, his tagged godwits traveled east to the Canadian prairies to feed, then made a five-day, 4,000-mile nonstop flight from central Canada east to the Atlantic Ocean, then south over the western Atlantic and Caribbean at the peak of hurricane season to the Amazon Basin in Colombia. From there they traveled down through the heart of South America to Argentina, then flew west over the southern Andes to finally reach Isla Chiloé on the coast of Chile, where virtually the entire Pacific population winters. A few months later, Senner's tagged birds departed Chiloé, but this time they flew north, paralleling the South American coast up the eastern Pacific to Central America, crossing the Gulf of Mexico and then continuing up through the Great Plains—a seven-day, 6,000-mile nonstop flight—before resting for a bit. From there they finally turned northwest back to their nesting grounds on the Beluga River in Alaska.

A journey of such epic measure should, by rights, tax any migrant, and Nathan expected to find ample evidence that carry-over effects from one season accumulated like falling dominos in subsequent stages of the godwits' lives. Instead, he was surprised to find that the godwits can somehow compensate for bad luck or challenging conditions on their migration. Godwits might straggle into Chiloé over a nearly two-month period, but when the flocks began departing again for their northbound return, they did so in a very compressed, seven-day time frame. Somehow, regardless of how stressed they were by the southbound leg, the godwits had recovered enough to leave en masse. Nathan could not find any correlation between their arrival date in Alaska and their breeding success there, nor between the timing of their migration stages and survival rates. Carry-over effects, which appear to be common in so many creatures, seem not to apply to this extreme migrant.

Why? Nathan suspects the godwits aren't really immune, just insulated by food. Throughout this entire nearly 19,000-mile migration loop, the godwits depend on only four major refueling sites, each of which is remarkably rich in the aquatic invertebrates, marine worms, or carbohydrate-laden plant tubers on which the birds feed. On the way south they use the prairie wetlands of central Saskatchewan, the Colombian Amazon, and the marshy pampas of Buenos Aires Province in Argentina. On the way north, the shallow playa lakes and marshes of the Great Plains are especially critical, coming as they do at the end of a 6,000-mile nonstop flight from Chiloé.

With such fertile, predictably stable stopover points, even a food-stressed godwit is apparently able to bounce back fully after each stage in a taxing migration. "Without such a high-quality nonbreeding site, it is easy to imagine the timing deviations that developed during southward migration growing unabated throughout the nonbreeding season and into the northward migration," Nathan concluded. Should any of those important stopover sites disappear or be degraded, the whole equation may change.

If any bird should suffer carry-over effects from a grueling migration, though, it would be the Hudsonian godwit's even more extreme cousin the bar-tailed godwit. This is the shorebird that makes the longest known nonstop migration of any species—that journey of up to nine days crossing more than 7,000 miles of empty Pacific Ocean from Alaska to New Zealand. Yet even this species seems able to compensate for bad luck and challenges on its marathon trip. Godwits that arrive late to New Zealand, where they must undergo the critical and energetically expensive task of replacing their flight feathers, can delay their molt a bit while they recover from their migration, then speed through it more quickly than normal and still be ready to leave with everyone else during a relatively brief period in March. Nor could scientists find any risk of heightened mortality among late-arriving godwits—they had the same fairly high survival rate from year to year as the rest.

Unlike Nathan Senner, the New Zealand team did not speculate on why the bar-tailed godwits seemed able to compensate for problems and delays in their southbound migration, though the rich food reserves on the tidal estuaries of the North Island may provide a similar resource cushion. If so, that's cold comfort once they leave New Zealand, since their next stop heading north is the Yellow Sea, which, as we've seen, is one of the most imperiled staging grounds on the planet. The New Zealand scientists also admit they had no way to measure whether carry-over effects might have degraded the godwits' reproductive potential once the birds reached Alaska—perhaps the single most important measure of a bird's success. "Therefore, the ultimate carry-over effects . . . may be subtle, and cannot truly be assessed without measures of individual fitness," they wrote.

Until now, those measures of individual fitness have been hard to come by, and teasing out carry-over effects has been squishy. But once the tagged Kirtland's warblers start arriving in Michigan, Nathan Cooper and his Smithsonian team would finally have a way to directly test what has, until now, been mostly theoretical.

Two months after leaving Cat Island, at the end of June, I reconnect with Cooper and his team near the small single-crossroad town of Luzerne, Michigan, about three hours north of Detroit in the middle of the Huron-Manistee National Forests. It's miles and miles of two-lane road to get there, through forests of mixed hardwoods and deep pines, across tannin-stained creeks that flow into the Au Sable River, legendary for its trout fishing.

Cooper looks infinitely more tired than the last time I saw him, and no wonder. After leaving Cat Island the third week of April, he'd had a single night back in Washington, DC (and not even in his own apartment, which he subleased to two other scientists while he was gone). The next morning he drove north in a truck stuffed with equipment, and as soon as he arrived in Michigan he began erecting a network of nearly a dozen 40-foot-high, antenna-topped receiver towers around the breeding grounds. That job was finished just in time to catch the arrival of the first tagged warbler on May 14, and eventually the team picked up the signals of 38 of the 63 birds they'd tagged in the Bahamas. Some of the missing Kirtland's may have evaded detection; even though 98 percent of the birds breed in this small 10-county area of Michigan around Grayling and Mio, a growing number have spilled over into new breeding sites in the Upper Peninsula, parts of northern Wisconsin, and southern Ontario. Chances are, though, that the ones they didn't detect hadn't survived their journey north. And since their arrival, at least four more warblers have disappeared, including one caught by a sharp-shinned hawk as the team watched, and another that had a severe parasitic infection when the crew captured it for a check-up.

Now I'm squirming through a dew-soaked stand of jack pine before sunrise, trying to follow Smithsonian interns Cassandra Waldrop and Justin Peel as they, in turn, follow a singing male warbler moving around his territory. It's chilly, with a smear of ground fog still clinging to the low-lying spots like the ghosts of old snowdrifts. Waldrop, her hair tucked up under a tan ballcap and a blue pack slung across her back, suspects the bird we're following is one of the tagged males from the Bahamas, if she can only get a glimpse of him. (The batteries on the tiny transmitters we deployed on Cat Island have a fairly short life, just enough to allow the crew to find the birds' breeding sites here in Michigan. By late June most of them have already stopped working, but we can still look for the bird's color bands and the wispy antenna sticking past its tail.)

Like almost every Kirtland's warbler in existence, this male is breeding in an artificially maintained habitat. Every year, managers clear-cut about 4,000 acres of mature jack pine forest in this part of Michigan, then replant some 5 million to 7 million jack pine seed-

Just before all hell broke loose: Laura Phillips, David Tomeo, and Iain Stenhouse relaxing on the tundra, unaware of an approaching grizzly.

Six million acres in extent, and holding North America's highest mountain, Denali National Park is also home to more than 160 species of birds, many of which migrate to wintering sites as far away as southern South America, east Africa, and southeast Asia.

Grizzly bears (and moose, which can be at least as dangerous as the bears) are a constant concern when conducting bird research in the Alaskan interior.

A male spoon-billed sandpiper, one of just a few hundred left in the world, calls repeatedly on the tundra of Russia's Far East, having just arrived from the Yellow Sea. (© GERRIT VIN/CORNELL)

Thousands of red-necked stints, dunlins, and other shorebirds swirl above the steel-plate mud of Dongling, just north of Shanghai—a small part of the 8 million shorebirds that depend upon the Yellow Sea tidelands every year.

Theunis Piersma and Bingrun Zhu scan for godwits—Zhu's doctoral study subject—along the coast of Bohai Bay.

At Tiaozini and other locations along China's Yellow Sea coast, the mudflats—full of marine invertebrates, a feast for migrant shorebirds—extend seaward for miles when the tide goes out.

Jing Li is one of a handful of Chinese conservationists whose work has demonstrated the critical importance of the Yellow Sea mudflats to spoon-billed sandpipers and dozens of other long-distance migrant shorebirds.

Coal-fired power plants, wind turbines, shrimp farms, and jammed highways crowd the shore of the Yellow Sea on what had, until a few years ago, been fertile mudflats.

Tiny semipalmated sandpipers like these can fly nonstop from the Bay of Fundy to Brazil each autumn because, in part, they feed on omega 3–rich invertebrates, like athletes using performance-enhancing drugs.

Great frigatebirds, like this male shading its single chick on Isla Genovesa in the Galapagos, almost completely forego sleep during foraging trips that can last up to 10 days—a trick also employed by migratory birds on long journeys.

On a hot summer's day, Dave Brinker assembles one of the directional antennas for a Motus receiver station in Pennsylvania.

Golden-winged warblers are one of several species that have surprised researchers by essentially swapping habitats in the weeks before autumn migration—a period in a bird's life about which very little is still known.

On a warm night in April, Doppler radar shows a diffuse blue-and-green blob over Jacksonville, Florida—not rain, but the radar echoes of millions of northbound songbirds, aloft in the night sky. Note that, unlike rain, the birds are not straying out over the Atlantic Ocean. (NATIONAL WEATHER SERVICE)

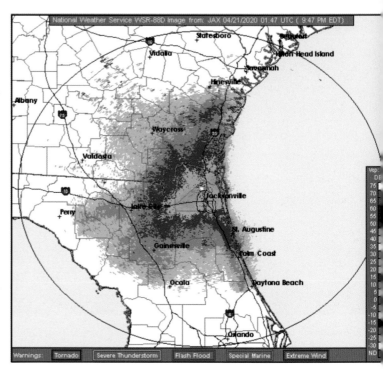

Built on hundreds of millions of observations from birders, eBird has proven to be one of the most powerful tools ever developed for understanding the abundance and movements of birds, like this map showing the year-round distribution of summer tanagers. (MAP: eBIRD/CORNELL LAB OF ORNITHOLOGY)

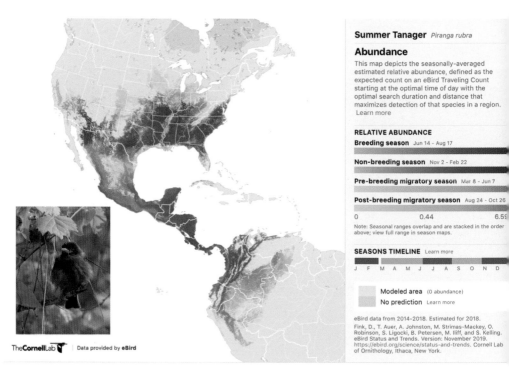

Nathan Cooper follows the beep of a nanotag transmitter, which reveals the location of a rare Kirtland's warbler in the impenetrable scrub of Cat Island in the Bahamas.

His mouth full of caterpillars for his chicks, a male Kirtland's warbler pauses to size up an intruder.

Eighty percent of the Bahamas—the only wintering site in the world for Kirtland's warblers— lies three feet or less above sea level, meaning that rising oceans pose an existential threat to this now-recovering species.

As the climate warms and springtime green arrives earlier and earlier each year, scientists are finding that migratory birds—especially long-distance migrants coming from the tropics—haven't been able to keep up with the rapidly advancing seasons.

A solar-powered transmitter, held in place by a backpack harness, peeks out from beneath the feathers of a snowy owl along the New Jersey coast. (© JIM VERHAGEN)

A new band glinting on its right leg, a female rufous hummingbird sits in the frigid sun of a January day in Pennsylvania—part of the vanguard of a remarkable expansion by several species of hummingbirds, made possible by changes in the climate and landscape. (© TOM JOHNSON)

Mount Shasta, more than 50 miles away, looms over the farm fields and grasslands of the Butte Valley in northern California.

His eyes peeled for an attack by its parents, Chris Vennum takes a Swainson's hawk chick from a nest high in a scraggly juniper tree, to which it will soon be returned.

Holding a freshly captured adult hawk— hooded to keep it calm—Brian Woodbridge waits with Karen Finely as the bird's mate is banded, so both birds can be released together.

Swainson's hawks have one of the longest migrations of any raptor in the world, traveling up to 8,000 miles each autumn from the grasslands of North America to the pampas of Argentina.

Great shearwaters, millions strong, nest on the tiny, remote Tristan da Cunha islands and Gough Island in the middle of the South Atlantic, migrating north from New England to Scotland in the boreal summer—a round trip of 12,000 miles each year.

Barely larger than swallows, Wilson's storm-petrels—like most pelagic birds—spend the vast majority of their lives far from land, moving with the seasons between the far northern and southern oceans.

The setting sun silhouettes a flock of greater flamingos on a salt lake in southern Cyprus—an island described as a "black hole" for migratory birds because of its intensive hunting and trapping.

Roger Little with a lime stick coated with powerfully sticky glue, a death trap for any songbird unlucky enough to encounter it.

Ending a night of antipoaching surveillance, a member of the Sovereign Base Area police force reaches up to catch a descending drone.

The roads in Nagaland are almost epically bad, making tourism away from the largest towns a proposition only for the most determined.

Slim and fast, Amur falcons nest from China and Mongolia to the Russian Far East, and make the longest overwater crossing of any raptor—2,400 miles across the Indian Ocean to southern Africa.

Bano Haralu (here with the author) was among those who first raised the alarm about the widespread slaughter of falcons in Nagaland. (© KEVIN LOUGHLIN/ WILDSIDE NATURE TOURS)

Tens of thousands of Amur falcons rise from one of many roost sites along the Doyang Reservoir—places that just a few years earlier had been the scene of great carnage.

The data stored in the small geolocator on this freshly recaptured blackpoll warbler will reveal the details of its migration from Alaska to the Amazon.

Like most Naga men, this fisherman carries a shotgun over his shoulder and slingshot at his belt—one reason why wildlife is scarce and wary near Nagaland villages.

Sunrise over Polychrome Pass in Denali National Park.

lings, creating huge expanses of dense young forest. These tracts are dotted with small grassy openings every 30 or 40 yards in all directions, and a scattering of dead, standing snags—the ideal recipe, tweaked and modified through decades of experimentation, for attracting this fussy bird. After just 15 years or so, the stands become too tall and overgrown for the warblers, which move into younger, more recently created habitat as the management cycle continues across nearly 150,000 acres of state and federal land. Where conditions are right the birds nest almost colonially, cheek by jowl. "Standing here, you probably have 50 percent of the global population of Kirtland's warbler within about 10 miles of you," Cooper tells me one day in Ogemaw County, the epicenter of the nesting range.

It's hard to overstate how confusingly featureless good Kirtland's habitat appears to a human encountering it for the first time. The trees are 6 to 12 feet high, planted with geometric precision five or six feet apart, and create a green wall through which you must force yourself—endless scratchy pines and head-high thickets of pin oak, all dripping with dew. Your arms and shoulders immediately become sodden and smeared with dirt from the clawing branches, and I soon understand why the scientists wear only their oldest, most tattered long-sleeved shirts. There are few landmarks, and it takes only a moment to lose your bearings. Shortly after following Waldrop and Peel into the pines, I pause briefly to jot a few notes—and then realize I have no idea where either of my companions have gone. The damp, lichen-covered ground muffles any footfalls, and I can only see a couple of yards in any direction. The warbler we are tracking is singing loudly, and I hesitate to call out, even quietly, because I'm not sure how disruptive that would be. I consider the embarrassment of getting lost within my first 10 minutes in the field, then make a best guess and force my way through the pines, stumbling over the deep furrows left by the mechanical tree planters—and almost crash into Waldrop.

"You lose us? It happens a lot—we play Marco Polo all the time in this stuff," she says. And as though on cue, a questioning "Marco?"

comes from Peel, 100 yards away in the brush. "Polo!" Waldrop yells back.

"Doesn't that scare the bird?" I ask.

"Nah, they're pretty tame," she says. The warbler sings again, and this time she cocks her head, listening intently. "Hear that muffled call? That usually means he has a mouthful of food. He's foraging for his chicks as he's singing."

But when we finally locate the target male, his legs are bare—not one of the tagged Cat Island birds. A short drive away, though, in a different plantation of pines, the pair leads me to a tagged male and his mate, whose first nest was lost to predators; as they do with many other pairs, the interns keep regular tabs on this family. The male chips and fusses at us from just a few yards away, while the female sits tight on her grass-lined cup, buried deep in the blueberries and thick sedges growing below a pine. Nearby, another male scolds me from barely arm's length away, several caterpillars clamped in his beak, as one of his fledglings balances uneasily on a branch, still unused to being out of the nest. "Tame" is hardly the word; most of the time the birds completely ignore us, even when we're mere feet from them.

The routine here is much the same as in the Bahamas. Soon after arrival in May, each tagged warbler was renetted, its body condition measured, a drop or two of blood and a little scoop of poop taken. Nests of both tagged and untagged birds are monitored and the number of eggs and successfully fledged chicks recorded. (The crew has to be very careful, because blue jays may learn to follow them, trying to puzzle out the location of nests they can raid for eggs or chicks. "I fake finding nests a lot," Waldrop says, "doing all the things down in the grass I would do if there was really a nest, so the jays don't know when I have a real one. It's kind of funny—I'll look back, and the jay will actually duck behind a tree trunk to hide from me. If I can, though, I'll lead him into a brown thrasher's territory. They don't call them 'thrashers' for nothing.")

Now, as the breeding season wanes and the families start to break up, Cooper and his interns are trying to net the tagged birds one

more time, for a final checkup and in some cases to deploy fresh
transmitters, which will tell the scientists when the warblers depart
the nesting grounds. The next morning finds me with Cooper and
Steve Caird, whom I'd met on Cat Island. We're trying to locate a
tagged male that eluded us the previous day, while keeping an eye
on the thick clouds that are rapidly building in the west. The bird is
ranging widely, singing as he forages, and my ear has become attuned
enough to distinguish when he's singing with his mouth full, carrying food for his fledglings. Caird and Cooper set up the mist net
and turn on the recorder, and in a couple of minutes the warbler is
in hand—tag number 86, pale blue and dark green plastic bands on
his left leg, a pale blue plastic and a numbered aluminum band on
his right.

At least three years old (based on subtle characteristics of his
plumage), Number 86 had been originally tagged April 5 at an old,
abandoned goat farm on the northeast end of Cat Island, about a
week before I'd arrived on the island. He left the Bahamas on May
2, and was picked up en route by one of the Motus towers in Florida.
He showed up here in Michigan on May 18, detected by the local
receiver array, and was trapped and given an exam on May 28. In
the weeks that followed he and his mate raised four chicks—a prime
example of a successful migrant.

With the bird in hand once again, Cooper goes through the process he's repeated endlessly since winter, and to which even Number 86 must becoming resigned. "Fourteen grams," Cooper says.
"Hmm—he's lost a lot of breast muscle compared with Cat Island,
which makes sense, since he doesn't need it here." Those pectoral
muscles, which power flight, beef up before migration, then slim
down again once the birds arrive, an endless seasonal cycle of expansion and contraction. Cooper snips the dead transmitter's harness
and fits on a new one.

"You're being a feisty little shit," he says to the bird, which is struggling more than usual, "but if you'd just chill it would go a lot faster.
I know it's not fair that you have to wear another one of these things,

but sometimes life isn't fair." Cooper does a final check and opens his hand. "See you in the Bahamas," he says as it flies away. The clouds, which have clotted the horizon, unleash a heavy downpour as we hurriedly pack up and trudge back to the truck. Number 86 is already singing again.

It was my last morning in the field with the Smithsonian crew. Almost three months later, I check in with Cooper to see how his first year's data are adding up. With a scientist's caution, he hems and haws a bit; he hasn't crunched the numbers on how body condition in the Bahamas affected breeding success, for instance, and the blood and fecal samples are at the Field Museum in Chicago, waiting for next-generation genetic testing to catalogue gut microbes and parasites. But several fascinating results are already clear. For one thing, the tracking data confirmed earlier evidence that the departure date from the Bahamas determines the arrival date in Michigan; late-leaving birds can't seem to make up lost time. And in a big surprise, the transmitter data suggest the most dangerous part of the Kirtland's migration is the beginning, as soon as the warblers leave Cat Island.

"We already know spring migration is the most dangerous time for songbirds, but with this new system we can get a much more complete picture of when these mortality events are occurring," he tells me. "If I'd had to guess, I would have thought that between Florida and Michigan is where most of them would die." After all, at 1,200 miles it's the longest part of the migration, and comes at the end when the birds might be most exhausted. Instead, all but two of the 25 tagged warblers that vanished en route disappeared between Cat Island and the US coast, a fairly short 300-mile flight. "Maybe they just weren't in good enough condition from winter—it's the dry season, fruit's dried up, the insects to some degree. It may just be a really tough time to leave. By the time they get to Florida and Georgia, where spring is in full gear, that may be a relief to them," Cooper speculates. It's still more evidence that conditions on the wintering grounds are critically important to migrants.

The rain that chased us out of the jack pines that last day in Mich-

igan was a fittingly ironic end, given water's outsized importance to
the Kirtland's warbler's fate. And the future promises not only drier
winter habitat for Kirtland's warbler, but a whole lot less of it. One
breezy afternoon back on Cat Island, we had climbed Mount Alver-
nia, all of 206 feet above the ocean—not only the highest point on
Cat, but in the entirety of the Bahamas. As a whole, these islands
barely emerge from the Atlantic, and 80 percent of the Bahamas lies
three feet or less above sea level. That means that even a modest (and
at this point, probably unavoidable) degree of sea level rise will inun-
date huge portions of this low-lying archipelago in this century.

"So you can imagine a lot of that habitat being lost," Cooper
said, looking down at the very flat island. "Nearly all of the focus
on Kirtland's warbler conservation so far has been on the breeding
grounds—and rightly so, it's been very successful. But that doesn't
mean [breeding habitat] is always going to be the limiting factor.
We have to start to think, are there ways we ameliorate some of
these effects on the wintering grounds? Can we do habitat management down here that promotes good quality habitat, even in a
changing Caribbean?"

Promoting habitat management that benefits the warblers—like
encouraging heavy goat grazing, something usually anathema to
conservationists—is one approach. Another intriguing possibility is
that Kirtland's warbler might naturally expand beyond its limited
Bahamian winter quarters. Just as the species is now breeding outside of its traditional northern Michigan range, perhaps as conditions
change in the south it will show similar flexibility there. At least a
few already go to Cuba, and they've been reported from Hispaniola.
Recently there was a Kirtland's photographed near Miami, the first
winter record for the United States (its favorite food plants, lantana
and black torch, grow in south Florida), and Pete Marra's redstart
crew caught the first Kirtland's warbler ever found in Jamaica.

Because migration in most birds, including warblers, is genetically
encoded and not learned, there are always a few individuals with odd
software that sends them in unexpected directions. When conditions

change, those pioneers may be perfectly positioned to exploit the new regime, and in some cases scientists have watched as new migration routes and wintering areas have emerged—including blackcap warblers in Europe and rufous hummingbirds in the southeastern United States.

"There have been some pretty cool modeling studies that looked at what sort of genetic variation you'd have to have for the whole population to switch to a new wintering location, and it's a pretty low level," Cooper told me, sounding decidedly unconvinced as he looked out at the low-lying forests of Cat Island stretching away from our feet, and we thought about how climate change is reordering so much of the planet and the calendar already.

"So it's possible. Or plausible, at least. But you know, if the Bahamas are underwater, a lot more Kirtland's warblers are going to be wintering in Cuba. They'll have to."

Six

TEARING UP THE CALENDAR

The final flocks of whimbrels were lifting off from the marshes of Virginia's Eastern Shore, heading north to the Arctic. It was the evening of my visit to the annual whimbrel watch with Bryan Watts and his team, and they were discussing their work tracking those big shorebirds over the years, the remarkable discoveries they've made about how whimbrels fly into the eyes of hurricanes, and the danger that continued hunting in the Caribbean poses to them. All around us, the soundscape of the tidal marsh was in full voice—the staccato grunting of clapper rails, the songs of seaside sparrows and marsh wrens, the guffawing of laughing gulls. My eyes, though, had been glued to the sky for the previous 30 or 40 minutes, as flock after flock of whimbrels took off and flew north past us. I hadn't been paying much mind to the scenery, which is why, when I finally glanced around, I was shocked to see how much my surroundings had changed. What had an hour or two before been an unbroken horizon of green tidal marsh to our east—stretching the better part of a mile offshore to Ramshorn Bay, and then across further big islands of marsh dotting the water to distant barrier islands—was rapidly disappearing as the tide surged in.

Every 12 hours and 25 minutes, at almost any given location along the edge of the sea, the high tide peaks under the forces of lunar gravity and terrestrial rotation. Along the Eastern Shore, the average tidal range is fairly modest, just a couple of feet between high and low extent—barely worth mentioning when compared with the 18-foot tides along parts of the Maine coast, or the whopping 43-foot range in the Bay of Fundy in eastern Canada. But in a flat, low land, even a little water makes a lot of difference, and the tides vary as well with the calendar, with the highest—the spring (or king) tides—coming with the new and full moons when the earth, moon, and sun are in alignment. This was a spring tide, and on this windy, cold May evening, the Atlantic was on the march.

"There's a rail," Bryan said, pointing to the edge of the lawn, above which our cars were parked. A clapper rail—its grayish, chicken-like body about the size of a grapefruit when viewed from the side but "thin as a rail," no more than a couple of inches wide when seen end on, as though squeezed in a vise—scrambled out of the flooding vegetation along the shoreline and skittered up onto the grass. Rails slink at the best of times, staying all but invisible in the densest cover they can find; this one, forced into the open by the rising water, took slinking to a whole new level, its tail held high and its head down low, neck scrunched, crouching on its gangly legs as though it could compress itself so completely that it would vanish entirely from sight in the clipped grass. Then it gave up on concealment, zigzagging across the lawn in a rush like a skittering mammal, and vanished into the brush a few yards away.

What had been a tidal creek beyond the dock, 15 or 20 yards wide when I arrived, was now a fast-moving river four or five times that width. Bryan's companions began moving their gear, knowing that a low spot in the driveway meant the dock would soon be an island; they moved a couple of cars, too, and a good thing, because within half an hour the area where they'd been parked was under a foot and a half of fast-moving salt water. More rails appeared along the edge, and Bryan pointed out a line of seven black dots 100 yards offshore—seven clapper rails, clinging grimly to some long, floating strand of vegetation, all that remained of tens of thousands of acres of sheltering cordgrass marsh just a few hours before. One by one, the rails struck out for shore—swimming, to my surprise. This is not something rails do with any panache; the birds rode so low in the water that only their curved necks and heads showed above the waves like delicate periscopes.

"Why don't they fly? Maybe they can't get airborne from so low in the water," someone said.

"Or maybe they don't want to be too obvious," Bryan replied, watching the little convoy of heads strung out toward shore. Only one rail still remained on its slender life raft, Noah without much of an ark. "If a peregrine comes through it may be better to be in the water," Bryan said.

Seconds later, a bird of prey vastly bigger than a falcon appeared— an adult bald eagle, flapping out from shore at treetop height. It passed over the squadron of half-submerged rails, swung left and away from the one bird still standing above the water, then swooped down and plucked from the waves a rail none of us had noticed. The eagle turned back to shore, the rail's twig-thin legs kicking a few times and then hanging limp.

"I'll be damned—he's done that before!" Bryan said with a snort of surprise. When we thought to look again, the final rail we'd been watching had abandoned ship and was making for shore, eventually scrambling out of the water almost at our feet, struggling through the half-submerged mesh of a straggly old fence that had, at my arrival, stood 50 yards from the water, and vanished into a tangle of weeds to await the turn of the tide.

Extreme tides like this are becoming more and more common along the Eastern Shore, where sea levels are rising three or four times faster than the global average, and at the highest rate anywhere along North America's Atlantic coast. This is a combination of geologic history and climatic change. The former involves something called glacial isostatic adjustment; 20,000 years ago, when a mile-thick sheet of glacial ice lay a few hundred miles to the north, this land bulged upwards, like a half-inflated air mattress with someone sitting on the other end. When the glaciers melted, the land beneath them to the north of us rose, and the bulge down in what is now Virginia subsided—and continues to do so, exacerbated by groundwater pumping that causes still more settling, a major issue around cities like Norfolk and Hampton Roads.

But land subsidence only accounts for part of the furious pace of relative sea level rise on the Eastern Shore. The rest comes from climate change effects. On a single day, in the midst of a record-breaking Arctic heat wave—August 1, 2019—the Greenland ice cap lost 12.5 billion *tons* of ice, enough to cover Florida five inches deep in water. More water obviously raises ocean levels on its own, but as sea temperatures rise, the existing volume of the water already in the oceans also expands. Sea levels in the lower Chesapeake Bay, which the Eastern Shore brackets, have risen 14 inches since 1950, and by 2100 are projected to rise another four and a half to seven feet.

The immense tidal flats I'd admired two hours earlier, covering God knows how many square miles, were now completely submerged—open water all the way to the barrier island chain eight miles to the east. The grunting of the clapper rails, which had been the constant background sound since I'd stepped out of the car, was gone. So were the bubbly songs of marsh wrens, and the buzzy *whup-weedle-BZZZ* songs of seaside sparrows; the wind was empty of all but the harsh calls of gulls. "Think about what an impact that a single extreme tide event like this has on those marsh birds," Bryan said. "Seaside sparrows, saltmarsh and Nelson's sparrows, marsh wrens, all those small birds that nest in the tidal marshes. Where do they go during a tide like this? What about their nests and their eggs? Their entire world is just *gone*."

Sea level rise—whether the current effects of an extreme tide in Virginia or the future inundation of Kirtland's warbler habitat in the Bahamas—is just one example of why climate change is the big enchilada when it comes to migratory bird conservation. From its effects on weather, precipitation, prevailing winds, habitat, food supplies—even what impact it will have on avian diseases and parasites—there isn't a corner of the globe, a cubic meter of the air column above it, or any moment in any migratory bird's annual cycle, that hasn't been (or soon will be) touched by the planetary fever that carbon emissions are producing. Winds are shifting, ocean levels are rising, glaciers are melting, sea ice is eroding. The great circulatory systems of the earth's atmosphere are wobbling and changing; this has meant not only more extreme summer heat waves in places like northern Europe and Alaska, but more extreme winter cold in areas like northeastern North America and (again) northern Europe as the polar vortex becomes unstable. Droughts are becoming more frequent and intense in the Sahel, the Mediterranean, the American Southwest, parts of southern Asia, and southern Africa—while storms and extreme rainfall events are intensifying in other places, like eastern North America, northern Asia, and parts of Europe. Temperatures are fluctuating—rising in most places, especially at the higher latitudes and certainly as a global average, but in some locations, spring is actually becoming later and colder, or winters are becoming harsher and snowier, because climate change plays out in weird ways and doesn't mean a universal warmth. Scientists fear that, during this century, many ecosystems will reach the breaking point beyond which natural flexibility will no longer be able to hold it all together. The days when any reasonable person could look at the evidence for climate change with a skeptical eye, perching on the fence about whether industrial emissions were really to blame, and staying cheerfully optimistic that it's a lot of fuss about nothing, are simply over.

Migrants—especially long-distance migrants, those that already exist in a fragile balance between distance, time, physiological ability, seasonal resources, and predictable weather—are among the species on which the hammer will fall first and hardest. Because birds are widespread and highly visible, and because so many people monitor their numbers and movements and have been doing so methodically for a long time, they have provided some of the earliest and most important evidence for how climate change is altering natural systems. For some species, the effects are already fairly grim, but the news is not entirely gloomy. While there is no doubt that climate change is going to pummel many migratory species—at least some, it seems all but certain, into oblivion—there are a few encouraging signs. Some species have shown an unexpected flexibility in the face of changes that are occurring at a faster pace than birds have ever experienced in the geologic past. Whether that will be enough remains to be seen.

Climate change is reshaping every single thing about migration. It is tearing up the calendar, altering the timetable on which birds must travel in order to find the food they need along their pathways, or accelerating the seasons in ways that increasingly leave them farther and farther behind during crucial periods like the nesting season. It is modifying the weather; not only are storms growing stronger, but continental winds are strengthening at some times and places and weakening at others, with unknown consequences for the many birds that depend on reliable tailwinds at critical steps in their migration—never mind the rising temperatures that alter when insects emerge, or simply make it too hot for a baby bird to survive. Climate change is reshaping the landscape—in dramatic ways, like the inundation of coastal wetlands during extreme tides, but also in more subtle but pervasive ways, as regions dry out or become seasonally sodden, as shorter winters and longer, hotter summers (or changing wet/dry seasons in the tropics) mangle once-stable plant and animal communities. Climate change, we now know, is even altering the physical size and shape of many migratory birds, as their bodies shrink in response to the rising warmth.

Let's start with the changing landscape. Coastal wetlands like those Eastern Shore tidal marshes, the thin rim of habitat on which so many species, especially migrant shorebirds, depend, are at grave risk from sea level rise. As oceans have climbed in the past, wetlands have been able to move inland in concert with the increasing depth. But today, in most areas, development along the coast will wall off any possibility of these ecosystems migrating inland, even assuming the marshes can keep up with the pace of rising water. I think about the mudflats of the Yellow Sea in China, and the last-ditch effort to save them, which seems to have been at least partially successful, in some measure because the Chinese government came to see those wetlands as a defense against the worst effects of sea level rise. But every coastal wetland I saw along the Yellow Sea was jammed up against high artificial seawalls and industrial development farther inland; as the waters rise, there will be no obvious route for gradual migration, just drowning.

In other parts of the world, higher temperatures and less precipitation will dry out the landscape. To take one example, in North America's intermountain West, millions of ducks, geese, swans, wading birds, shorebirds, rails, and other water-dependent migrants on the Pacific Flyway pass through the Great Basin. This region's unique lakes and marshes have no outlets to the sea, and the basin's interlocking, progressively saltier wetlands—from fresh to saline to hypersaline—are packed with invertebrates like brine flies and brine shrimp, making this a critical stopover and staging area for waterbirds that breed in or migrate through the basin. Millions of eared grebes, for instance, leave their breeding grounds in the northern United States and western Canada and migrate in autumn to hypersaline lakes, especially Great Salt Lake in Utah and Mono Lake in northern California, where they undergo molt and become flightless, packing on so much weight that even when their wing feathers regrow they are unable to fly for several months. To complete their migration they must first fast for several weeks, losing two-thirds of their weight until they can again become airborne. Then at night they leave in waves, sometimes hundreds of thousands at a time, and fly to the Pacific coast.

A research collaboration led by the US Geological Survey found that the Great Basin wetlands on which the grebes and millions of other birds depend have already dried out significantly since 1980, owing to falling river flows and snowpack depths; the amount and timing of water coming into the wetlands is changing, and bird populations there are already reflecting that. Shorebird numbers have fallen by 70 percent; Wilson's snipe, black terns, and western and

Clark's grebes, among others, are in serious decline. Climate change is exacerbating what had already been serious impacts from water diversions in the basin. "Even [the] loss of a small amount of habitat, food resource or a key site in this critical region could trigger dispro-portionate population declines, particularly because nearby options are becoming limited," the study's authors warned.

When it comes to another critical migratory link—the Sahel in Africa, on which many palearctic migrants to Europe depend—there is no clear consensus about what a warmer future may mean. The Sahel is warming faster than most of the planet, and that trend, at least, seems certain to continue. But the Sahel has been an especially tricky region to forecast, dependent as it is on monsoon rains that are tough to capture accurately in climate models. Some projections sug-gest the region will continue to dry out, as it did during devastating droughts in the 1970s and '80s. Other and more recent models seem to suggest a wetter future, with more intense monsoon rains drifting farther north, while still others split the difference, predicting more moisture in the eastern and central Sahel and less in the west. What-ever happens will profoundly affect the fate of millions of migratory birds that depend upon the Sahel as they migrate from Africa to Europe and back.

At best, though, the overall picture for migratory birds is alarming. Combining avian data from the Breeding Bird Survey and Christmas Bird Count, along with low-, medium-, and high-emissions climate projections, the National Audubon Society found that more than half of nearly 600 North American bird species would lose more than half their current geographic ranges by the end of this century. For about a third of the species, there would theoretically be some range expansion into new areas to offset the loss, but 126 species would have no such escape hatch; Baird's sparrow, for example, appears poised to lose virtually its entire breeding habitat in the northern prairies, as well as all of its wintering habitat in the arid grasslands of northern Mexico. Even where the study found that a bird's range could expand north, based on climate models, the realities of plant migration may limit any benefit. Scarlet tanagers are projected to expand almost 1,000 miles north into central Canada by 2080—but the tanager needs mature hardwood forests, and no one expects fully grown oak or maple forests to appear essentially overnight in what is now boreal spruce woods around James Bay.

Of course, one of the most important "landscapes" for migratory birds is the airspace above the land itself—and how climate change is and soon will alter the wind and weather patterns is going to have far-reaching consequences for migrants. Few have tried to look as deeply into this question as a team at the Cornell Lab of Ornithology headed by research ecologist Frank La Sorte, which is mining eBird, radar, climate data, and other sources to understand how birds are using the skies today, and how conditions in the future are likely to help or hinder migration. La Sorte and his colleagues have found that springtime migrants, especially insect-eaters like warblers, carefully track the emerging "green wave" of new vegetation that spreads north from the Gulf of Mexico beginning in March. They've shown that many migrants in eastern North America make a clockwise loop migration, up through the middle of the continent in spring—a longer path but one along which a low-level jet stream of southerly tail winds is strongest—and then a shorter and more direct route south over the western Atlantic in autumn with prevailing northwesterlies behind them. In the West, migrants appear to follow river valleys and other greenways, and pay less attention to prevailing winds—but may therefore be at greater risk if their migration timing and local food resources become disconnected because of climate change.

What does the future hold? Predicting how something as complex as a planetary climate system will respond is insanely difficult, but as all-time-hottest year follows all-time-hottest year, and polar ice melt keeps smashing old records, it appears that, if anything, models up to this point have low-balled the consequences. Warming may be twice as bad as predicted, climatologists warned in 2018, even if we somehow manage to keep the average global temperature rise to 2 degrees Celsius, which itself increasingly seems like a pipe dream. The Cornell team has modeled how changing weather conditions are likely to affect migration in the Western Hemisphere. Climate extremes are expected to increase with global warming, so the team used a bizarrely warm spring in March 2012 to see what effect such episodes have on birds. They found that such warm surges initially accelerate ecological productivity, but result in a corresponding sag later in the summer just as birds are preparing to migrate—a situation that disproportionately impacts long-distance migrants that need abundant food at that time. The team used historical records and climate projections, along with eBird data for almost 80 species of migrants, to see when new climate regimes would emerge in different portions of the birds' annual cycles, and concluded that by the second half of this century, migrants will begin to experience novel climate conditions both on their tropical wintering grounds, and in late summer on their temperate breeding areas. By 2300, all 80 of the migrant species will face novel climates throughout the year. Prevailing winds, which are especially critical for long-haul migrants, will also change. La Sorte and his colleagues used data from the 143 contiguous US Doppler radar sites, along with wind projections and information about where, when, and at what altitude most migrants fly, to determine that through this century, tail winds during spring migration will increase by about 10 percent—a boost to northbound birds— but westerly winds will decrease by twice as much in autumn, making nocturnal migration at that time of year less efficient. But it's an ill wind that blows no good; those diminishing westerlies mean that birds flying south over the western Atlantic, which now must fight strong westerly crosswinds as they migrate between North America and the Caribbean or South America, won't have to work as hard to stay on course.

The biggest threat from climate change, though, may be occurring on the nesting grounds, as the seasons shift beneath the feet of migrant birds. The poster child for the potentially devastating ways climate change can wreck a migratory bird's world is the European pied flycatcher. This active, pot-bellied songbird—the males black and white, females brown and white, but both with large, flashy white wing patches—nests from the British Isles to southern Russia, and winters in western Africa south of the Sahara. It is one of the most-studied songbirds in Europe, one of those "model species" partly because it conveniently accepts man-made nest boxes, making it handy for researchers, and also because of its unusual, polygynous breeding system, with males often having more than one mate. More recently, though, it hasn't been the pied flycatcher's sex life that has attracted the most attention.

As the name suggests, pied flycatchers are largely insectivorous during the summer, and while the adults spend a lot of time snatching flying insects from the air, when they feed their chicks they depend to a remarkable degree on caterpillars—a trait shared by many, perhaps most, migratory songbirds across the temperate and boreal zones of the Northern Hemisphere. Caterpillars are soft and easy for young birds to digest, and across northern forests—especially in oak-dominated woodlands like those in which many pied flycatchers nest—there is an explosion of them a month or so after the new spring leaves emerge. A single pair of songbirds may have to provide more than 6,000 caterpillars to its brood of four chicks in the weeks it takes to raise and fledge them, so hitting the caterpillar crescendo isn't a luxury—timing arrival, nest-building, and incubation just right to hit the peak is a necessity.

That system worked, presumably over many millennia, for pied flycatchers and other songbirds migrating back north from the tropics, but in the era of rapidly warming global temperatures the connection has increasingly unraveled. Spring is coming earlier and earlier in the Northern Hemisphere, with leafout and its subsequent caterpillar peak advancing in lockstep with rising temperatures. Migratory birds are also arriving somewhat earlier, but not at the same pace. One early study of the phenomenon in eastern North America, by the Smithsonian's Pete Marra and several colleagues, found that for every one degree Celsius rise in average spring temperature, migratory songbirds were coming back, on average, one day earlier—but leafout for plants was advancing three times as fast, leaving the birds further and further behind the curve. This is known as a seasonal (or phenological) mismatch, and it has migrants like the pied flycatcher caught in its tightening grip. Over and over again, scientists have found that spring is outrunning bird migration, with long-distance fliers that must return from the tropics in particular trouble. A pied flycatcher, wintering in the forests of western Africa, has no way of knowing whether it's an unseasonably chilly or unusually warm spring in the Mediterranean or central Europe; the triggers for its body to begin to lay on fat, its pectoral muscles to increase in mass, and all the other physiological changes it must undergo in order to make a 2,500-mile flight across the Sahara and Mediterranean are coded into its genes. It's the photoperiod, the subtly changing ratio of daylight and darkness that sets off the premigratory changes, as well as internal circadian rhythms in the bird's own body, that determine when any given migrant heads north.

That's been a disaster for pied flycatchers. Between 1980 and 2000, spring in the Netherlands advanced significantly, but ornithologists Christiaan Both and Marcel Visser found that the flycatchers didn't alter their migration timing at all. They did, however, accelerate how quickly they built their nests and laid their eggs once they got back to Europe, managing to advance the start of their breeding by about 10 days to try to compensate for the warmer weather and onrushing season. But that only gets them so far. "Owing to . . . their relatively inflexible arrival date, however, this window has become too narrow, and a significant part of the population is now laying too late to exploit the peak in insect abundance," the two researchers noted in 2001. Although other scientists had observed this trend toward earlier breeding in some birds and its link with climate change, Both and Visser were the first to raise a warning flag about how the growing mismatch was going to steal food from the mouths of hungry chicks. The results have been dire. Pied flycatchers have suffered enormous population declines, down more than 50 percent in the UK since 1995 and off by 90 percent in parts of the Netherlands.*

Interestingly, although growing seasonal mismatches have been documented in North America, and although migrant numbers in general have been falling for decades, there have as yet been no correspondingly catastrophic declines in particular species like those of the pied flycatcher in parts of Europe. One reason may be that neotropical migrants coming back from Latin America and the Caribbean face fewer obstacles along the way than do palearctic migrants from Africa that must cross both the Sahara and the Mediterranean. But another reason may lie in the far greater diversity of insect life in at least some North American regions. At Hubbard Brook Experimental Forest in New Hampshire's White Mountains, where over the past few decades some of the most important migration research has been carried out, there is no caterpillar peak; with hundreds of species of moths and butterflies feeding in the mixed hardwood-conifer woodlands, there is an ever-changing buffet for insect-eating birds throughout the breeding season. Insect diversity is much lower in oak-dominated forests, and this appears to be particularly true in parts of Europe, where the limited number of moth species accentuates the seasonal caterpillar peak.

The altered calendar produces winners and losers. In the UK, arrival dates for 11 of 14 species have advanced by up to 10 days since the 1960s, with the biggest change among short-distance migrants

---

* The nonmigratory great tits that share the European forests with pied flycatchers haven't shifted their egg-laying dates—but then, they were nesting several weeks earlier than the flycatchers to begin with, so they didn't face the same pressure from a climatological mismatch. What has happened is an increased—and, for the flycatchers, often fatal—degree of conflict with the tits over nesting cavities. As male flycatchers have played catch-up with spring, they have found themselves, more and more frequently, prospecting for nest cavities at the peak egg-laying period for great tits. This leads to ferocious battles, often to the death, which the flycatchers rarely win. Especially after mild winters (another product of a warming planet), when tit numbers are particularly high, up to 9 percent of male flycatchers may die in such conflicts, pecked to death by the slightly larger and heavier tits.

that winter in southern Europe or northern Africa; many have also extended their stay in autumn so that their time in Britain has lengthened significantly. Perhaps not coincidentally, these same species, like blackcaps and chiffchaffs, have also shown increasing population trends, possibly because they can now produce two broods in a summer instead of one. In 2008, scientists from France, Italy, and Finland analyzed bird observatory records from across Europe, looking at how 100 species of migratory birds fared during two periods, 1970–1990 and 1990–2000. They found that during the earlier period, a species' success was tied to where and in what habitat it nested and wintered; species that wintered in the African Sahel, where prolonged drought was a persistent problem during that period, did poorly. After 1990, though, the only factor that accounted for population change was migration timing. All the species that were declining during that decade were those (mostly long-distance migrants) whose arrival dates hadn't budged over time, while those with stable or increasing populations (mostly short-distance migrants) were keeping up with the earlier seasonal timing by arriving earlier themselves, staying in synch with insect populations and giving them time to produce additional broods of chicks as a result. Short-distance migrants also have the advantage of being able to monitor the weather much closer to their breeding grounds. A bird wintering in the southern United States, like an eastern phoebe or a hermit thrush, or one that migrates only as far south as the Iberian Peninsula, like a blackcap or a chiffchaff, is in a position to test the winds—to leave a couple of weeks early if there are warm, sustained southerly winds to carry it along, or to hunker and wait for better conditions if cold northers keep blowing day after day. A blackburnian warbler in the foothills of the Andes, or a willow warbler in the Congo basin, hasn't a clue what's going on, meteorologically, thousands of miles to the north.

Scientists have an increasingly clear sense of how changing climate is reshaping migration timing across Europe and North America, the regions with the longest and most extensive data sets on bird movements. The information is much spottier elsewhere in the world, though it paints a very similar picture of galloping seasonal change and a parallel but lagging response from birds. In Japan, wintering birds were found to be arriving nine days later in autumn and leaving a full three weeks earlier in spring, shortening their time on the wintering grounds by more than a month. A large analysis combining the results of almost 90 studies and more than 1,000 datasets from across the Southern Hemisphere found that there, too, austral spring was advancing far more rapidly than were bird movements. In Australia (where changes in seasonal rainfall, rather than temperature, seemed to be the driving force), plants had advanced the time of flowering and fruiting by almost 10 days per decade, while bird migration had changed by only two and a half days. Conversely, a meta-analysis of multiple long-term datasets in China showed a strong spring advance for trees and shrubs, but actually found a slight delay in springtime bird arrivals—though the study's own authors cautioned that the number of bird species involved, and avian datasets in general from China, were too limited to draw firm conclusions.

There are fewer detailed studies, like those involving pied flycatchers, exploring the effects of climate change on individual species, but those that have been undertaken show how complex the question can become—and how, by solving one problem posed by the shifting calendar, a bird may stumble into another. Take barnacle geese, for example, a smallish species with a black head and neck and a white face, which winters in western Europe and breeds in eastern Greenland, and on Arctic islands from Svalbard to northwestern Russia. The geese time their arrival in the Arctic to coincide with snowmelt, which has been advancing in spring by almost a day per year for several decades. The barnacle geese that winter in the Netherlands have managed to keep pace, but only by eliminating their traditional northbound stopover sessions along the Baltic and Barents Seas, where they were once able to graze for up to three weeks on emerging vegetation, building up their reserves of fat and protein necessary to produce eggs before continuing to the nesting grounds. Thus, they now arrive in the Russian Arctic up to 13 days early to catch the snowmelt, but aren't in condition to lay eggs until they've rebuilt their energy stores. Egg-laying dates have crept forward a bit, and average clutch size has increased slightly, but not enough to make up the difference. As in the oak forests of Europe, there is an insect peak in the Arctic, and in years with early snowmelt the goslings—which feed themselves from the moment of birth on mosquitoes, midges, and other bugs—miss it, suffering much higher juvenile mortality rates as a result. Barnacle geese that winter in Scotland have bene-fited from milder winters there, and despite a seasonal mismatch with the insect peak on their nesting grounds in Svalbard are producing larger broods—but fewer chicks are surviving to adulthood because of growing predation by Arctic foxes. (Barnacle and snow geese, com-mon eiders, common murres, and other colonial Arctic-nesting birds face an even bigger furry threat because of climate change. As Arctic sea ice disappears, and with it the opportunity to hunt seals as they have traditionally done, desperate polar bears have turned increas-ingly to bird nesting colonies for food, sometimes consuming 90 per-cent of the eggs and chicks.)

One of the most fascinating case studies, because it shows how the effects of climate change can differ dramatically even within a single species of migratory bird, involves Hudsonian godwits, the species that Nathan Senner studied that appear to be immune to carry-over effects despite one of the longest nonstop migrations on earth. Shore-birds are almost in Nathan's DNA. When he was eight years old, he had what he describes as a "conversion" when his father, my old friend Stan Senner, took the boy to the Copper River Delta Shore-bird Festival on Prince William Sound in Alaska. Nathan saw tens of thousands of western sandpipers and dunlin crowding the mudflats below the snow-capped Chugach Mountains and was hooked imme-diately and permanently on birds. Now based at the University of South Carolina, he and his students continue to study godwits and other Arctic-nesting shorebirds.

Nathan has been comparing the dramatically different ways two populations of Hudsonian godwits are reacting to climate change.

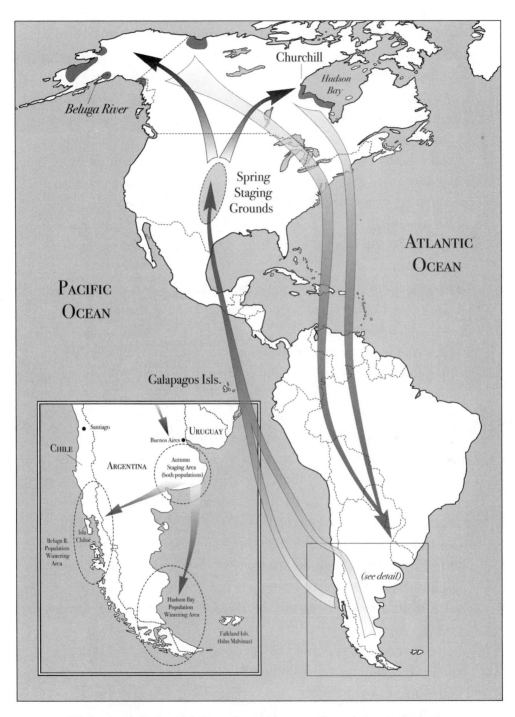

Hudsonian godwits, which form a few widely scattered populations in the Arctic and subarctic, follow similar routes to and from South America, but diverge in spring in the Great Plains—where they diverge as well in how they are faring in the face of climate change.

One group (those that have dodged carry-over effects) spends the winter in Chile, the other in Tierra del Fuego in Argentina, and as the austral summer ends in March and April all head north, the Argentine birds a few weeks later than the Chilean godwits. Both sets fly far out into the Pacific Ocean, as far west as the longitude of the Galapagos Islands, catching tailwinds associated with the cold Humboldt Current there. They cross Mesoamerica and the Gulf of Mexico, sometimes making landfall on the Gulf Coast but more often continuing to the central Great Plains of Kansas, Nebraska, or the Dakotas, in all a 6,600-mile nonstop flight. Here, after resting and feeding, the two populations diverge—not only in migratory paths, but in fortune.

The Chilean birds swing northwest, making a largely unbroken flight to south central and western Alaska, arriving the last week of April or the first week of May. The godwits that Nathan studied in this group, breeding along the Beluga River west of Anchorage, now arrive about nine days earlier than they did 40 years ago. But they suffer no mismatch, because the climate all along the route on which they fly, and at their end point in Alaska, has warmed at a fairly smooth, consistent rate, allowing them to accelerate their timing to keep pace with the seasonal changes. Their chicks hatch just in time to catch the insect wave, and the population is thriving.

Not so the godwits that started in Argentina. They arrive on their staging grounds in the Plains a few weeks behind the Alaskan birds, and after feeding and resting there they fly another 2,000 miles north to Hudson Bay, arriving the last week of May or early June—about 10 days *later* than they once did. Why? Because climate changes occur in paradoxical ways, and the Hudson Bay godwits are experiencing seasonal whiplash.

"On the northern portion of their migration route, the Dakotas and prairie provinces up to Hudson Bay, conditions are actually cooling," Nathan told me. That's part of the global weirding that is climate change; some parts of the planet are getting colder, at least for now, and at certain times of the year. Snow and ice now linger later in this final section of the godwits' route, preventing the migrants from pushing north as early as they once did. And it gets worse. "The unfortunate thing is that those cooling conditions in May are counteracted by really warming conditions the rest of the summer— June and July are experiencing some of the most drastic warming" anywhere in the North, he said. As a result, the godwits have to delay their arrival on their breeding grounds and so get a later start at nesting—but the rapid warming still brings out the insect swarms too early. Instead of reaching a peak just as the growing chicks need the most food, the bugs top out too soon, leaving the older chicks underfed right when their energy demands are highest, a seasonal mismatch reminiscent of the pied flycatcher's.

"So the godwits are between a rock and a hard place," he said. "They can't arrive any earlier or else they're going to encounter a lot of snow. But if they arrive any later, they're going to be even more mismatched with the emergence of the bugs their chicks rely on for food." The result? Many years the godwits are unable to breed suc- cessfully, with just 6 percent of their chicks making it to adulthood. Recent research, by a US-Canadian team that included Nathan, found that across the North American Arctic mismatches tied to ear- lier snowmelt are increasing, and are especially severe in the eastern Arctic—which is, perhaps not coincidentally, where the biggest pop- ulation declines in species like red-necked phalaropes and semipal- mated sandpipers have occurred. Still, not every shorebird seems to be suffering equally. The semipalmated plover, a plump little brown- and-white bird with a single black band across its neck, also nests at Churchill with the Hudsonian godwits. Even though insect emer- gence in 2010 peaked a full month before the plover chicks hatched, and a week and a half *after* they hatched in 2011, researchers found no impact on the young birds' growth in either case—suggesting that for this small shorebird, at least, there is enough arthropod food available even when the calendar is out of whack.

It's not that shorebirds are rigidly inflexible. An Old World god- wit, the black-tailed, which Nathan and his former advisor Theunis

Piersma have studied together, has shown remarkable adaptability. "Outside the breeding season, pretty much any way we look, these birds are able to respond to the changes that are thrown at them," Nathan said. Where the entire population once migrated from the Netherlands to sub-Saharan Africa, many of the godwits now detour to the southern coast of the Iberian Peninsula, where rice fields provide superb winter habitat, and the godwits move effortlessly from Spain to Portugal as conditions in either one change. The godwits are also incredibly flexible in the timing of their northbound migration, depending on the weather in Europe; an individual may fly north at the beginning of January during an especially mild year, he said, or wait until the middle of March the next if the cold lingers.

Even in the nesting season, Nathan and Theunis's research showed that black-tailed godwits can be more adaptable than expected, extending their nesting season as summers in northern Europe have become warmer and wetter. When an unusual spring snowstorm struck the Netherlands in March 2013, the godwits, having just arrived, made an unexpected reverse migration south to escape it. All of which is great news—but there is one obstacle even these ultra-adaptable godwits can't overcome. Intensive monocrop agriculture has largely replaced the flower-spangled natural meadows that once covered the dairy farms where most of the godwits breed. Early mowing destroys many nests, and those chicks that hatch find few insects to eat in what little foraging habitat remains—which in turn is also taller and denser because of warming summer temperatures. The result has been a steep decline in the population of godwits, which despite their adaptability at other times in their annual cycle haven't been able to adjust to the changes on the breeding grounds.

"That's the nub of it," Nathan told me. "They're flexible up until the point where they're not flexible."

Scientists have found other examples of behavioral flexibility in the face of climate change. One of the most unusual—in part because it builds on a foundation of meticulous data collected more than a century ago—has come to light in California. Between 1908 and

1929, pioneering field biologist Joseph Grinnell, who founded the Museum of Vertebrate Zoology at the University of California, conducted exhaustive surveys of many of California's most biodiverse regions, documented in voluminous records totaling 74,000 pages of notes. Beginning in 2003, his successors at the museum began resurveying those same sites, a similarly herculean task. In the process they discovered that many of the 202 species of birds found during both surveys had not shifted their range upslope to cooler climates, as one would have expected in a warming world—but the birds were nesting an average of seven to ten days earlier than they did in Grinnell's era. The average temperature in this earlier summer time frame is 2°F cooler than the later period, allowing the birds to exactly cancel out the two degree rise in average temperature that the state has experienced over the past century, without moving.

The ability of an organism to change in the face of different environmental conditions is one aspect of a grab bag of possible modifications to a plant's or animal's physical form, physiology, or behavior that biologists call *phenotypic plasticity*. This differs from evolutionary change, which is wholly genetic and passed down through generations; an organism's phenotype is a combination of its genetic background (its genotype) and how the expression of that genotype is affected by its environment. So far, evidence for true climate-driven evolutionary change among migratory birds is sparse, and differentiating between evolution and phenotypic plasticity has been difficult—though the difference is important. Ornithologists assume that birds will evolve in response to climate change, but will they be able to do so quickly enough to ward off its worst impacts? Or will phenotypic changes stretch only so far, eventually snapping as climatic change accelerates? When scientists in Sweden observed great reed warblers returning from Africa six days earlier than they had two decades previously, was that evolution or plasticity? After a great many calculations, they concluded it was the latter—but admitted they could not be certain. However, researchers working with pied flycatchers in Germany were able to exactly replicate a classic experiment first carried out in 1981, in which chicks were raised in captivity and shielded from all external seasonal cues to determine the genetically based timing of migration. In 2002, newly hatched chicks from the same region were raised in exactly the same conditions—even in the same cages and housing used 21 years earlier—to see if their migration timing had changed. It had, advancing in spring by more than nine days, close to the 11-day advance observed in a nearby wild population. (Although the experiment was carried out almost two decades ago, the results, which confirmed an evolutionary change in the face of climatic warming, were only recently published.)

The ever-busy Theunis Piersma was part of a team that found a surprising climate-induced change in red knots that nest in the Russian Arctic, where—as in so much of the world—the seasons are running amok, with spring advancing half a day a year over three decades. In that time, red knots have begun to physically shrink; especially in years with early snowmelt, juvenile knots weigh less, and have shorter bills, legs, and wings. This could be evolutionary or it could be (and likely is) phenotypic plasticity, since knots that are missing the Arctic insect peak and are undernourished may not grow as well as they otherwise would. But this change has life-or-death consequences for the knots once they reach their wintering grounds on the mudflats of the west African coast. Those with typically long bills can more easily reach a larger, more abundant type of clam that lives well below the surface, but the stunted knots with their shorter bills are forced to subsist on smaller, rarer clams closer to the top, and on low-quality seagrass rhizomes. It may not be a surprise that the climate-dwarfed knots have a significantly lower survival rate.

A reduction in body size actually appears to be a near-universal response to climate warming, and has been shown in a wide variety of animals, from salmon to squid and salamanders to ground squirrels. Red knots are not the only examples among migratory birds. Researchers from the University of Michigan and Chicago's Field Museum examined more than 70,000 bird skins, representing more than 50 North American species collected over a four-decade period, and found that as average temperatures on the birds' respective breeding grounds rose, body size across almost all of the species declined—with one notable exception. Wing length actually increased, perhaps because with less muscle mass to power them, the wings must become more energy-efficient, and also because as breeding ranges expand to the north, migrants must fly farther. (Long-distance migrants have always averaged much longer, more tapered wings than short-distance travelers.)

For some migratory birds, the biggest effects from climate change aren't on body weight or wing length, but the food they need. A warming climate in the Arctic is already altering food webs, including those on which one of the world's most dramatic migrants depends.

I like peace and quiet, and there is not a lot of either to be found out on the tarmac at a major airport. Terminal E of Philadelphia International was some distance behind me in the bustling, noisy chaos of this January night—planes taxiing back and forth, support vehicles crisscrossing each other or backing up with alarms beeping, some sort of siren sounding somewhere in the distance adding to the cacophony. I was leaning out the open passenger-side window of a pickup truck with half a fishing rod in my cold hands, the line disappearing into the darkness in the direction of runway 09L/27R, onto which a parade of 737s, A321s, and other big passenger jets was landing with metronomic, earth-shaking regularity. Each time one of the planes landed I could feel the *whump* of touchdown through the ground, even before I heard the squeal of tires and the roar of its thrust reversers slowing the massive machine down the runway.

I was not engaged in some lunatic style of fishing—well, not exactly, anyway, because the whole thing did feel more than a little crazy. This was workaday stuff for Jenny Martin, though, sitting behind the wheel of the truck. She was one of the federal wildlife biologists assigned to the airport to keep planes and animals—mostly birds—safely apart, and she was helping me with what has become one of the most exciting research projects I've ever been involved with, studying one of the migratory birds that may be at most immediate and serious risk from climate change.

Somewhere out in the darkness was a snowy owl, and if I could catch it, we would affix a new generation of high-tech GPS transmitter to its back and take it far from the dangers of the airport, releasing it in distant farm country in a bid to learn more about its behavior and wintering ecology. Catching it was proving to be the problem, though. Earlier in the evening I'd flubbed my first—and, it seemed more and more likely, perhaps only—chance to get the bird. The fishing line led to the trigger on a spring-loaded bow net, a bit like a giant mousetrap about three feet wide, in the center of which sat a pigeon wearing a custom-made double-layer protective leather flak jacket. It's the kind of rig I've used for 30 years to catch raptors as large as bald and golden eagles, though nothing in my experience had prepared me for working in the middle of an airport next to an active runway. Maybe it was the noise and the planes, a case of sensory overload, but I'm inclined to blame garden-variety buck fever. In any event, when the huge white owl ghosted out of the twilight, hovered low over the fluttering pigeon, and landed just to the side of the bow net, almost involuntarily I'd given the half-rod in my hand a convulsive jerk, tripping the net and startling the owl—which was, of course, several feet outside the trap—off into the dark.

Cursing doesn't help, but it does make you feel better. I sprinted out, reset the net, and climbed back into the truck with Jenny. Fortunately, snowy owls have little fear of anything human—which is one reason they find airports uncommonly attractive. Coming from the most remote parts of the Arctic, usually with little or no experience of anything human, they find trees and jetliners equally alien—and because airports like the one in Philadelphia are often the only flat, treeless expanses in an urban environment, when snowy owls migrate south in winter they often end up at airports, which despite the giant, thunderous "birds" must feel a little bit like home.

My interest in snowy owls had always been fairly casual. They're gloriously beautiful, of course, riveting in the way that only an immense, white bird with a killer stare from incongruously daffodil-yellow eyes can be. I'd seen many of them over the years, along the Atlantic coast or on inland farmland, and had tried to catch a few to band them on the rare occasion when one showed up in my area. But my real interest in owls focused on a much smaller migratory species, the northern saw-whet owl, a bird the size of my fist, which a crew of volunteers and I have been studying for more than two decades in the mountains of Pennsylvania. Snowies were just an occasional diversion.

That changed on a single day in early December 2013 when my phone rang. My good friend and longtime colleague Dave Brinker, a wildlife biologist with Maryland's Natural Heritage Department (and with whom I also worked on the Motus tracking system) was on the other end. "Have you been watching what's happening with snowy owls?" he asked.

I had. For the previous two weeks, online forums and birding listserves in the Northeast had been lighting up with increasing reports of snowy owls. That in itself wasn't unusual; snowies are irruptive migrants, their numbers fluctuating dramatically from year to year. Each winter there are always at least a few along the Great Lakes or New England coast, but every three to five years there will be a big surge, when instead of dozens of owls coming south from the Arctic there will be hundreds, even a few thousand. What we were seeing at the end of 2013, though, was something very different, and it was dawning on many of us that an invasion of perhaps historic significance was underway. A few days earlier, for instance, birders in Newfoundland found some 300 snowy owls at Cape Race, the easternmost spot on the continent; one fellow counted more than 75 of them from a single spot there as he panned his binoculars across the tundra landscape. It's hard to precisely gauge across the decades the magnitude of snowy owl irruptions, as such incursions are known, because no one (even today) conducts standardized surveys, but as far as anyone could tell, this was the largest irruption since at least the winter of 1926–27, and perhaps as far back as the 1890s. From a research perspective, this was literally a once-in-a-lifetime event. "None of us are going to live long enough to see something like this again," Dave said.

And so Project SNOWstorm was born—what's grown into a collaboration of 40-some researchers, banders, wildlife veterinarians, and pathologists, all volunteering their time and expertise as we tag and track snowy owls—more than 90 to date—from the prairies of North Dakota to the islands and peninsulas of the Great Lakes, the St. Lawrence River valley in Quebec, the Atlantic beaches of Maryland and New Jersey, the farmland of Pennsylvania, and the coast of New England. We've followed some individuals for years, back and forth between their Arctic breeding grounds and their wintering areas down here in the south. (All of our tracking data are available via interactive maps online at www.projectsnowstorm.org.) But although I've trapped and tagged many snowies since that night at the Philadelphia airport, just a few weeks after Dave's initial call, I've never again experienced quite the waves of mingled exhilaration and sheer relief that I did when that owl came back out of the darkness and into the beams of the truck's headlights, feet out and talons ready, into my net and then our hands.

An hour later we'd given the pigeon (unharmed, as usual) a well-earned meal in its cage, banded the owl and fitted it with a transmitter, then put the raptor into a large pet carrier and drove it out of the city toward Amish country farmland 50 miles away, where dozens of other snowy owls were already wintering. It was nearly midnight when we opened the door on the crate and watched the owl, a juvenile male, fly into the night across wide Arctic-flat farm fields, knowing we'd be able to follow his every move.

Because snowy owls are large and strong, they can carry a matchbox-sized, solar-powered transmitter that weighs less than two ounces and packs a lot of horsepower under its hood, allowing us to learn more than ever before about a tagged owl's movements. The transmitter sends us a fire-hose blast of information. Around the clock, as frequently as every six seconds, it records the bird's latitude, longitude, altitude, and flight speed. We get the local air temperature from an on-board sensor, and a tiny accelerometer can even record every wing beat and every thump of a hunting strike. Each day or two, a modem in the unit dials up through the cellular network and sends us its data; I don't know about you, but I get text messages from giant owls. Thanks to this technology, we've been able to document, in unprecedented detail, new or poorly known aspects of snowy owls' winter lives—the way many of them along the coast hunt waterbirds like loons and ducks far offshore, for example, or spend weeks at a time in the middle of the frozen Great Lakes (hunting waterfowl there, as well, in the small leads and openings that prevailing winds create in the otherwise unbroken ice). We've started to see patterns linking where and how the owls feed with what kinds of environmental contaminants they pick up, like mercury and rodenticides. By tagging owls in close proximity to one another, we've had insights into their social behavior (they are not, for the most part, very chummy, and females tend to dominate the smaller males), especially at night when they're most active. And by working with airport authorities, we're looking for better ways to keep owls and planes safely apart.

So part of what drives us—as with any research project of this sort—is simple curiosity. But what really motivates us to learn as much as we can about snowy owls, as quickly as we can, are the twin realizations that there are far fewer of them than we once thought, and that they are squarely in the crosshairs of climate change. Because they nest in the most remote, most northerly areas on the planet, snowy owls have always been hard to census, but the best estimates—combining breeding-season tallies from Alaska, Canada, Greenland, Scandinavia, and Russia—suggested that the global population was about 300,000; not a huge number, but a comfortable one. However, not long before we launched Project SNOWstorm, scientists realized that those estimates had overlooked a critical, newly discovered aspect of snowy owl biology. Tracking projects conducted by colleagues of ours in Canada, Alaska, and Russia showed that these birds are highly nomadic, moving hundreds or thousands of miles

from one year to the next; an owl nesting in the central Canadian
Arctic this summer may be in Greenland the next; one nesting now
in Alaska may be in Siberia the following year. Those add-'em-all-up
estimates had grossly overstated the global population, which more
careful analysis puts at no more than 30,000, and perhaps as few as
10,000, individual owls.

At the same time, ornithologists have also recognized how immediately perilous climate change may be for this raptor. A question we
get a lot, both from reporters and the public, is whether irruptions
like the one that made headlines in 2013–14 are caused by climate
change. The answer is no; snowy owl irruptions have been going on
for as long as anyone has been paying attention, at least since the mid-
nineteenth century. Although snowy owls will eat almost anything
during the winter—from muskrats to geese, mice to ducks, rabbits
to gulls; we even trapped one that was defending a rotting dolphin
carcass from vultures—during the nesting season their fortunes are
inextricably linked with lemmings, the small, hamster-like rodents of
the Arctic. Lemming populations in many parts of the Arctic boom
and bust on fairly regular, roughly four-year cycles, which in turn
produce the cyclical irruptions of snowy owls—which, common
belief aside, are not comprised of starving owls flying south looking for food, but rather usually plump, well-fed juveniles born a few
months earlier during the salad days of a lemming peak. Part of the
nomadic nature of these birds seems to be an adaptation to find and
exploit regional lemming highs, though how they do this across vast
Arctic distances remains a mystery. But without a lemming boom,
snowy owls rarely even attempt to nest—and in order to peak, lemmings must breed through the winter, protected from the bitter cold
beneath a thick, insulating layer of snow.

That's where climate change actually does come in, because the climate is changing faster in the Arctic than anywhere else on earth, and
as it does, it can alter or eliminate the conditions lemmings need to
hit their reproductive stride. Warmer, wetter winters mean less (or less
fluffy) snow cover, and more thaw-freeze cycles or freezing rain that

create ground-level ice—the worst conditions for lemmings. This isn't theoretical. On the Scandinavian peninsula, lemming cycles broke down beginning in 1994 and didn't recover for two decades, during which both snowy owls and Arctic foxes largely disappeared there; in northeastern Greenland, the same thing happened to lemmings starting in 1998 and remains so. In other parts of the Arctic, like Wrangell Island in the Russian Far East, lemming cycles haven't collapsed, but they have stretched—from every four years to every eight, meaning snowy owls in those regions can breed less often. Interestingly, one area that runs counter to this trend is the central and eastern Canadian Arctic and subarctic, which is where most of the snowy owls we see in eastern North America breed. In that region, although summer and especially autumn temperatures have risen (causing, as we've seen, major problems for nesting shorebirds), winters remain extremely cold, and snowpack depth has been increasing since 1995, perhaps because of higher autumn humidity. That actually means improving conditions for lemmings, and thus for snowy owls, too—but perhaps only in the near term. Climate models suggest that here, too, snow cover will eventually erode as winter temperatures rise.

What will that mean for snowy owls, especially given that there's an order of magnitude fewer of these raptors than we realized just a decade ago? One can make a strong case that snowy owls are among the handful of species, along with walruses, polar bears, and ivory gulls, at most immediate and direct risk from climate change. If lemming cycles begin to break down more widely across the Arctic—if snowy owls cannot breed as frequently, as successfully, and in as many places as they traditionally have—the population will fall rapidly. While our research project can't on its own change the trajectory of climatic warming in the Arctic, we can learn about the other threats that face snowy owls when they're on their wintering grounds, from airplanes and vehicles to chemical contaminants, and make that part of their annual cycle a little safer. Every owl we save down here is one more that has a chance of weathering the coming changes in the Arctic.

————————

You'd expect a snowy owl on a frigid winter's day. Other migrants are considerably more surprising.

It was New Year's Eve day, and the temperature at daybreak, when I left my home in central Pennsylvania, was 10°F. It warmed through the teens as the sun rose and I drove south, following the directions I'd scribbled down while on the phone the evening before. In the back of the car, a couple of cylindrical wire-mesh cages jiggled and rattled with every pothole. Sunlight shone off a fresh covering of snow that had fallen two days earlier. An hour and a half later I turned off the interstate, followed a few smaller roads to a quiet development, and pulled into a driveway. A knock on the door was answered by a fellow in his eighties, I would guess, bustling with energy and excitement for so early on a cold morning.

"Come in, come in! She's been here three or four times already this morning, but right now she's in her favorite spot," he said. He showed me to a window facing his backyard, where a gnarled apple tree stood, dozens of wrinkled old fruit still clinging to the branches, a few still capped with fresh snow. "Right there, near the top."

Glinting metallic green in the light, a hummingbird sat in the morning sun, her head swiveling constantly to glare at the parade of chickadees and finches swarming a seed feeder. As we watched, she roared out at a titmouse that had dared fly too close, tail-chasing the larger bird and scattering angry, high-pitched chips we could hear through the window. After the intruder was vanquished, she buzzed over to a small porch along the side of the house, and sipped at a bright red feeder hanging just below a utility lamp, whose floodlamp bulb kept the sugar water inside from freezing.

"Okay," I said, satisfied with what I'd seen. "Let's get to work." In a few minutes I'd set up one of the cage traps so the feeder, placed inside, was positioned almost exactly where it usually hung. I connected the trap's sliding door to a radio-controlled tripper, thumbed the remote switch to make sure it worked—*swoosh!*—then reset it and went into the house. Within five minutes the hummer had zoomed down from her perch, hovering cautiously as she examined from several angles this new thing that had appeared around her favorite watering hole. Suspicious, she instead flew to the tree, probing her bill into a number of the wizened apples. (I assumed she was finding syrupy juice, but when I later examined one of the apples, it was also full of dormant fruit fly larvae, left from warmer days.) Back she came to the feeder—slowly, carefully edging her way into the trap for a quick sip and a dash back out. I left the trigger alone; only after she'd come in several times, and taken a couple of long drinks, did I push the button and drop the door.

If the notion of a hummingbird in the middle of a frigid, snowy winter landscape strikes you as strange, you're not alone. Every year, a small group of fellow hummingbird banders and I (there are only about 200 of us in North America) get calls like the one I'd answered the night before, from homeowners baffled and worried because, long after the local hummers have migrated south, they still had at least one hummingbird hanging around as the mercury dropped and autumn transitioned to winter. Is it lost? they ask. Hurt? In need of rescue?

Fact is, the rapidly increasing number of hummingbirds showing up in fall and winter in the Midwest, East, and South are not lost or hurt, but rather the leading edge of a newly evolving migration route for several normally western species—one underpinned by the kind of normal mutation that drives evolution, and facilitated by human-altered landscapes and a warming climate. The bird I'd caught in that snowy Pennsylvania yard was a female rufous hummingbird, a species that nests from northern California and Idaho up into south central Alaska—the most northerly hummer in the world. Traditionally, rufous hummers migrate down the Rockies and into the mountains of western and central Mexico for the winter. But starting in the 1970s, birders and backyard hummingbird enthusiasts along the Gulf Coast started reporting more and more overwintering rufous hummers, and pioneering researchers like Nancy Newfield in Louisiana began banding them, confirming that these weren't lost, doomed vagrants, as ornithologists long assumed, but regular, reliable migrants, some of which returned for half a dozen or more winters. We now know that thousands—perhaps tens of thousands—of hummingbirds of a dozen or more western species, most notably rufous hummers, show up in the Midwest and East every year from late summer to early winter. Most eventually migrate south to the Gulf Coast for the winter, though a few—especially in mild years—may linger as far north as New England and southern Canada, because despite their appearance, these hummingbirds are incredibly hardy little beasts.

Migration, remember, is a genetically coded behavior, not the result of a conscious decision. Within any population of migratory birds there will always be a few individuals with a hiccup in their software; instead of orienting south from British Columbia to Michoacan in Mexico, a particular rufous hummingbird may instinctively fly, say, west, out into the Pacific, or even north in autumn, toward the Arctic. If so, obviously, those birds are culled from the gene pool. But what about a hummer that flies east? Centuries ago, when the Eastern Seaboard was largely forested and the climate was a lot colder, that may have been a death sentence, too. Today, however, humans have reshaped the land with farms and flower-laden backyards, while the climate has become steadily warmer. Combine that with a rufous hummingbird's innate cold-hardiness (this is a species that arrives in Alaska in April when the ground is still snow-covered, and which can drop into hibernation-like torpor every night to save energy) and you have a new world ready for colonization. Instead of dying off, these trailblazing birds return in spring to their nesting grounds thousands of miles away, and pass on those once-deleterious genes to fresh generations.

My late Alabama friend Bob Sargent, whom I met at Fort Morgan in the 1990s, was among the earliest banders studying this phenomenon, and he trained me in the delicate work of banding these small but feisty birds so I could expand that research to more northern areas like my then home state of Pennsylvania. In the nearly two decades since, I've banded more than 100 western hummingbirds—rufous, Allen's, black-chinned, Anna's, and calliope hummers among them—and seen them shrug off deep snow and bitter cold. (Seriously bitter: one rufous hummingbird that a colleague of mine banded survived air temperatures of −9°F and wind chills of more than −30°F.) The banding process is much the same as with any bird, though the bands are so small I must make them myself, using a precision jeweler's shear brake to cut a sheet of thin metal I get from the federal Bird Banding Lab on which are printed 100 tiny, unique alphanumeric codes. Trimmed to proper dimensions—exactly 5.6 millimeters long and 1.4 millimeters high for a female like this one—and then shaped in a specially made jig, the band goes onto the hummer's small leg using a custom pair of pliers that forms it into a perfect ring. Proportionately, the finished band on the bird weighs about as much as a fancy metal wristwatch weighs on a male human.

Reaching into the cage trap, I gently removed the hummingbird, which cussed me as energetically as it had that titmouse, and wrapped it in the toe of a nylon stocking—an easy way to keep it controlled and calm. Affixing the band and measuring its wing, bill, and tail took just a few moments. Using a short straw, I blew away the feathers on its throat and body, noting the heavy fat deposits that showed yellowish beneath its skin; this bird, I told the homeowner, was getting ready to head south. Using a magnifying loupe, I checked the upper mandible of her bill, which was smooth—the sign of an adult, since juvenile hummers have fine grooves on the bill surface. "This isn't the first time she's made this trip," I said. "She's had her passport stamped a few times already." Last, I weighed her on a sensitive digital scale: a porky 4.42 grams, 0.16 of an ounce, well above a lean nonmigratory weight of perhaps 3 grams. I was even more confident she'd be heading for the Gulf in a few days—and with such a full fuel load, I explained, she could easily fly 600 miles nonstop, getting to central Georgia, perhaps, in one 24-hour flight. After all, ruby-throated hummingbirds are able to fly that distance across the Gulf of Mexico.

We stepped off the porch, and I had the old gentleman hold out his hand, asking him to remain very still. I placed the hummer on his palm and slowly withdrew my fingers; she lay there for about 30 seconds, her tail quivering in time with her normal four-breaths-a-second respiration rate—then exploded into buzzing flight, zooming back to her perch in the apple tree. The look on the man's face was one of stunned elation.

Hummingbirds are not the only birds evolving new migratory routes in the face of a changing world. The most notable example is the Eurasian blackcap, a gray Old World warbler with an inky crown (rusty in females), and a complex suite of migratory behaviors depending on regional population, from highly migratory to completely sedentary. Like hummingbirds, a few blackcaps have always been born with a faulty genetic code, sending them in the wrong direction, or for too long or short a distance, or at an inappropriate time of the year. This is nature's way of throwing spaghetti at the wall and seeing what sticks; what is maladaptive for tens of thousands of generations may, because of changing environmental conditions, suddenly confer an advantage. Thus, a few blackcaps from the central European breeding population have likely always migrated northwest, into the UK, instead of southwest into the Iberian Peninsula where much of the population winters, a phenomenon called mirror-image migration. The UK-bound birds, like hummingbirds in eastern North America, rarely survived in Britain until conditions changed in the twentieth century. With warmer winters and an abundance of back-garden bird feeders, they're now a common and increasing wintering species, especially in southern Britain.

What's more, scientists used stable chemical isotopes in the feathers of blackcaps nesting in Germany and Austria to determine their wintering areas and found that, perhaps because they don't have as far to migrate, the UK population arrives earlier than Iberian blackcaps each spring, and produces bigger broods when they nest. Arriving before the Iberian birds, the British blackcaps not surprisingly also pair up preferentially with each other, a behavior known as assortative mating—and that created a buzz in the ornithological world when the findings were published in 2005, because assortative mating can be the first tentative step toward speciation. Indeed, later research confirmed that after just a few decades, there are now weak but significant genetic differences between the UK and Iberian populations. Biologists once assumed that geographic isolation was the main driver of evolution, but temporal isolation can work, too, and in this case, climate change is nudging the two populations of black-caps apart.

There has as yet been no similar research to see if something like that is also happening with rufous hummingbirds, but each banded bird is another data point on the road to understanding. Two days after my banding trip, the phone rang; it was the elderly gent hosting the hummingbird. "She came in at daybreak, and kept coming back again and again all morning, just sitting on the feeder when she wasn't drinking," he said. "I thought something was wrong with her, but about an hour before lunch she took off and flew straight up in the air. I ran outside with my binoculars and saw her as she leveled off, so high I could barely see her." Then she aimed due south, he said, and was gone.

Seven

AGUILUCHOS REDUX

An hour after dawn, the Butte Valley still lay in a basin of shadow, the July sunrise bright on the rugged, juniper-clad hills to the west, and splashing light on the high, snow-covered double cone of Mount Shasta 40 miles to the south. My companions and I stood on a steep, sparsely wooded slope of pale brown grass, our footing uncertain; loose lava rock rolled and shifted beneath our boots with every move, stirring up a low haze of dust. Below us lay an orderly gridwork of agricultural fields—rows of crops in immense rectangles, circles of green alfalfa surrounding center-pivot irrigation rigs. In the distance was the little town of Dorris, California, just a couple of miles from the Oregon border, where an American flag hung limp in the still air atop what the chamber of commerce brags is, at 200 feet, the tallest flagpole west of the Mississippi. Beyond that lay patches of sagebrush and bleached grass, part of the 18,000-acre Butte Valley National Grassland.

A PhD student named Chris Vennum had a great view, near the top of a tall juniper along one of the hills that enclose the valley, but his mind wasn't on the scenery. A big female Swainson's hawk, screaming her anger in high, piercing wails, was dive-bombing the scientist as he reached into the bird's bulky stick nest for its lone chick; even with a bright orange climbing helmet for protection, Vennum ducked each time the hawk turned to attack and we shouted a warning. Pummeled in the past, he'd learned from experience not to take the threat lightly.

It's the sort of scene I've experienced countless times as a birder, a writer, and a raptor researcher; I've been strafed more than once myself in similar situations. But that's not where my mind was wandering on this morning. The sight of that Swainson's hawk overhead— each sooty wing coming to a candle-flame point, the morning light glinting off her dark chestnut-colored breast and head, her smaller mate circling and calling a bit farther out—brought up potent memories of a time, more than two decades ago, when this species faced a terrifyingly imminent danger, and a future that was anything but secure. At the far end of its migration route, on the grassy pampas of Argentina, pesticides were decimating these hawks, with only a small group of scientists and conservationists trying to stave off an ecological catastrophe. I found myself in those days with a frontline view of that ultimately successful battle, working beside that team on the Argentinian plains as they struggled to understand and counteract the danger. Although that threat, and my involvement, lay half a world away, the story actually began here, among the sagebrush flats and farm fields of the Butte Valley in northern California. Twenty years later, I was back to close the loop—and also, perhaps, reassure myself that in a world of multiplying dangers, we actually can change things for the better.

Few raptors on earth travel as far as Swainson's hawks, which nest across the grasslands of the American West—from northern Mexico up through the Great Plains to southwestern Canada, and west to the Coast Ranges and Central Valley of California; a few even breed in the northern Yukon. In autumn all but the Central Valley hawks migrate to the pampas, the austral analog of the North American plains. It is an annual journey of up to 18,000 miles from one grassy sea to another, but for most of the twentieth century no one had ever traced it in any detail because tracking devices were too bulky and heavy for even a two-pound hawk. That changed in 1993 with the advent of the first small satellite transmitters. There was a rush among raptor biologists to tag peregrine falcons, the sexiest and most glamorous birds of prey, but a biologist in northern California had

other ideas. "Everyone jumped on peregrine falcons," Brian Woodbridge told me a few years later, "but the first thing that came to my
mind was, 'I need to find enough money to put some on some Swainson's hawks.'"

He was driven by more than just simple curiosity. For almost
15 years Brian had been studying Swainson's hawks in the Butte
Valley—climbing the prickly, resinous juniper trees to band their
chicks, trapping and color-banding the adults, getting to know
the lives and lineages of the dozens of pairs of hawks that nested
in this 130-square-mile valley a few miles from the Oregon border.
The hawks had a sweet gig; the valley's alfalfa fields brimmed over
with ground squirrels and voles, easy pickings for such adept hunters, which raised lots of healthy chicks on the bounty. But the Butte
Valley was only part of their world, and every few years Brian would
see a big dip in the number of marked adults returning in the spring
from their migration. He worried that some unrecognized problem,
somewhere far to the south along the unknown migration route or on
the poorly studied wintering grounds in Argentina, was killing the
Butte Valley's hawks. That worry peaked in 1993, when the valley's
Swainson's hawk population took an especially heavy hit.

That summer, the local national forest supervisor found Brian
enough money to fit two adult female hawks with the new satellite
transmitters, following their autumn journey via regular data downloads as they headed south. One of the birds winked out in Arizona—
a transmitter failure. The remaining hawk traveled down the Gulf
coastal plain of eastern Mexico, through the narrow waist of Central America and south along the eastern slope of the Andes. Once
in Argentina, it continued on to La Pampa Province, a board-flat
land that bears a remarkable resemblance to Kansas, with squared-off
grids of farm fields, pastures and shelterbelts at rigorous right-angles,
and small planted groves, or *montes*, of exotic eucalyptus trees. That
winter, Brian and two of his former field assistants traveled down
there, expecting to conduct some basic life-history research—nothing
earthshaking, just some ground-floor biology, since almost no one

had ever studied Swainson's hawks in their nonbreeding range. The trio followed the transmitter coordinates deep into the pampas, and were jubilant to find the hawks coming in to roost in the *montes* at dusk in flocks numbering in the thousands.

But jubilation was quickly replaced by shock and horror when they discovered equally large numbers of dead Swainson's hawks littering the fields and woodlots. Frantically questioning local farmers, Brian and his friends soon learned that the birds, which feed on large insects during their time in Argentina, were likely being killed by a powerful pesticide, an organophosphate called monocrotophos, which landowners—who were converting grazing land to row crops like sunflowers and soybeans—were spraying to control a grasshopper outbreak. Quickly shifting their focus to forensics, the Americans began to make methodical surveys of the devastation, finding dead hawks with poisoned grasshoppers still in their mouths. Sifting among the fetid carcasses and insect-stripped skeletons, trying to tabulate the losses, they also found leg bands, including one that Woodbridge himself had placed on a nestling in the Butte Valley years earlier, and whose breeding success as an adult he'd documented. Returning the next winter with a larger team comprised of both American scientists and Argentinian researchers and government officials, they found even greater numbers of dead and dying hawks, as many as 3,000 lying in a single field. The team estimated that up to 20,000 Swainson's hawks, mostly breeding-age adults, had died in this one relatively small area of the vast pampas—and there was every indication that the carnage extended across most, perhaps all, of the hawks' wintering range in Argentina. The true numbers were impossible to gauge, but it was clear that just a few more years like these would quickly plunge the entire species—long among the most common and widespread raptors in North America—toward extinction.

That did not happen. By January 1997, in the middle of the austral summer when I joined Brian and his team back on the pampas for their third season, things were moving in a positive direction. Bird conservation groups, led by the American Bird Conservancy, had negotiated a deal with the manufacturer to pull monocrotophos off the market. The Argentine government had moved quickly to ban the chemical for use against grasshoppers and to buy back existing stocks from farmers, and had launched a major education campaign about the danger to the raptors. Everyone was holding their breath, waiting to see if the destruction might at last abate.

Our base of operations was Estancia La Chanilao, a lovely old ranch owned by a middle-aged man named Agustín Lanusse—tall and thin, penetrating eyes, a dark beard, and thinning hair. I initially took him for a local rancher, but soon learned that his background was far more tangled and unusual. Agustín's uncle, Alejandro Agustín Lanusse, was an Argentine general, part of a military junta that took control of the country in a 1970 coup. For two years General Lanusse served as president—but in 1973 he reestablished free, direct elections in which he was soundly trounced, and after which he facilitated a peaceful transition of power. In his final years, General Lanusse testified against the military's "Dirty War" from the mid-1970s through the mid-1980s. His nephew, Agustín, turned his back on politics and the family legacy entirely, working as a shepherd and a park ranger in remote Patagonia before marrying into a ranching family in La Pampa. His wife had died shortly before Brian and his colleagues first showed up at Chanilao in 1995, but despite his grief Agustín worked tirelessly with them over the subsequent years, as concerned as the Americans were with the deadly toll the pesticide was taking.

My time in La Pampa was at once magical and exhausting. The monotony of soybean fields and pastures was broken by shimmering lakes with flocks of pink-white flamingos, and clouds of shorebirds that had migrated there from Arctic Canada. More than once we had to hit the brakes when gangs of rheas, the South American relative of the ostrich, sprinted across the dirt road in front of us, long-necked and leggy like dinosaurs dressed up in feather boas. I slept in a tent pitched in the shade of a small copse of trees near the ranch house, rising at three in the morning to help set bal-chatri traps, little wire-mesh cages festooned with monofilament nooses. These were baited with live mice and set out in the fields near the *monte*; at daybreak, the air heavy with the smell of eucalyptus, the hawks would glide down from the trees and, by the thousands, gather in these fields, waiting for the sun to warm the ground and generate the thermal air currents that would carry them skyward. Although in Argentina Swainson's hawks feed mostly on large insects (grasshoppers and dragonflies in particular), a meal is a meal, and many would grab at the mice, snaring themselves by a foot. These birds were banded, their blood and feathers sampled to see what kinds of toxins they were exposed to—we even washed their feet with alcohol that would later be run through a gas chromatograph to assess the chemicals they had picked up.

In the evening, after showers in the bunkhouse, we usually gathered on the lawn in front of the ranch house, sharing a few tall bottles of cold local cerveza beaded with condensation in the damp heat, to watch the show. The hawks would sail in from every point on the compass, gathering in ever-increasing numbers over the eucalyptus grove—sometimes hanging almost motionless, all facing into the wind in vast sheets and layers, other times swirling in majestic pillars that reached, vertiginously, high above our back-craned heads. Unlike most raptors, Swainson's are highly gregarious outside the breeding season; some years earlier, when I had helped document the largest migration choke point for the species at Veracuz, in eastern Mexico, we would count thousands per hour passing high overhead on their way south. What we witnessed at Chanilao was orders of magnitude more dramatic. On a good night, when most of the Swainson's hawks wintering in the vicinity gathered at Chanilao, there might be 10,000 raptors in the air, still one of the most awe-inspiring sights I've ever witnessed. The summer sun would sink, the muggy sky would go orange, and the hawks would slide down en masse to roost in the trees. Slipping between the trunks in the dim light of the forest, I could hear the rustling, slapping, rattling sound of wings against leaves and branches as the buteos settled in for the night—and the crunch of dry bones underfoot, where beneath the leaves still lay the remains of hundreds of hawks that had died from monocrotophos two years earlier.

We took our meals with Agustín, his sister-in-law, and his three teenaged daughters from his previous marriage, slender young women who had thrown themselves enthusiastically into our field work. The table groaned under platters of beef, potatoes, and boiled vegetables, while cigarette smoke swirled around us. But because the Lanusse family did not begin to eat until 10:00 or 10:30 at night, and because our field work precluded us napping during the traditional afternoon siesta, we became steadily more sleep-deprived as the weeks went on, until by the end we all felt as though we were swimming in mental molasses.

That month in the pampas went quickly, though, each day a different task—joining Argentine grad student Sonia Canavelli tracking radio-tagged hawks to learn how far they traveled by day to hunt; collecting regurgitated pellets—pink and crumbly wads the size of walnuts, made of chitinous insect bits that would be analyzed to determine the birds' diet; driving the squares of empty dirt roads that stretched for miles across the flat landscape, searching for concentrations of wintering Swainson's hawks on other estancias. In the dusty little pampas towns, we saw posters and signs urging farmers to protect the hawks, to avoid using monocrotophos. We met people wearing pro-hawk buttons that Argentine conservation groups and the government had distributed; we did interviews with TV stations and newspapers, and developed such a weird degree of local celebrity that we were often stopped on the street, and at one point even presented, by its beaming *abuelita*, with a squalling baby to kiss. One day after six or seven hot, dusty hours conducting road surveys, one of the scientists and I stopped at a local ranch to ask about hawks. The worker we met brightened immediately; *sí, sí*, big smiles as he nodded enthusiastically at Mike Goldstein's questions. Lots of rapid-fire, Argentine-accented Spanish that I couldn't follow, though I caught a lot about *aguiluchos*—"hawks"—and some gestured references to the leg bands we'd been deploying. How wonderful, I thought; another example of the far-reaching conservation message. It wasn't until we were back in the car that Mike sourly explained that the guy was cheerfully explaining how he liked to shoot the hawks for the "bracelets" they wear, which he was collecting. There were more dangers in this far land, I realized, than just chemicals.

Each day, we waited for news of some big chemical kill, a call from Agustín's network of fellow ranchers saying that the hawks were dying again. But that call never came; it was a wet year, the grasshoppers weren't much of a threat, and the farmers had absorbed the lessons of the PR campaigns and were switching to less noxious chemicals when they did spray. As the years passed, and the news from Argentina remained positive, we began to relax. Back in California, those winters when a hefty chunk of the hawk population would vanish became a thing of the past. I kept in touch with Brian Woodbridge; a few years later I heard that he'd left the Butte Valley for a job with the US Fish and Wildlife Service, working with northern spotted owls—a subject far more politically charged and challenging than Swainson's hawks. Still later, he was part of the agency's team working on golden eagle conservation. But although Brian was gone, others picked up the torch, grad students who maintained and expanded his long-running Swainson's hawk study. At a raptor research conference a couple of years ago, a decade and a half since my time in Argentina, I ran into one of them, a muscular PhD student with buzz-cut blond hair named Chris Vennum. "You should come out next summer when we're banding chicks," he said. "Brian will be there, Pete Bloom, Karen, all the old crew are coming in to help. It'll be a reunion." Best of all, he said, the Butte Valley's hawks had continued their slow recovery from that pesticide-induced nadir, effectively doubling their numbers. The next season, he said, should set a new record.

Good news, obviously—and good news can be scarce these days on the migration front. So the next summer, I picked up a rental car at the airport in Medford, Oregon, and drove south through the hazy smoke of wildfires, looking forward to a victory lap with the scientists who have championed this isolated valley's raptors for decades. And it was, sort of. But it was also a chance to gaze with them into an uncertain future, wondering how far they could stretch their optimism, the kind of worries that bedevil anyone trying to shepherd migratory birds on a fast-changing planet. Twenty years ago the world saved Swainson's hawks from a single, sharply defined threat and celebrated a well-earned conservation victory. Today, the hawk's most ardent advocates admire the bird's adaptability, and take comfort from its ability to thrive in many human-altered landscapes. But also I heard concerns about new issues—some immense and diffuse, like climate change and a drying landscape, some much more specific and near at hand, weirdly unexpected. I mean, honestly—what raptor biologist ever expected to lose sleep over America's hunger for year-round strawberries?

The story of the Butte Valley's Swainson's hawks—both their near-death experience in Argentina, and their wider role in understanding the ecology of migratory raptors—is an interlocking tale of scientific and avian lineages stretching across decades. Although Brian Woodbridge is most closely associated with the valley's hawks, he was neither the first nor the last scientist to study them; over the past 40-plus years there have been, in a sense, four linked generations of researchers and their crews working in this corner of California. Many of those folks were now coming together for several days to help Chris Vennum, the latest in that line, with his field work. It would be, as Chris had said, something of a reunion.

The elder statesman of the bunch, and a legend among western raptor biologists, was Pete Bloom. Pete has worked with virtually every species of bird of prey in the West, from tiny American kestrels to (most famously) California condors. In the 1980s, when he was part of the condor recovery program, the massive vultures were careening to extinction because of lead poisoning and DDT-related

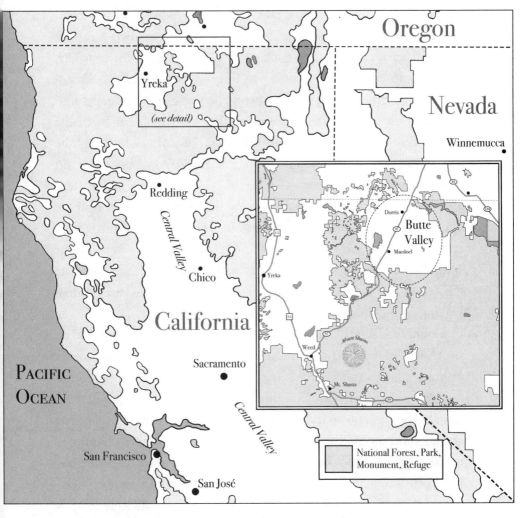

Butte Valley and environs, northern California.

eggshell thinning. He was one of the biologists charged with cap-
turing the last few wild condors in the mountains of Southern California, which he did by crouching for days or weeks at a stretch in a
camouflaged pit beside a calf carcass, ready to reach through a narrow slit and grab the legs of a condor coming down to feed—more
or less exactly the way American Indians had once trapped golden
eagles for their feathers. On Easter Sunday in 1987, Bloom caught

the last free-flying condor, a male known as AC9, though in that case he and his colleagues used a cannon-propelled net over a carcass rather than the old Native trick. AC9 spent 15 years in captivity, fathering 15 chicks that became part of the successful captive breeding and reintroduction program, before he was himself released into his old territory in 2002. After producing still more chicks in the wild, AC9 finally disappeared in 2016—presumed dead, though he was never found—by which point there were 270 condors in the wild and nearly 200 more in captivity. Grainy video shot that day in 1987 shows the net exploding up and arcing down over dozens of ravens and the single, immense vulture, and Bloom—wiry and lean, bushy dark hair and beard—bursting from the pit and racing to the great bird. The camera cuts to a close shot as he and two other biologists cradle its black wings and orange-and-purple head, then ease it into a big dog kennel for the trip to the zoo.

Bloom first started studying Swainson's hawks in the Butte Valley in the late 1970s, part of exhaustive statewide surveys for California Fish and Game, and he still helps to keep an eye on them. Although Swainson's hawks remain common across most of their range, there have been huge, inexplicable collapses in some regional populations several times in the past century. In the 1890s they were ubiquitous on the prairies of Saskatchewan, but gone from there two decades later; similar declines or local extinctions were noted in parts of Montana, Alberta, and Manitoba in the years after white settlement, variously ascribed to persecution by settlers or just put down as a mystery. In many of those areas, Swainson's hawk numbers later rebounded, but in few places were the declines as pronounced or as permanent as in California, where the hawks vanished entirely from all of the state except the Central Valley and the northeastern Great Basin country, including the Butte Valley. Bloom's surveys, which led California to list Swainson's hawks as threatened in 1983, suggested that the state's population had dropped more than 90 percent from historical levels, from more than 17,000 pairs to just 400. In a prediction that proved prescient, Bloom speculated that pesticide use on the wintering grounds might have been a factor.

Brian Woodbridge came on the scene in 1981, hired to study peregrine falcons in the region and basically loaned to Bloom at the end of the peregrine season to help with Swainson's hawk research. "Pete told me, 'Learn to climb nests, and try not to kill yourself,' and pretty much cut me loose," Brian recalled. A lot of his job entailed rappelling down cliffs into falcon and eagle eyries. "And when I was finished with peregrine monitoring for the season I'd wander around, checking on the Swainies. I became friends with the Forest Service biologist here, and the next year they offered me a job monitoring northern goshawks and Swainson's hawks." He continued to work for the Forest Service, studying raptors for the next 21 years. Toward the end of his tenure, Brian hired a University of Nevada master's student named Chris Briggs as a research tech, who as part of his later doctoral research started color-banding all the hawk chicks they could find—a dramatic new twist to the research, as we'll see. Briggs took over the Swainson's hawk project entirely in 2003 when Woodbridge left for the US Fish and Wildlife Service, and Briggs in turn hired another master's student, Chris Vennum. By the time of my visit, Briggs was teaching at a college in New York, and Vennum was leading the project as he worked toward his own doctorate.

It's fitting that from this lineage of scientists has grown a study that is all about the lineages of hawks. Because Brian Woodbridge (and Pete Bloom before him) had such a long, stable record of the Butte Valley's hawks, many of which Brian had trapped and marked with colored, numbered bands that can be easily read at a distance, he was able to detect the damage those pesticide kills in Argentina were inflicting in the 1990s. But when Briggs (and now Vennum) started color-banding nestlings as well, it created a population in which almost every hawk is, at a glance, a recognizable individual, its family traceable back through many generations. Almost nowhere else on earth has this been done on such a scale, and it's allowed the researchers to ask and answer many previously puzzling questions about the ecology of migrant raptors.

When I arrived, Briggs—his beard speckled with some gray, lamenting the extra pounds he'd added since he was no longer in the field all summer—was already on hand to help, with Bloom, Woodbridge, and several others due to come in that evening. That first morning Briggs and I kept cautious watch as Vennum scrambled up the tangled branches of the tall juniper overlooking the Butte Valley. Briggs shouted warnings as the mother hawk dove, while I was positioned some distance from the nest in case the young bird—unable to fly well but liable to glide a long way—should panic and bail out of the nest. Watching from the other side was the fourth member of our field crew, one of Briggs's students from Hamilton College in New York named Amelia Boyd—the latest twig on this ever-branching scientific family tree. After Vennum lowered a blue canvas shoulder bag containing the chick, Amelia cradled the youngster in her lap to band, weigh, and measure it. Nearly full size, but with wing and tail feathers that were still growing in, the chick was remarkably calm, lying quietly and unrestrained on its back on the digital scale, or crouching patiently next to Amelia as she finished the paperwork.

Done at the hillside nest, we drove a bit, then hiked out across a flat expanse of rabbitbrush and sage to another nest about 40 or 50 feet up in a dense, bulbous juniper. The adults circled and squealed their displeasure, the noticeably larger female, as usual, hanging closest. Amelia buttoned up a long-sleeved shirt despite the high desert heat—the temperature would top 104°F by late morning—strapped on a helmet and pulled on a pair of gloves, then scrambled up through the tree. "These junipers are like climbing a ladder," Vennum told me. "Pretty easy, but messy." His field clothes bore that out—heavy canvas jeans shredded and frayed, his socks showing through rents in his tattered boots, the sap from the trees smelling like cat piss. When Amelia's head appeared in a gap in the canopy, the big, dark-morph female hawk saw her chance, flipping instantly into a slanting attack.

"Here she comes—*duck*!" Vennum shouted, as the hawk missed by a foot or two, screaming as she passed.

"Where is she?" Amelia hollered, looking around frantically. A pre-med undergrad who signed up to work as a research tech because she likes to climb, Amelia had earlier admitted she was pretty freaked out by the prospect of getting clocked on the head, five stories above the ground, by a large bird with eight knives on its feet. "She's high above you and circling," Vennum yelled up. "You're good." A minute or two later, Amelia had bagged the single chick in the nest and started quickly down to safety.

Once the chick was on the ground, the parents drifted off, and the three processed the youngster quickly. Amelia drew a little blood from a vein in the wing, which she would later examine for parasites and to analyze the ratios of immune cells, which in some birds can help predict whether they'll survive to reproduce. But the green color band on its left leg, with a white two-character code that's easily readable from a distance, was the key to the unusual value of this long-running study. In the early years, Bloom and Woodbridge color-banded any adult Swainson's hawks they could catch, but when Briggs came on in 2008 he expanded that work to include almost every chick in every nest they could find, every year, across the 10-by-20-mile valley—a prodigious task, totaling almost 1,100 nestlings to date. As a result, more than three-quarters of all the adult Swainson's hawks in the immense valley are now color-banded, one of the highest such rates in the world. The scientists can thus tell which individuals have returned and which have disappeared, whether they've shifted territories, who is paired up with whom or has split from their long-time mates. Most of all, Briggs and Vennum can now map the valley's Swainson's hawks back through generations and across bloodlines, an ability that has offered surprising insights into the hawks' biology and behavior. For one thing, they know this small breeding population is intensely loyal to their homes; color-banded birds from the Butte Valley rarely show up anywhere else except the neighboring

Klamath Valley a few miles away, and pairs will keep returning, year after year, to nest sites that are no longer suitable because the habitat around them has changed. Such incredible tenacity, Briggs thinks, may be a result of the hawk's extraordinary migration. "Perhaps the irony of being so mobile, and migrating so far, is that by the time they get back to the breeding grounds they don't have time" to look for new places, he told me. Whatever the cause, Swainson's hawks are unusually slow to respond to new habitat and new opportunities. Even today, the species is only slowly colonizing parts of southern Butte Valley that were flood-irrigated until the 1990s, but which now hold plenty of ideal but underused habitat.

The researchers have also learned that while most pairs contribute few offspring that live long enough to breed, the remaining one-third of the nesting females produce virtually all the chicks that survive to adulthood and return to nest. "It's really just a few individuals that are driving this population," Vennum said, pulling off his helmet and rubbing his short hair. "They're the supermoms, just a few nests that produce most of the recruits"—the biological term for youngsters that survive to enter the adult population. "Then you have a majority of duds that really aren't contributing much. It's interesting; some territories are occupied every year for decades, while others come and go with time. Some pairs produce a ton of chicks but very few of them return as breeders. Still, we don't know what's creating these supermoms. Is it that they're just able to attract more experienced males? Is it just because they're able to get a good territory, that they were just lucky? There was a paper that came out earlier this year titled, 'Pluck or Luck,' trying to get at this whole question of why some birds are good parents and some aren't. Their conclusion was that the majority of it was luck. But I don't know."

Unlike populations farther east, Swainson's hawks in California are predominantly dark-morph birds, "morph" being the term for a palette of color variations within a single species. In most of the Swainson's hawk's wide range, light-morph individuals are the norm—creamy undersides, a dark brown back and hangman's hood with a white chin, and charcoal-gray wing feathers. In the Butte Valley, though, such birds make up less than 10 percent of the population, the rest being dark- or rufous-morph hawks that exhibit lovely and highly variable tones of cinnamon, chestnut, and mahogany. Why this is so—and why so many other related buteos, like red-tailed, ferruginous, and even broad-winged hawks, also exhibit an east-to-west light-to-dark morph gradient—is something that Briggs has obsessed over for years, without yet having an answer. But because he knows everyone's family history in the Butte Valley, Briggs can show that while female Swainson's hawks don't seem to care about the coloration of their mates, males very much do. They tend to pick females that share the morph pattern of their mothers—a so-called Oedipal complex. Intriguingly, males that pick mates that *don't* match their mother's morph produce significantly fewer fledglings over their lifetimes, which are also shorter than average.

"Whether that means those males are terrible quality and they just can't attract the females they think are attractive, or [are forced to] the periphery of the valley into some crappy territory, or whether they're just not willing to put in the effort to a female that they don't find attractive, we don't know," Briggs said. One possibility, though, is that the morph colors may be a visible expression of some aspect of the bird's immune system. "It could be some MHC compatibility thing," Briggs said, then noticed my puzzlement. "MHC—major histocompatibility complex," he explained. The MHC is a set of genes that code for proteins in a cell, which in turn allow the body's acquired immune system to recognize foreign invaders and trigger a response. The diversity among MHC genes is tremendous, and studies in a wide variety of organisms, from fish to mice to humans, have shown that mating is most common between those individuals with the most dissimilar MHC sets. (When, for example, college women are asked to sniff the used tee-shirts of college men and choose the ones they find most attractive, the women almost always pick those belonging to men whose MHCs are very different from their own.) "Maybe color morph in Swainson's hawks is a visible indication of their MHC," Briggs explained. It's just speculation right now, but the blood samples that Amelia was taking could explain why, if you're a male Swainson's hawk, you want a girl just like the girl who married dear old dad. Choosing a mate with the wrong immunity complex might compromise your breeding success. Because very few raptor populations anywhere in the world are so intensely tracked, the Butte Valley may be one of the only places where such questions—why supermoms exist, whether color morphs are signaling something critical about health, and much more—have a chance of being answered.

Vennum and his team were nearing 100 nests for the season, a new record, but it was far from a banner year in terms of productivity; it had been a dry summer, which appears to have driven down rodent numbers, especially on native grasslands in the valley where most of the nest failures occurred. Such ups and downs are normal for rodents and raptors, but in the Butte Valley, Swainson's hawks have a lifeline—alfalfa fields surrounding center-pivot irrigation rigs. Those luminously green circles in the otherwise sere landscape are the perfect habitat for ground squirrels, which exist here in astounding densities of up to 133 per acre. Because they can eat as much as 45 percent of the alfalfa crop, the squirrels are considered a plague by farmers, who welcome hunters who use the rodents for target practice, shooting thousands of "squeaks" (Belding's ground squirrels, the most common species) and "longtails" (California ground squirrels) every spring. But the rodents also provide the prey foundation for many raptors, including Swainson's hawks, which explains why this part of northeastern California and neighboring Oregon has one of the densest concentrations of birds of prey in the United States.*

---

* Lead poisoning is a serious threat to raptors, but this is one literal bullet that Swainson's hawks have largely ducked. While red-tailed hawks, bald and golden eagles, ravens, and other birds scavenge the shot and discarded ground squirrel carcasses—and have been shown to suffer very high lead levels as a result— the main squirrel-shooting season occurs before the Swainson's hawks return in spring. And fortunately for all raptors in California, a first-in-the-nation statewide ban on lead ammo went into effect in 2019.

Knocking off for the day, we drove up Rt. 97, the busy truck route that runs through the valley on its way to the Canadian border, and Vennum started detailing the local raptor neighborhood. "That tree right there is a Swainson's hawk pair, the one beyond it is redtails, and"—he pointed across the two-lane road—"there's a Swainie territory over there known as BLM Pine One. The scrubby little tree to the west of it is another Swainie nest called Captain Kangaroo— I have to change the name of that territory. But we haven't even gone 100 yards. You'd have to say the Swainson's and the redtails are doing okay in the Butte Valley." A Swainson's hawk territory in the Butte Valley averages about 1,000 acres, versus up to 10 times that in other parts of California, a testament to the abundance of food.

If 40-plus years of research in the Butte Valley has demonstrated anything, it is the fundamental link between hawks, rodents, and alfalfa, but the valley wasn't always a paradise for squirrels or their predators. Back when the Modoc, Klamath, and Shasta peoples lived here, the Butte Valley's 130 square miles formed an expansive wetland dotted with sagebrush islands, great for migratory waterfowl (which Natives hunted during the birds' flightless molt) but marginal habitat for a grassland raptor like the Swainson's hawk. Beginning in the 1860s, though, white settlers drained the tule marshes and converted the land to farming and ranching, forcing out the Indians but opening a door for Swainson's hawks. Potato and hay farming (along with lumber from the surrounding mountains) was an economic mainstay for much of the early twentieth century, but when Brian Woodbridge arrived in the valley in the early 1980s that market was waning, and the weedy old potato fields produced little in the way of prey or hawks. Federal subsidies for milk production had ended, and with them much of the alfalfa production that had boomed in the valley in the 1960s and '70s—and alfalfa is the key.

"Any increase in alfalfa productivity is an increase in Swainson's hawk populations," Brian told me that evening, settling himself in a lawn chair outside the house the field crew was renting. "Everywhere they live, Swainies use ag habitats. It's amazing, the degree of reliance on ag habitats in this bird—it's pretty much 100 percent in the Butte Valley. These two communities, the hawks and the farmers, fit together for better or worse." Pete Bloom slid a chair over and joined us; although his thick hair and beard have gone silvery-white, Pete still has the sinewy build of an active field biologist. Brian, too, had a lot more gray in his goatee than the last time I'd seen him; he was 61 and retirement, he told us, was looking better and better. A car door slammed, there were more hugs and greetings—Karen Finley, Brian's old field assistant who had been with him on that first trip to Argentina, had driven down from the Willamette Valley in Oregon, where she and her husband run a huge bee-keeping and honey operation.

The next morning I was riding in Chris Vennum's truck with Brian and Karen, looking for a pair of unbanded adults that Vennum badly wanted to trap. We found both birds perched on adjacent utility poles, so we drove slowly but steadily up the gravel road. As we passed them, Karen opened the passenger side door and carefully dropped a bal-chatri trap with a mouse as Vennum drove on, Brian craning his neck to watch the hawks. We'd gone barely 50 yards when the smaller male dropped off the perch and slanted down, out of sight below the tall grass on the roadside, so we did a quick turn-around in a barnyard as a couple of Australian cattle dogs barked at us, and sped back. The hawk was flopping on the ground, snared by one foot, and Brian was out of the truck in a fluid rush, enveloped in our dust cloud, before Vennum even had it stopped. Throwing his shirt over the bird, Brian grabbed its legs, then he and Chris undid the nooses and slipped an elegant leather falconer's hood on the male to keep it quiet.

We were discussing how to set the trap for the female when a four-wheeler raced up behind us, an older, sun-weathered man riding with one of the bluish cattle dogs perched behind. His name was Tom, a yellow feed cap on his head and a large Jack Daniels belt buckle at his midsection. "I wanted to make sure you weren't going into the alfalfa—I've had a lot of trouble with people shooting squirrels," he said. "They hit those pipes with ricochets, and if you get a single hole in one it's 300 dollars to replace it." Vennum explained what we were doing, and they chatted a little about Swainson's hawks. Chris has an unfortunate tendency to lapse into bio-speak on occasions like this, framing the hawks' epic migration in kilometers instead of more easily understood miles, and talking about how young ones "recruit" into the population, but he got across the main point—these raptors eat a lot of the ground squirrels that are an alfalfa farmer's bane.

"You're okay," Tom said, climbing back on his four-wheeler. "I just wanted to know who was up here."

We quickly caught the female, and both birds got metal bands on their left legs and green-and-white color bands on their right. Karen, holding the female, didn't hesitate when I asked how the study had changed since her days as a field tech 20 years earlier.

"Color-banding the chicks—that's been transformative. Years ago we would have said, 'Oh, there's a metal-banded bird on that pole,' and that would have been it. But now they can read the color band and say, that's a second-year male from such-and-such a nest. Amelia can go back and read the genetics for that bird, and see who is its family."

"Well, these birds *are* like family," Brian said, checking the fit on the female's band.

Karen nodded. "Yes! These could be the grandkids or great-grandkids of the birds I used to know. And I guess they probably are."

Late in the day I was in Vennum's truck with Melissa Hunt, a bio-technician for the US Geological Survey who studies goshawks in the area, and who was helping out with the Swainson's work when she had time. We'd dropped a bal-chatri trap for a pair of hawks sitting on a center-pivot rig as the sky to the southeast darkened, and sheets of rain started to drape the view of Mount Shasta. Wind began licking the tall grass, and the temperature plummeted into the fifties. Lightning bolts—which lingered on the retina for what seemed like long, long seconds—crackled to the valley floor. The male dropped twice, but we couldn't see the trap for the high grass, and the first time we eased up in the truck he was only sitting on the ground nearby and flushed. Finally he dropped a third time, and as the first rain started splatting the windshield Chris and Melissa grabbed him and got him hooded and into the vehicle.

The rain was pounding now, and mixing with hailstones. "We need to find a pole barn!" Vennum shouted over the din as we bucked and bumped down the dirt road, past some parked equipment, and skidded under the shelter of a high, open-sided hay barn. The rest of the crew piled in right behind us, as blueberry-sized hail whitened the ground. It was almost impossible to talk over the noise on the metal roof—some of our communication was by gesture—but the hawk got his band, and by the time we were done, the rain and hail had passed.

"His mate's gonna say, 'Hmm, where were you that you stayed so dry?'" Brian joked, then got serious. "With such heavy hail, we may have lost some chicks." With their thin, hollow bones, even large birds like raptors can be crippled or killed by large hail, and if the female doesn't shield them with her own body, the chicks can easily die. Fortunately, when we released the hawk back on his territory, the female flushed from their nearby nest and Vennum, standing in the bed of his truck and peering with binoculars, was able to see that at least one of their chicks was moving and looked fine.

Twenty years after their near-brush with catastrophe in Argentina, Swainson's hawks are riding high not just in the Butte Valley, but across most of their range. Unlike so many highly migratory birds, their numbers seem stable or perhaps even increasing, since by evolutionary luck they can thrive in some human-altered landscapes. "This is a species that has adapted really well to agriculture, even becoming urban in some areas," Vennum said. "If the human population goes to nine billion like they predict, then we'll have to convert a lot of the remaining land to agriculture. And this is one species that could do really well in that situation." Which is true, but the lessons of the 1990s remain clear—not all land, or all agriculture, is created equal. A farmland raptor is at particular risk not only from the wrong chemicals, but from changes in cultivation practices, development pressures, or swings in the market; after all, it was the change from traditional grazing to row crops like soybeans and sunflowers that led to the pesticide poisonings in Argentina. In California, a quarter of all urban development in the Sacramento and San Joaquin Valleys was on formerly irrigated farmland, while acreage devoted to grape production has more than doubled since 1990, and the demand for almonds and olives has exploded; consequently, vineyard or orchard planting has destroyed a lot of former Swainson's hawk habitat in the Central Valley. At the same time, overgrazing and a drying climate are reducing the quality of the remaining rangeland habitat, giving the hawks fewer options off farmland.

They don't grow grapes in the high desert of northeastern California, but in the Butte Valley where the alfalfa boom once supercharged the recovery of the valley's Swainson's hawks, the new hot crop is strawberries. The valley is only one link, albeit a critical one, in a strawberry cultivation chain stretching the length of California—a dizzyingly complex, labor-, transport-, and chemical-intensive process that keeps ripe berries on American tables most of the year. Plants grown from clusters of cloned cells and raised first in greenhouses and then in fields in the hot Central Valley are dug up and trucked north to the much cooler Butte Valley. Their offshoots, known as "daughter" plants, are picked and planted, primed by the chilly nights and warm days here to flower. Before they do so, however, they are hauled as much as 400 miles back south and replanted, producing the big, if rather insipid, berries that fill supermarkets.

Strawberry-growing started in the Butte Valley about 15 years ago, Vennum said, but has really taken off in the past few years. I saw quarter-section fields, 120 acres each, covered in plastic mulch, beneath which was pumped fungicides that literally sterilize the soil in preparation for planting. (I also saw thousands of disintegrating, slumping gray-white bales of old, tightly wrapped plastic from previous seasons, piled up in small mountains on some farms, looking like dirty ice cream melting in the sun.) Hundreds of farm laborers worked the berry fields. Many of them were riding, belly-down and a foot from the ground, on winged contraptions affixed to tractors, six workers to a side as the machines rumbled slowly down the rows in formation and the workers quickly and expertly nipped off buds so the plants wouldn't bloom prematurely.

The immediate concern to hawk biologists is the sterility of the fields that strawberry farming requires. "Nothing uses them," Vennum said. "Not squirrels, not hawks. Nothing." At the moment, the berry production is mostly confined to a few thousand acres of the valley, and the hawks still have other options. But if strawberry cultivation continues to expand, that may change—and strawberries depend on pumped water to an even greater extent than alfalfa. Brian Woodbridge told me that the current level of pumping is thought to be more or less sustainable with the recharge from the winter snows, but with a warming climate, climatologists forecast a dramatically reduced snowpack in California. "If we run out of water, I'd expect to see a big hit," Vennum said. The history of Swainson's hawks, from the Butte Valley to Argentina, is a reminder that with every mile a migratory bird has to travel, the chances that something crucial will change along its route increase. Although the biologists are optimistic about the Swainson's hawk's chances, they know the hawks' destiny is, like farming, bound up with water in an increasingly warm and often arid world.

And while we know a lot about the lives and travels of Swainson's hawks, thanks in significant measure to the 40 years of work in this remote valley, there are still gaps in our knowledge. One that intrigues Vennum and Briggs is exactly where the juveniles go when they leave the breeding grounds, and how they get there. Most of the satellite-tagged birds over the past quarter-century have been adults, for two reasons. Most juveniles hawks (like most juvenile birds of any sort) die within weeks or months of leaving the nest, falling victim to other predators, starvation, exhaustion, or accidents. If you are investing several thousand dollars in a satellite transmitter, you want to put it on a bird with the best chance of surviving and paying you back in data—and that means an adult. But in a less selfish sense, researchers often avoid tagging juveniles because those birds already have the odds stacked against them; the weight of a backpack transmitter, though small, is another burden that may tip the survival balance for an inexperienced youngster. For all those reasons, very few juvenile Swainson's hawks have ever been tracked.

"We know a lot about Swainies, especially what's going on here in the valley, but the juveniles are the vexing mystery," Vennum told me that evening. "I mean two, three, four years of their life is just a big question mark." What little evidence there is suggests the young birds make a return trip in the spring from Argentina, but Vennum said he wouldn't be entirely surprised to learn that, like young ospreys, some juvenile Swainson's hawks remained in South America for a year or two until they're old enough to breed. Vennum was able to deploy satellite tags on six young Butte Valley hawks, only two of which lived long enough to migrate south and return; two of them never even made it out of the valley. But the data from the young survivors, along with those from tagged adults, show that as the hawks are passing through the Southwest into Mexico, they are keying in on places with center-pivot irrigation, just like back home—an alfalfa-and-rodent highway leading toward the pampas.

But then the conversation, which had been fairly technical, took a turn that struck me like a gut punch. Chris Briggs was explaining how 10 years earlier, when he was working on his PhD, he'd traveled to Argentina, hoping to trap Swainson's hawks. His goal was to test their blood for chemical isotopes and stress hormones, which might shed light on important aspects of their biology, but under the scientific veneer was the excitement of a raptor nut who would finally get to see those great tornadoes of hawks that we'd witnessed in the late 1990s. Trouble was, he couldn't find the hawks—at least, not many of them, not like he'd expected. "We just weren't finding birds there. The biggest flock we saw in almost two weeks was probably a couple hundred hawks. We went to all the places you guys did, Chanilao and the rest. Because the flocks weren't big, we trapped three birds that entire trip, and I'll tell you, it was tough sledding to get those three."

"My God," I said, shocked. The memories of those evenings at the ranch, watching the *aguiluchos* gather in almost uncountable numbers, are among my most precious out of a lifetime of travel and bird adventures. Part of me had always harbored the hope I would get back there again, to stand in the humid summer dusk and watch the spectacle once more. That it might be gone left me with a hollow ache, like news of an unexpected death.

"I'm so sorry, Chris. It was such an amazing thing to see."

"Oh, I can't tell you how excited I was to go down there and finally see that for myself," Briggs said, his shoulders slumping. "I wouldn't want to say it was disappointing to see a few hundred birds, but. . . ." His voice trailed off, then he shook himself. "Who knows—maybe there are still big flocks and we just don't know where they were. The sense down there is that the hawks are just more dispersed, and they're not all in that core area in northern La Pampa where you saw them." There's been no sign of a major die-off, he said, no evidence that the population has fallen. Perhaps my companions and I back in the 1990s were just lucky; maybe some confluence of circumstance, like the grasshopper outbreak that spurred the deadly pesticide use, had concentrated the hawks in that part of the pampas. What we took for typical may have been exceptional, maybe even unprecedented.

We don't know. That's the joy and exasperation of studying migratory birds; there is still so much that eludes us. Even with whiz-bang technological advances and sci-fi remote sensing, Big Data number-crunching and radar and sat tags and all the rest, what we don't know about the global journeys of birds still vastly outstrips what we do know. It's a big world, and while humans are omnipresent, we're not omniscient. There is a lot of land out there, a lot of miles passing under their wings, and still a lot of secrets that the *aguiluchos* keep to themselves.

Eight

OFF THE SHELF

L ook at a map, and you'll see that the Outer Banks form the jut-
ting jaw of the Eastern Seaboard, like that of a cocky boxer inviting every passing hurricane to take a swing at it. And over the years
countless storms have taken their shot at this slender necklace of barrier islands, stretching more than 200 miles down from southeastern
Virginia into North Carolina, enclosing Albemarle and Currituck
Sounds to the north and immense Pamlico Sound, the largest lagoon
on the East Coast, farther south, which together cover more than
3,000 square miles of water. You can see the evidence of those storms
on the long drive down Highway 12, past Corolla and Nags Head,
Rodanthe and Buxton. In many places the islands of the Outer
Banks are already barely half a mile wide and not more than a few
yards above sea level; as the sea rises and storms intensify, salt water
pushes inland, abetted by ditches and canals built to drain uplands
for agriculture, killing maritime woodlands and leaving stark gray
"ghost forests" in their wake. (And frequently washing out sections
of Route 12 during hurricanes, until this one road that serves as the
tourism lifeline to much of the Banks can be hastily rebuilt.)

But the same geography that makes this crooked finger of barrier
islands so vulnerable to hurricanes and tropical storms also makes
it a mecca for anyone interested in seabird migration. For it is here,
especially at the far southern end of the Outer Banks near the village
of Hatteras, that the edge of the continental shelf and the deep water
beyond come closest to dry land, and the sweep of the Gulf Stream

is most easily accessible from shore. From here is the shortest route to an utterly different world, that of the pelagic migrants—the birds that, much as those European swifts that spend eight or nine months a year on the wing, have all but completely severed their connections with dry land. The annual movements of such birds—shearwaters, albatrosses, storm-petrels, and others—are among the most profound mysteries remaining about migration. In some cases, we're not sure exactly which hemispheres they inhabit; to a surprising degree, we don't even know what species are out there, still unknown.

It is a long drive down the Outer Banks in high tourist season in August, even on a midweek day, a slow crawl through the tourist towns and stoplights, condos and restaurants and beach shops that crowd the northern Banks, with some relief when you hit the boundaries of Pea Island National Wildlife Refuge and, a bit farther south, the start of the 70-mile stretch of Cape Hatteras National Seashore, where high dunes threaten to engulf the macadam from the ocean side, and vast salt marshes and estuaries run to the horizon on the other. Darkness was closing in by the time I finally reached the south end of Hatteras Island, the literal end of the road, and got myself checked into the hotel. I unloaded my gear—some food and a cooler, an overnight bag, a backpack with binoculars and cameras—as three college-age guys unloaded their own necessities, a car trunk full of beer, crate after crate of which they toted to their room. I was relieved to see they were a floor down and half a unit away from mine, since I had a feeling I'd be getting up before they finally went to sleep.

I was up by four, and the night had been quiet, thanks to earplugs. By five I'd driven the few miles to a marina beside the ferry terminal, just shy of Hatteras Inlet. A dozen other cars were already in the parking lot. The stars were hidden behind thick clouds, but the temperature was already in the low eighties, and despite a gusty breeze the air was so close and damp that it felt as though I was breathing through a wet washcloth. Lightning flickered to the north, far enough away that the thunder was lost to the distance. "Welcome to humid Hatteras," Brian Patteson said as he gathered the 10 of us birders—all male, most a fair bit older than I—on the forward deck of his boat, the 61-foot *Stormy Petrel II*. We stood in the red light from its pilothouse as Patteson, wearing shorts and a faded, sleeveless tee shirt, ran a hand through his hair and began a safety briefing, checking his clipboard occasionally with a flashlight as he discussed life jackets and man-overboard drills. "Never had someone go over the side, but it's better to be ready, especially on a day like this," he said in a Virginia accent seasoned by decades on the Outer Banks, where he runs a fishing charter business but has made a specialty of pelagic seabirding in the Gulf Stream. Patteson attracts birders from around the world who want to glimpse the avian life usually hidden beyond the reach of anyone not signing up for an extended deep-water cruise.

"It's gonna be a little active out there, once we cross the bar, a little lively," Patteson said. "With the low tide, we'll have to take a longer way out of the inlet, maybe 40 minutes, and then it's another, oh, maybe two hours out to the shelf break. We'll start in a few minutes when there's enough light to see where we're going."

We stowed our bags under the benches in the small cabin, and arrayed ourselves around the deck as Patteson cast off with the help of Kate Sutherland, his longtime trip leader, and Ed Corey, a bearded redhead who comes down to help guide every few weeks from Raleigh, where he's a state wildlife biologist. The boat traced a complicated, meandering course, threading between buoys that marked the serpentine channel among hidden shoals and sandbars, some of which showed white as the wind and waves picked up the farther out we traveled. This part of the coast is known, famously, as the Graveyard of the Atlantic, where more than 5,000 ships have foundered and sunk over the centuries, some 600 of them off Hatteras alone. I looked behind us to see an immense thunderstorm over Pamlico Sound—the dark blue-gray water of the inlet cut by our wake, a small white fishing boat following hundreds of yards back, and rearing over it all the almost sculpted undersides of the storm's skirt of clouds, while pouring from its center onto the distant bay, miles wide, was a thick and gauzy trunk of rain.

To my right, fragments of the newly risen sun—a deep, angry red-orange—peeked through ragged clouds just about the time the boat started to rock in the growing swell, taking spray over the bow. Patteson's voice came over the PA system. "Okay, we're coming up on the bar—better if everyone got in the cabin. The wind's sou'west, so we'll close up the starboard side because of the spray." I was the last off the deck and the benches were full, so I reached up to grab a line of wooden handholds that ran the length of the cabin ceiling, just as the boat began to roll and buck, taking the waves on its quarter with increasingly violent lurches. I was standing a few feet from the open portside door, through which I had a great view of the rising sun before it disappeared into the clouds, but mostly I was aware of how hot and oppressively humid it now was inside the mostly closed, tightly packed cabin. My hands were slimy, and trickles of sweat ran down my forearms, coalescing into streams dripping off my elbows; I could feel other rivulets snaking down my back and sides beneath my quickly sodden shirt. Releasing the grip of even one hand was risky, though, and despite hanging on with both I almost lost my balance a couple of times when the boat hit a deep trough, spray cascading over the cabin and running in rivers down the port deck.

This went on for a very long time. My arms and hands began to ache, my calf muscles complained, and sweat gathered in the wrinkled corners of my eyes where I tried, with minimal success, to blink it away. How the hell are we supposed to use binoculars in this, I wondered? I wasn't sure I could even move out of the cabin when the time came, although the lure of marginally cooler air outside was tempting. About two-and-a-half hours after we left the dock, though, the seas calmed dramatically—we had crossed the shelf break and were into the Gulf Stream. One by one, we cautiously left the cabin and blinked in the brightening sun, which was breaking through as we left the clouds behind us. I felt a little like Dorothy stepping out of her black-and-white tornado into a technicolor Oz. The water was a vivid, clear cerulean blue, dramatically different from the dark, grayish inshore seas, and spangled with long wind-ordered rafts of golden sargassum, the floating seaweed of tropical oceans, which formed intensely yellow lines stretching for miles. Flying fish, silvery and iridescent blue, launched themselves from our bow wave, skittering along the surface like skipped stones until they caught the wind beneath their outstretched, winglike fins and sailed for hundreds of yards.

I was more than a little stunned by the change. "Is this typical?" I asked Kate, who had donned a pair of heavy rubber gloves and was carefully decanting smelly, yellow fish oil into a jug that would hang off the stern and drip to create a slick to attract seabirds, which use their exceptional sense of smell to find food on the open ocean.

"With these winds it is," she said, tightening the top of the jug and then fetching a block of frozen chum from the cooler, placing it in a wire mesh cage that trailed in the wake. "They're from the southwest today, and the Gulf Stream current flows northeast, the same direction, so the waves just lay down under these winds. It's a big improvement, isn't it?"

That was an understatement, and with a little care we could now move easily around the ample deck and get to know one another in a way that the earlier passage made difficult. There were, among others, a fellow visiting from England; the director of a national estuarine center in Florida; a colorfully tattooed environmental educator from Long Island; and a 75-year-old University of Michigan professor of English and poetics who was happy to explain the four-beat similarities between fourteenth-century European poetry and his current field of study, rap—and who proceeded to quote several modestly profane lyrics from the latter to make his point.

That professor, Macklin Smith, is also a celebrity in certain birding circles, having seen more than 900 species of birds in continental North America, making him the top lister in the continent by the American Birding Association's count, a position he's held for many years. Smith is a regular on Patteson's trips. "I started coming down here in the seventies, before anyone knew what was out there," he told

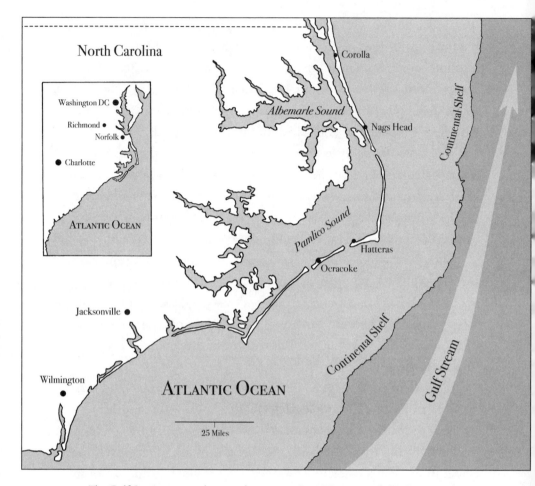

North Carolina

Corolla

Washington DC ●

Richmond ●
Norfolk ●

● Charlotte

ATLANTIC OCEAN

Albemarle Sound

Nags Head

Pamlico Sound

Hatteras

Ocracoke

Continental Shelf

Continental Shelf

Gulf Stream

Jacksonville ●

Wilmington
●

ATLANTIC OCEAN

25 Miles

The Gulf Stream comes closer to the coast at Cape Hatteras and the Outer Banks of North Carolina than anywhere else along the Atlantic, making the usually distant pelagic world unusually accessible to birders and researchers.

me as we stood swaying on the rear deck, scanning the sky for birds. "I'm not expecting to see anything new today, but you never know."

You never know, indeed. Brian Patteson's operation out of Hatteras, like several other boundary-pushing pelagic birding companies around the world, is constantly rewriting what we think we know about seabirds and their migrations. More than any other aspect of migration study, pelagic trips are the venue where gobsmacking discoveries and mind-bending rarities are the norm—as I'd learned all too well, and all too painfully, myself. When I'd contacted Patteson the previous winter, explaining my interest in the unknowns of pelagic bird migration and wondering when might be the best time to encounter something unexpected, he immediately suggested the end of May, when oddities are most common—a time when, unfortunately, I needed to be on my way to Alaska for my annual Denali field work. So I booked a couple of August trips instead—and thus was not entirely surprised to hear that, at the end of May, one of Patteson's cruises had turned up a Tahiti petrel, a large, narrow-winged dark-gray species with a white belly. Even by seabird standards, this was dumbfounding. As the name suggests, the Tahiti petrel breeds on the islands of the southwest Pacific like the Marquesas, the Society Islands, and French Polynesia, coming to shore to lay eggs in burrows dug beneath rain forests. It has been recorded on a few occasions off Hawaii, and is rare along the Pacific coast of Baja Mexico and Costa Rica. Not only had it never been seen off North Carolina, but none had ever been recorded within hailing distance of the entire freaking Atlantic Ocean.

"Someone said they think maybe it got blown across Panama by a storm," said Smith, who also missed this rarest of rarities, but how a Pacific seabird would end up in the wrong ocean is really anyone's guess. Or even whether the Atlantic is, in fact, the "wrong" ocean for it at all. The Atlantic covers 41 million square miles, and the number of people out on it who would (first) bother to notice a rare petrel and (second) be able to identify it are ridiculously, vanishingly small. That tiny number would also be largely restricted to a fairly short boat ride from land, and except for guides like Brian and his team, few of them spend more than a couple of days a year at sea. Swinhoe's storm-petrel, a small, all-dark seabird that's been seen fewer than half a dozen times off the East Coast, was thought to breed only in the western Pacific around Japan and the Yellow Sea, and range no farther west than the Indian Ocean—until a small number were found breeding on the Selvagem Islands off Morocco. More recently, at least one short-tailed shearwater—a sooty-gray bird the size of a large tern that nests on small islands off Australia and Tasmania, and normally spends the nonbreeding season in the north Pacific and Bering Sea—was photographed off Cape Cod, only the third record of that species in the Atlantic. Unexpected as that was, it perhaps made a little more sense in the context of another discovery—a nesting colony of the same shearwaters on Bouvet Island in the Southern Ocean, more or less equidistant between Antarctica and the tips of South America and Africa, and more than 1,200 miles west of the closest colony.

So Macklin and I had both missed the Tahiti petrel, but there were other rarities to watch for on the bathwater-warm current, which Brian's sensors pegged at 85 degrees Fahrenheit, only fractionally cooler than the sultry summer air. We were over 3,000 feet of water, and moving steadily deeper. Along much of the Southeast coast, the continental shelf is actually a series of steps and terraces for hundreds of miles beyond land, dropping by stages from the shallow inshore waters to the deep abyss. Off Hatteras, though, the bottom simply falls away; by the time we were 30 miles or so from shore, the depth was nearly 6,000 feet, and Kate was shouting, "*Black-capped petrel!* Black-capped petrel at two o'clock, moving right!" We swarmed the starboard side of the boat, crowding the rail, as a bird the size of a small hawk with long, tapered wings flared high against the wind, flashing its white underside, then stalled and entered a fast, shallow glide toward the water, its white nape and rump shining against the dark sea. It flared up again, and repeated the glide—a style of flight known as dynamic soaring, and a slick way to use the difference in wind speeds close to the water and higher above it. It works like this: The petrel turns into the wind, tipping its wings to catch it like a sail, slowing but rapidly gaining altitude like a kite pulled against the breeze. After climbing for a few seconds, the petrel rolls away from the wind, which is now at its tail as it slashes down into the slower surface layer of air, losing altitude but gaining speed until it turns back into the wind once more. This cycle is repeated endlessly as the bird zigzags its way across the ocean, rarely flapping its wings and using almost none of its own energy.*

This last point is critical, because pelagic seabirds like petrels or albatrosses travel farther annually than any other group of migrants, crossing many tens of thousands of wind-raked miles of ocean every year. Shorebirds like godwits are the grand masters of the do-or-die sprint, expending extraordinary amounts of energy to cross utterly hostile environments like the open sea in a matter of days. Pelagics take it easy, shaped to an environment that provides all their needs for months or even years at a time, without ever approaching even the rumor of dry land. This is especially true of the group known as tubenoses, named for their odd, tubular nostrils, out of which drips a highly saline solution extracted from the seawater they drink. Tube-noses include petrels like this one, as well as the very similar shearwa-ters, albatrosses with wingspans up to 11 feet, and storm-petrels the size and delicacy of swallows.

The black-capped petrel we jammed the rail to see belongs to the genus *Pterodroma*, its name meaning "winged runner," which seemed fitting as we watched the bird vanish over the horizon in moments. Fortunately, a second soon appeared, which hung around a bit longer. Black-capped petrels are an endangered species, but for much of the twentieth century they were little more than a cipher. They once nested abundantly on half a dozen islands in the Caribbean, where Spanish-speaking colonists called them *diablotíns*, little devils, for their weird midnight cries as they came and went at their nest burrows under the cover of darkness. By the middle of the nineteenth century, though, they appeared to be extinct, done in by hunting and introduced predators; the only hint that they survived was occa-

---

* My friend Rob Bierregaard, who has used highly precise GPS tracking units to follow young ospreys, found to his surprise that they use dynamic soaring as well when making 2,000-mile nonstop flights across the western Atlantic from the New England coast to South America—the first record of this flight style among raptors.

sional reports at sea of petrels fitting their description. Not until 1963
were a handful discovered nesting in the highlands of Hispaniola;
the entire population today is likely fewer than 2,000. Based on radar
studies a tiny remnant population was recently found on the island
of Dominica, where despite many previous searches it had last been
seen in 1862, and they may nest as well on Cuba and Jamaica. (Ironically, the same young biologist that found the *diablotín* in Hispaniola, David Wingate, had in 1951 also rediscovered the only other
seabird that bests it for both rarity and Lazarus tendencies—the Bermuda petrel or cahow, which had been thought extinct since the mid-
1600s, and which today still numbers fewer than 120 pairs.)

While cahows are only occasionally spotted off the Outer Banks,
black-capped petrels are, despite their rarity, predictable from May
through October, when Patteson's trips sometimes spot hundreds.
They are known to range in the Gulf Stream as far north as Nova
Scotia, although the handful of black-caps that had been fitted with
satellite tags spent the majority of their time at sea between Colombia
and the coast of New Jersey, staying in the deep water off the continental shelf. That may seem like an immense area, and it is—almost
1.5 million square miles—but among pelagic migrants such movements could almost be considered loafing around the neighborhood.
As they soar effortlessly on the perpetual sea wind, distance is essentially meaningless to the tubenoses, whose travels connect incredibly
distant pockets of tremendously abundant food. A wandering albatross, with 11-foot-wide wings the largest of them all, may during
its so-called "sabbatical year" between biennial breeding attempts fly
74,000 miles, circumnavigating Antarctica two or three times without ever seeing land. The pelagic seabirds chase oceanic riches that
swell and diminish with the changing seasons, while returning once
every year or two to land—almost always on some flyspeck island or
remote archipelago insulated from predators by distance from any
mainland—where they spend the bare minimum required by biology
to lay an egg (always just one) and raise a chick. They compensate for
this extremely low reproductive rate by living a long time; the oldest

known wild bird of any species is a Laysan albatross named Wisdom, banded in 1956 as an adult and, at the age of at least 69, still returning to Midway Atoll in the Hawaiian chain each year to nest.

Understanding seabirds means becoming fluent with a largely unfamiliar geography; few people have heard of, let alone can find on a map, the Juan Fernandez Islands, Trindade and Martin Vaz, the Desertas Islands, Lord Howe, Marion, or the Antipodes Islands, to name a few centers of global seabird breeding. The three volcanic islands of Tristan da Cunha, a small chain that lies in the very empty middle of the South Atlantic, are typical of the lot. Tristan is the most remote inhabited archipelago on earth, its several hundred British-citizen residents some 1,500 miles from South Africa and more than 2,000 miles from Brazil. Two of the islands, as well as even lonelier Gough Island, 250 miles to the south, are given over to birds, including rockhopper penguins, several species of albatrosses, and a variety of storm-petrels, shearwaters, petrels, and diving-petrels. Gough is often called the greatest seabird island in the world, with more than 5 million nesting pairs in all, including the endemic (and critically endangered) Tristan albatross, and the surpassingly lovely sooty albatross, a bird the color of wood smoke, its eyes rimmed with white and a slender yellow streak that curves up like a shy smile along each side of its dark bill. The numbers are staggering—2 million pairs of prions (a type of small, largely nonmigratory petrel) nest on Gough, including a million McGillivary's prions, which weren't even recognized as a distinct species until 2014. There are between 1 million and 1.5 million pairs of Atlantic petrels that migrate between Brazil and southern Africa, and 1 million pairs of great shearwaters, which during the austral winter migrate north, foraging the North Atlantic from Cape Cod to northwestern Scotland. But even on so remote and seemingly secure a place as Gough Island, seabirds face new and bizarre threats.

Gough (pronounced "Goff") Island is closed to all visitors, and the only outpost there is a small South African meteorological station. Fortunately, not all seabird islands are quite so inaccessible. A few

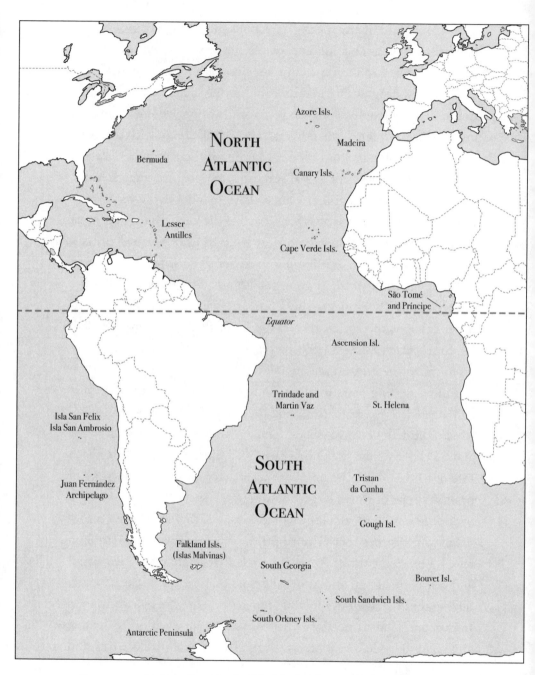

Important seabird breeding islands of the North Atlantic and South Atlantic Oceans.

years before my Hatteras trip I loaded my gear into a Zodiac and, with two companions, carefully nosed our way onto a cobblestone beach on Kidney Island, an 80-acre lump in the Falkland Islands off the southern tip of Argentina (which has tried to claim the archipelago as Las Malvinas). Treeless, the island is blanketed in what's locally known as tussac—what can only be described as old-growth grass, a towering nine- or ten-foot-tall species that forms a dense canopy through which we pushed our way. I was following Craig Dockrill, then the director of Falklands Conservation, and trusting him not to let us stumble into the angry, 600-pound sea lion bulls that roared and bellowed all around us, the air rank with their wet-dog scent.

We raced to set up our tents in a small clearing as twilight gathered, then followed Craig to the far end of the island, where the air overhead was a cyclone of birds, a splendid chaos moving against the fast-dimming sky—tens of thousands of sooty shearwaters, roughly 100,000 of which nest among the roots of the great tussac stands. The birds bided their time, whirling endlessly in a clockwise funnel as darkness—and their numbers—steadily grew; should they land too soon, while there was still enough faint light to see, they would be snatched by brown skuas that patrolled the edges of the gyre, heavy-bodied and compact gull-like birds, or by giant petrels the size of albatrosses—two seabirds that have evolved into highly aggressive predators, ready to pounce on any careless shearwater. (A few days earlier, I'd watched two giant petrels, their heads and necks bathed in gore, kill and rapidly dismember a gentoo penguin the size of a beagle, so I could appreciate the shearwaters' caution.) Fog streamed in, and in minutes we could see nothing beyond the hazy outlines of the towering grass clumps—and that was the moment for which the shearwaters had been waiting. Around us in the damp darkness I could hear heavy thumps, like someone pelting us with potatoes, as the birds—the embodiment of grace in the air, but clumsy on earth—landed awkwardly. One barely missed my head, and I caught it in my headlamp beam, a duck-sized bird, glossy gray-brown with a slender, hooked bill, wings still akimbo, staring back at me with shiny black eyes before scrabbling down its burrow.

Sooty shearwaters are among the most abundant seabirds in the world, and the only shearwater known to occupy both the Atlantic and Pacific basins. A few years before my visit, scientists had fitted a few sooty shearwaters on Kidney Island with satellite transmitters, and found that when they leave the Falklands they travel very rapidly, covering about 12,000 miles north in about three weeks—first out into the middle of the South Atlantic, then making a 90-degree left as they pick up the southeasterly trade winds, almost clipping the eastern bulge of Brazil. Then they track up the western slope of the Mid-Atlantic Ridge, the massive deep-sea rift that splits the Atlantic down the middle, eventually spending the northern summer along the Grand Banks east of Newfoundland. They do not linger along the way—there's no stopover rests for these birds, as with so many landbird migrants, but neither do they put on large fat reserves before they leave the Falklands. Instead, the tracking data suggest they use a "fly and forage" approach, opportunistically taking what prey they can along the way, but primarily rebuilding their reserves on fatty capelin, squid, and krill once they reach the productive northern waters—a strategy that can only work across such staggering distances because their dynamic soaring flight allows them to save so much energy.

The Kidney Island shearwaters showed very strong migratory connectivity—all the tagged birds had very similar flight routes, and all spent the northern summer in the same fairly small area off Newfoundland. Interestingly, that is not the case with sooties from the Pacific. While a few of the Pacific birds nest in southeastern Australia, and others along the coast of Chile, most—an estimated 21 million—breed in New Zealand, where the native Maoris take about 360,000 chicks a year in a traditional harvest. Those that survive undertake annual figure-eight migrations around the Pacific that total as much as 46,000 miles of travel, almost twice the journey of the Falkland population—and they take highly individual routes to do so, as researchers found when they put geolocators on mated pairs from two New Zealand colonies. One pair flew due east to the southern coast of Chile, then split up—one hugging the coast of South America and spending the austral winter (the northern summer) off Baja and California, while its mate used prevailing easterly trade winds to fly northwest to Japan, then to the waters off the Kamchatka Peninsula in Siberia. One member of another pair took a more direct route to Japan and Kamchatka, while its mate spent the northern summer along the Gulf of Alaska. In the words of the researchers who deployed the geolocators, the shearwaters "used the entire Pacific Ocean . . . ranging from Antarctic waters to the Bering Sea and . . . from Japan to Chile."

It was too late in the season for me to expect sooty shearwaters off Hatteras; they're a spring species here, not one you find in the dog days of August. And it was, to be honest, a fairly slow day by Hatteras standards; reports from the waters off New Jersey, Long Island, and Massachusetts suggested that a lot of the birds that normally haunt the Outer Banks had shifted hundreds of miles north that summer, for unknown reasons. But we had enough to keep us busy. An Audubon's shearwater, one of the smaller species in that group, whizzed past the bow, its willow-leaf wingtips slicing the surface of the sea as it glided and turned, then plunged headfirst into the floating bright yellow-brown mats of sargassum to catch—well, no one is quite sure what this species of seabird eats, though one was once observed eating bits of squid vomited up by dolphinfish, and it is presumed to mostly hunt for fish, squid, and pelagic crustaceans. In any event, sargassum clumps are natural buffets, floating islands of biodiversity in the Gulf Stream that shelter a bewildering array of tiny crustaceans, worms, mollusks, and the larval young of hundreds of species of fish, including some like marlin that eventually grow to be some of the largest in the sea. The plunging shearwater flushed several small flying fish, and as the day went on, I found myself increasingly charmed by these little skittering marvels, with their kaleidoscopic colors and intricate "wing" patterns—and by the whimsical names that observers have coined for them. There were sargassum midgets and purple bandwings, Atlantic necromancers and rosy-veined clearwings, diablos and patchwings and berrywings. The smallest, unidentifiable juveniles have been dubbed "smurfs," and that seemed to me to be perfect.

The sargassum also often held small flocks of red-necked phalaropes, delicate shorebirds that—despite being little bigger than sparrows—spend their winter on the open ocean, and were already deep into their "autumn" migration even though by the human calendar it was still summer. Phalaropes are odd birds on a number of fronts; they are (take a deep breath) reverse sexually dimorphic, primarily monogamous but routinely sequentially polyandrous and rarely sequentially polygynous—which, when translated from ornithologese, means that the females are larger and more colorful than the males and will sometimes take multiple males as mates during the course of the same breeding season in the Arctic, while the males will less often pair up with second females. In late May and June, when the phalaropes arrive on the breeding grounds, the females are gunmetal gray with splashes of bright chestnut on their necks, while the males are similar but considerably duller and more camouflaged. That is because phalaropes also practice sexual role reversal—the females compete intensely for males and defend potential mates against other females, and once they have laid their clutches of four eggs in small nest cups concealed by tundra grasses, they abandon their care to the males and will seek other mates. The male phalarope—whose levels of prolactin, the ostensibly female hormone normally associated with maternal care, is off the charts when compared with most male birds—has sole responsibility for incubation and rearing the chicks.

Red-necked phalaropes breed across the entirety of the Northern Hemisphere in subarctic and Arctic regions, with most of the Eurasian population wintering in the Arabian Sea, among the islands of the East Indies, and east to the Bismarck Archipelago. The North American birds—well, that's more complicated. Those from the western Arctic migrate to the Pacific off South America, wintering primarily where the cold Humboldt Current collides with Peru and brings nutrient-laden water to the surface, producing a riot of plankton. Those from the eastern Canadian Arctic used to gather every autumn, up to 3 million strong, at the mouth of the Bay of Fundy (more specifically, in Passamaquoddy Bay on the Maine–New Brunswick border), a phenomenon ornithologists had noted as early as 1907. The draw, it seems, was a zooplankton called *Calanus finmarchicus*, pushed out of deep water at the mouth of the bay by Fundy's mighty tides. Where the phalaropes went from there was a mystery, however—no large concentrations have ever been found in the Atlantic in winter. Some biologists speculated that the phalaropes must cross Central America and join the western birds in the Pacific, while others thought it unlikely, given the almost complete lack of records for the species from Mesoamerica. But by the late 1980s this question was supplanted by an even bigger and more immediately worrisome puzzle, for the multitudes of phalaropes in the Bay of Fundy were melting away like snow in the sun.

My friend Charles Duncan, later the director of the Western Hemisphere Shorebird Reserve Network but at the time a chemistry professor at the University of Maine in remote Machias, had summers off and had befriended a local captain, so he witnessed the vanishing firsthand. "He let me tag along on his whale-watch and sightseeing trips in Passamaquoddy Bay and off Grand Manan," Charles recalled. "I kept daily checklists for no particular purpose other than organized record-keeping." The realization that something was wrong dawned slowly. In 1985, Charles was still counting up to 20,000 phalaropes a day on Passamaquoddy Bay, tallying up flocks ranging from a few dozen to a few thousand. But the next year his maximum counts dropped by an order of magnitude to 2,000. The next year, another order of magnitude drop to 200, then just 20 the following year, and finally zero.

"At first we just thought we were missing things—we were there on the wrong tide, or the wind was wrong, or some such. By about the third such summer, with things getting worse each time, we

knew something was up. Were we seeing an entire species or population collapse and no one else was in a position to notice? We had
the feeling we should tell someone, but who should we tell?" Charles
drafted letters to the US Fish and Wildlife Service, the Canadian
Wildlife Service, and the Maine Department of Inland Fisheries and
Wildlife, "none of whom I knew," he told me. "I was just an amateur
birdwatcher, and professionally, a chemistry professor. We were baffled, frankly, about what was going on and what we—or anyone—
could do about it."

To this day, no one can say with certainty what happened. Some
biologists contend that a string of powerful El Niño events in the
Pacific in the early and mid-1980s—which dramatically raised sea
surface temperatures in the Pacific, causing the collapse of marine
food chains and the deaths of millions of other seabirds—might have
decimated the Fundy birds if they did, in fact, winter in that ocean.
Others, including Charles, don't buy that explanation; among other
things, the timing fails to explain why the collapse happened in stages
over a five-year period. They point to a more fundamental cause—
the equally unexplained disappearance of the *Calanus* zooplankton
from the surface waters of Passamaquoddy Bay, where the phalaropes
once gathered. As Charles puts it, "There's no food available at this
restaurant any more." As the years passed and no similarly immense
concentrations were found elsewhere, it became clear this wasn't a
temporary aberration. Charles notes that a couple hundred thousand
phalaropes still use a different part of the Bay of Fundy where some
Calanus still concentrate, in the company of closely related red phalaropes that likely winter off the west coast of Africa. A few hundred
have been found to winter off the Southeast coast, close to where I
was cruising in the *Stormy Petrel II*. But for so many migratory birds
to just vanish in such a sudden and inexplicable manner remains
essentially unprecedented anywhere in the world.

One mystery about the Bay of Fundy birds seems to have been
solved, however—though in a bizarre twist, the answer came from

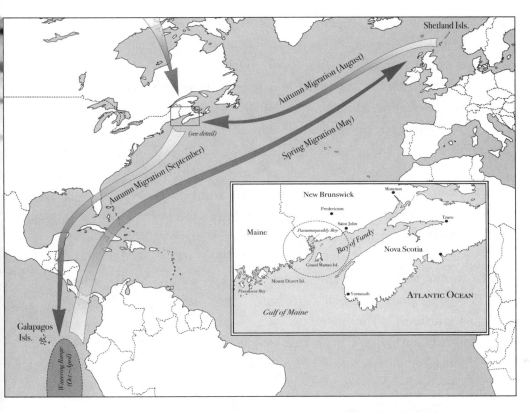

Geolocator tracking has revealed that red-necked phalaropes from the Shetland Islands cross the North Atlantic and Central America to winter off the northwestern coast of South America, staging up in the Bay of Fundy along the way. While the wintering area of the eastern Canadian Arctic population remains a mystery, it seems likely that they, too, follow a similar route to the Pacific.

nearly 3,000 miles away, in the middle of the North Sea. While the bulk of the red-necked phalaropes that breed in Europe winter in the Arabian Sea, those that nest in Scandinavia, Scotland, and Ireland do not—and no one knew where they went. So in 2012, scientists in the Shetland Islands, north of Scotland, affixed geolocators to nine phalaropes. The next year they recovered only one—but that bird had flown almost 7,000 miles across the Atlantic, down the US East Coast, across the Gulf of Mexico and Central America, and wintered off northwestern South America. Additional tagged Shetland phalaropes in subsequent years did the same, as did birds tagged in Green-land and Iceland. If birds from those areas were swapping oceans, it seems reasonable to believe the Bay of Fundy phalaropes were, too.

Now that we were deep into pure Gulf Stream water, Patteson started turning the *Stormy Petrel II* back on itself, paralleling our previous wake and watching for birds attracted to the slick of drip-ping fish oil, which we could see and smell in the damp air. Bridled and sooty terns—both Caribbean species that wander north along the Gulf Stream—flew past purposefully in twos and threes; unlike the tubenoses, they hunt by sight, and the whiff of the slick was lost on them. But the odor had drawn in small flocks of Wilson's storm-petrels, which gathered to hover and dabble their long, twiggy legs and webbed feet on the water as they nibbled up the oil. Wilson's storm-petrels are often described as the most abundant wild bird in the world. They aren't (that title likely belongs to the red-billed queala, an African finch that may number 1.5 billion) but with a global population of up to 10 million pairs breeding largely on the islands, archipelagos, and peninsulas of the Southern Ocean, and at least that many again in nonnesting adults and juveniles, the storm-petrels are anything but rare. Wilson's storm-petrels are also among the most wide-ranging birds in the world, found in the nonbreeding season in almost every major ocean, sea, and gulf except the north Pacific. Just seven inches long and the color of black coffee, with a bright white patch that wraps around the body above the tail, and legs as thin as uncooked spaghetti, a storm-petrel seems far too dainty to bear a life at sea. More than almost any other marine migrant, these pixieish seabirds have been the focus of maritime superstition and awe; the word *petrel* is a diminutive of Peter, and a reference to the way in which they—like the apostle Peter, striding with faith on the Sea of Galilee—appear to walk on water as they patter and flit along the surface. Along the Canadian coast they're called "Careys," a trun-cation of the old British nickname "Mother Carey's chickens," which in turn is a corruption of *Mater Cara*, "dear mother," a reference (perhaps originally in Italian) to the Virgin Mary. A sailor in rough seas will look for divine help wherever he can—and should that help fail, well, some of the old mariners believed that storm-petrels were the reincarnated souls of their drowned comrades.

"*Whoa!* Everyone, get on this bird! Band-rumped storm-petrel!" Kate shouted. "Dead astern, the one that's bigger and has longer wings." The bird looked as though a Wilson's had been juicing on steroids, a third bigger than the rest of the flock, its rump patch much wider and brighter, its flight loose-limbed and floppy compared with the stiff-winged Wilson's. It made a few passes, then was quickly lost amid the lapis swells, and I jotted a few quick notes about it, along with the other species we'd been seeing. But here's the thing: even as I looked at my growing list, I couldn't really be sure exactly *what* I'd seen. Almost all of the seabirds we'd found so far are (if you'll pardon a maritime analogy) adrift in poorly charted taxonomic waters. Hang on, it gets confusing.

Let's take band-rumped storm-petrels. These birds nest in the eastern Atlantic along a 4,500-mile arc of islands running from the Azores in the north to the Berlengas Islands off Portugal, south through Madeira and the Canary Islands off northwestern Africa, and another 2,400 miles to Ascension and St. Helena in the middle of the South Atlantic, as well as (possibly) on small islets near São Tomé in the Gulf of Guinea. Other populations breed in the Galapagos, Hawaii, and Japan. Such a wide geographic spread of isolated colonies is the first hint that lumping them all as one species might be questionable, but across their range they differ only in relatively minor ways, like the shape of their tails—and remember, these are birds to whom distance means little. Yet as scientists have plumbed the DNA of each population, they have found almost no evidence of gene flow between colonies; wide-ranging as they may be in the off-season, the storm-petrels appear to be utterly faithful to their birthplaces.

What's more, on many breeding islands there are two distinct nesting populations, which use the same burrows but at different times of the year, one in the hot season and the other in the cool—avian timeshares, if you will. They differ physically, with hot-season populations generally having more deeply forked tails and their cool-season counterparts much squarer tail feathers, and some have unusual vocalizations. The hot-season birds on the Azores have been formally split into a distinct species, Monteiro's storm-petrel, named for the Portuguese ornithologist who first detected the seasonal differences, and who died in a plane crash shortly thereafter. Genetic work suggests there are still other species to be described in this complex—as many as three other species in the northern Atlantic alone and perhaps 10 worldwide, hiding within what is now considered the band-rumped storm-petrel.

"Was that a Grant's?" I asked Ed and Kate, after the band-rumped had disappeared among the waves. Grant's storm-petrel is one of those maybe-it-is, maybe-it-isn't possible species in the band-rumped complex, a cool-season breeder in the Azores, Canaries, Madeira, and other islands off northwestern Africa and Portugal—though no one has yet taken the step of formally describing it and giving it a scientific name.

"I'd need a better look," Kate said, "but it looked big, and Grant's would be one we'd expect at this time of year." Grant's storm-petrel is fairly common in the Gulf Stream from late spring through August, occasionally making it as far north as Cape Cod, and west into the Gulf of Mexico. It has a bigger head and heavier bill than the hot-season forms, and its vocalizations are different, but to be honest, even the experts are trying to figure out the identification clues, and the best I could do was put a big question mark next to its name in my notes.

And it doesn't end there. That Audubon's shearwater I watched plunging into the sargassum for food? Like the band-rumped storm-petrel, this is best thought of as a species complex, in which DNA evidence shows varying degrees of relatedness. Some experts have already split Audubon's shearwater three ways, with those breeding in the Caribbean considered "true" Audubon's shearwaters; those nesting on the Cape Verde Islands (and in the past, Bermuda) as a species

dubbed "Boyd's shearwater"; and those breeding from the Azores to
the Canary Islands as "Barolo shearwaters," which have been recorded
a handful of times off North America. And within each of these pro-
spective species there are finer divisions—three discrete forms within
the Caribbean population that could be full species, and several more
under the Barolo shearwater umbrella. Several new species have been
split off from Cory's shearwater, a large, pale brown seabird that's
common off the East Coast in summer—the Cape Verde shearwater,
which breeds in those islands off the western bulge of Africa, and
Scopoli's shearwater from the Mediterranean, both of which show
up on trips out of Hatteras. In 2017, a petrel that looked like nothing in any field guide showed up on one of Patteson's trips; whether
it was a hybrid, a weird and previously unknown plumage of a routine species, or an extraordinarily rare seabird that has never been
described for science was anyone's guess. The same bird or another
like it appeared two years later, leading Patteson's team to dub it the
Whiskey Tango Foxtrot petrel. Steve N. G. Howell, an ornithologist
and seabird expert from California who guides Patteson's trips a few
times each year, and who has written the definitive guide to pelagic
birds in North American waters, understates the situation when he
describes the taxonomy of many seabirds as "vexed"—but of course,
the birds have no problem with the situation. It's we pigeonholing
humans, trying to sift out the relationships between (and thus, the
evolutionary histories of) these birds, that find square pegs and round
holes so frustrating that we ask, "WTF?"

Understanding the taxonomy is critical for conserving migratory
seabirds, though, many of which have miniscule populations and
a host of threats, both at sea and on their breeding islands. Hence
the urgency in figuring out whether a species is widespread and thus
fairly secure, or actually a welter of highly localized, possibly quite
rare cryptic species. The black-capped petrels that Brian Patteson and
his team see come in two forms, a dark-faced variety and another
with a whiter face. The former is slightly smaller and is thought to
nest on Hispaniola and possibly Jamaica, while the white-faced form

is bigger, has a flight feather molt schedule that suggests it may breed earlier in the year than the dark-faced birds, and may nest in the Lesser Antilles—to the extent that anyone knows where black-capped petrels breed outside of Hispaniola. Not surprisingly, some experts think these two forms could well be distinct species. There are also intermediately colored birds, which could represent other unknown populations, or just immature stages of the two main varieties. Estimates based on at-sea surveys suggest the black-capped petrel population (including all forms) totals between 1,000 and 2,000 birds—but scientists know of only about 50 nest sites in Hispaniola. Where are the rest?

Scientists are trying to puzzle out all of this, but working with seabirds presents its own unique challenges. "Last year the American Bird Conservancy came out with us to try to trap some black-capped petrels and put trackers on them, to find out where they're breeding," Kate told me. I must have looked shocked—how would you begin to trap a bird with the limitless sea at its command?—because she answered my unasked question. "They had floating mist nets, and kayaks to man them and get the birds. We came out at night, in the full moon, so we could see where the petrels are feeding—but we didn't even see one. We really didn't think it was going to work; the birds are pretty smart."

As it turned out, the researchers—including folks from the conservancy, the US Geological Survey's South Carolina Cooperative Fish and Wildlife Unit, Clemson University, the National Fish and Wildlife Foundation, and BirdsCaribbean—finally succeeded the following year, working again with Patteson and his crew. This time they brought in an expert from the Northern New Zealand Seabird Trust, who used a specially designed gun that fired a net, snagging the petrels out of midair when they came in to check out the smell of floating chum. The team hopes that the 10 petrels they fitted with satellite tags—a mix of white-, dark-, and intermediate-faced birds—will eventually lead scientists to previously unknown nesting sites in the Caribbean, and allow them to begin to unravel some of the many mysteries surrounding them.

I was once party to a scheme to raise money—a *lot* of money—to kill rats. To save seabirds—a *lot* of seabirds.

Not to be indelicate, but just as one uses bait to catch rats, so does one need a lure to attract those whose money you wish to use to kill rats, which in this case were destroying seabird colonies. The lure, for people with an interest in birds and what the development professionals euphemistically call "significant capacity," was the most famous birder in North America, field guide author David Allen Sibley. The Nature Conservancy's Alaska chapter hatched the plan in concert with the US Fish and Wildlife Service; they would invite a dozen or so exceedingly wealthy people with a passion for birds and bird conservation, take them to one of the most remote places in the Northern Hemisphere—the western Aleutian Islands—and hope to convince them to pony up a million dollars or so for rat eradication on critical seabird nesting islands. In case a trip to a stunning oceanic wilderness seething with millions of birds, whales, sea lions, and sea otters wasn't enough, the chance to spend a week or so birding there with David Sibley would, they hoped, seal the deal. The lure for Sibley, in turn, would be the chance for him to finally see the whiskered auklet, a crazy-cute roly-poly the size of an orange with flamboyant facial plumes and white-button eyes, impossible to find anywhere in the world but this isolated corner of the north Pacific, and one of the only North American birds he'd still never seen in the wild. Along the way, the rich folks would be shown both a rat-ravaged island, where the survivors of long-ago shipwrecks had eaten once-thriving seabird colonies into extinction, as well as islands like Kiska, where rats only just achieved a foothold during World War II, and eradication could save tens of millions of nesting birds. They would also visit islands where foxes, which had been introduced by Russian and

American fur traders, had been eradicated at great labor and expense. (My job: stay out of the way, take a lot of pictures, and write it all up for the conservancy's magazine.)

Which is why I found myself on a noisy turboprop flying west toward Shemya, the penultimate island in the Aleutians, where we would meet a federal research vessel. Shemya lies some 1,800 miles west of Anchorage, our departure point—the distance from Nashville to Los Angeles, far closer to Asia than the rest of Alaska, and almost all of that long way over frigid, empty seas and uninhabited islands. All of which took on pressing importance when the pilot announced that the weather in Shemya was going from marginal to God-awful, so we were turning around and rerouting to Adak. And later, that we were rerouting from Adak to Dutch Harbor on Unalaska Island, for the same reason. And later still, word that conditions at Dutch were going south, too—but that we were landing, one way or the other, because we were almost out of fuel. We did.

We never made it to Shemya, though, nor to the worst of the rat-infested islands. After some logistical gymnastics, we eventually met the Fish and Wildlife Service's research ship, the *R/V Tiglax*, and spent several days exploring the central Aleutians. I helped biologists conduct nest surveys hundreds of feet up the sides of steep-sloped islands, where countless auklets, murrelets, and storm-petrels incubated their eggs and tended their chicks in burrows among the rank beach grass beneath our feet. In the long twilight before the brief midsummer darkness at one or two in the morning, we watched astounding numbers of seabirds—maybe half a million—swirling in incomprehensibly large flocks around a dormant volcano, waiting for the safety of darkness in which to land. (The volcano, called Kasatochi, wasn't as dormant as we thought; three years later it exploded, almost incinerating the two biologists working there.) And we found David his whiskered auklets, including one that tumbled onto the deck at night, befuddled by even the dim lights from the pilothouse, its curlicue black crest and long, filamentous white "whiskers" bobbing comically as, cradled in a biologist's hands, it looked from one to the other of us.

And I thought a lot about rats and birds. "It's not oil spills that really keep me awake at night," a biologist friend had told me once a few years earlier, as we watched hundreds of thousands of seabirds crowding the pristine cliffs of St. George in the Pribilof Islands, far out in the Bering Sea. "Even with a bad oil spill, there's a chance at recovery. A rat spill, though? Rats are forever."

It might seem that the riskiest part of a migratory seabird's life would be its time on the open sea, at the mercy of winds and storms, but the ocean holds few terrors for them. For many species, the greatest dangers today await them on land, because there are now no places so remote that humans—and our hangers-on like rats and mice, cats and dogs, goats and sheep—haven't found them. Even after we leave, our commensals remain behind to wreak havoc. What had for millennia been islands of security, guarded by distance and isolation, can now be death traps for birds with no innate sense of self-protection when they come to shore. One of the biologists with us on the *Tiglax*, who studies the interactions between rats and birds on Kiska, described in horrifying detail how the rats are able to eat into the brains of the living birds, whose instinct to sit tight and protect their single egg overwhelms even pain and death.

Nowhere has this played out in a more bizarre and dramatic fashion than on Gough Island, part of the Tristan da Cunha group in the empty South Atlantic. As mentioned earlier, Gough is one of the most important seabird nesting islands on the planet, home to millions of seabirds of 22 species, including several endemics that breed nowhere else. One of these is the Tristan albatross, which numbers barely more than 5,000 pairs. One of the largest birds in the world, weighing more than 15 pounds and with a wingspan of 10 feet, the albatross would seem to be impervious to most dangers, much less the one that threatens its existence on Gough: mice.

Nineteenth-century sealers apparently introduced house mice to

Gough Island, where the rodents found a predator-free world. They also experienced a lean period each winter, when insects and seeds were hard to come by—but that's when many of Gough's seabirds return to the island to nest. With time, Gough's house mice evolved not just an increasingly carnivorous diet, but a larger and larger body size, so that today they are half again as big as normal house mice— and kill some 3 million seabird chicks each year on Gough. The carnage includes four out of five of all albatross chicks, two-thirds of the chicks of endangered Atlantic petrels, and virtually every egg and baby of an endangered burrow-nesting seabird known as McGillivary's prion, which was only described for science in 2014 and will, at current rates, be extinct in a few decades. (Many other small, burrow-nesting petrels, prions, and storm-petrels, once incredibly abundant on Gough, are now so scarce that scientists were unable to get much data about them.) It is, for the birds, an especially gruesome death, since the mice can only nibble. They open multiple wounds in the chicks that, lacking any instinctive defense against land predators, simply sit there stoically as many mice eat away at them over the course of days, losing blood and strength until they die. In the case of the pigeon-sized prions, few chicks survived more than a couple of days after hatching. Because Tristan albatrosses lay only one egg every other year, the loss of any chick is a serious blow to the species, which is also under threat from longline fisheries that have devastated albatross populations in general. More recently, scientists have documented the first cases of mice attacking the adult albatrosses on Gough, an especially worrisome development that could push these magnificent birds to extinction by 2030.

Surprisingly, Gough is not the only seabird-nesting island where mice have grown into an existential threat. Marion Island, off the coast of South Africa in the Indian Ocean, has a variety of invasive mammals, including mice, sheep, and goats. In the 1950s five house cats were released to control the mice, but instead, their ever more abundant progeny began killing almost half a million petrels per year. The cats were eventually exterminated over a 16-year period starting in the late 1970s, but while that effort succeeded, it also freed the mice from any control. As on Gough, they began killing large numbers of seabird chicks and adults—in this case by scalping them. Mice were also a problem on the Antipodes Islands, a volcanic, sub-Antarctic archipelago 470 miles south of New Zealand, home to nearly two dozen threatened seabirds, including the Antipodes albatross, as well as endemic snipe, pipits, and parakeets, all of which were threatened by the impact the rodents had on the island ecosystem. But my old friend in Alaska all those years ago was wrong about one important thing: rats (and mice) are not forever, not any more. New Zealand has been a pioneer in island restoration, figuring out ways to eliminate rats, mice, and other nonnative species from endangered island ecosystems. Those techniques are going global, and on a larger and larger scale. In 2016, experts used helicopters to methodically spread 70 tons of rodenticide over every square yard of the main island and surrounding sea stacks on the Antipodes (work paid for by a successful "Million Dollar Mouse" fund-raising campaign); trained dogs later sniffed out the few survivors, and two years later the islands were declared mouse-free. Macquarie, another sub-Antarctic island almost 1,000 miles south of Tasmania, was declared pest-free after a seven-year, $25 million Aus. ($17 million US) effort to rid it of rats, mice, and rabbits.

The most ambitious pest-eradication campaign of all just wrapped up in South Georgia, the extraordinarily beautiful, ruggedly mountainous, and biologically rich island between South America and Antarctica. A few years ago when I was in the Falkland Islands—the inhabited land nearest to South Georgia—I saw helicopters being positioned, along with more than 100 tons of blue-green bait pellets, for the first of three summer seasons dropping poison across the 380-square-mile island—an area eight times the size of Macquarie, which until then had been the largest eradication project ever attempted. Two years after the poisoning campaign ended in 2015, teams returned to South Georgia, blanketing the island this time with wax tags and "chew cards" baited with peanut butter or vegetable oil, which would record any nibbling attempts by rats. Baited tubes, coated inside to record rat footprints, were placed in likely areas, and as on the Antipodes, trained dogs (muzzled so they wouldn't nip curious penguins) scoured the land. Not a rat or a sign of a rat was found, and conservationists were delightfully shocked by how quickly the island's imperiled birds responded to the release from predatory pressure.

After years of preparation, and some delays caused by its incredibly remote location, a similar eradication campaign was poised to begin on Gough Island in February 2020 when the coronavirus pandemic struck. The team of 12 Royal Society for the Protection of Birds conservationists had to be evacuated, and the eradication program was delayed by at least a year. Assuming its success, however, the same folks will then shift their attention to Marion Island, home of the bird-scalping mice. A recent assessment of the benefits of mammal eradication on islands ranked Gough third (after Socorro Island and San José Island in Mexico) in terms of the greatest benefit to the largest number of imperiled species should the pests be eliminated. Some people, though, are thinking *really* big. New Zealand originally had no native mammalian predators, and the introduction of mammals like stoats, rats, and opossums has been devastating to the island's birds. Although New Zealanders have made great strides clearing mammals from many of its offshore islands, in 2016 then Prime Minister John Key set a goal of eliminating all nonnative predators from the entire country by 2050. It's a pipedream—but then, so was tackling an area the size of South Georgia not that long ago. That same global assessment of island pests calculated that humans could save one-tenth of all the vertebrates now threatened with extinction— birds, endemic mammals, reptiles, and amphibians—by clearing nonnative mammals from 169 islands around the world.

And that rat-elimination scheme that David Sibley and I were part of, luring those rich folks to the Aleutian Islands—how did that go? Well, if the logistics didn't come off quite as intended, it did achieve its ultimate goal. The money was raised, and a few years later the US

Fish and Wildlife Service used that and a lot more—almost $2.5 million, in all—to eradicate Norway rats on the most appropriate target imaginable, a 10-square-mile island in the western Aleutians named Rat, where the rodents first washed ashore following a Japanese shipwreck in the 1780s. It worked; puffins, auklets, and other seabirds have returned for the first time in centuries. With the pests gone, and at the urging of local Aleut leaders, the US Board on Geographic Names in 2012 officially restored the island's traditional name, Hawadax. Rats remain on 16 other Aleutian islands, and budget cuts have stalled any further attempts to remove rats, foxes, rabbits, and even cows from islands in the chain where they don't belong—but it's a start.

Nine

TO HIDE FROM GOD

It was two in the morning, a warm, somewhat muggy night, and Andreas was pushing the old, battered truck hard, taking curves fast enough to make me tense up each time in the left-hand passenger's seat, waiting for a slip or a skid, bracing my legs unconsciously until the vehicle settled back again onto its loose shocks while he accelerated down the straightaways. We had a long way to go, as the autumn-dry hills and dusty olive groves of Cyprus flashed by in our headlights, and it wouldn't do to keep the police waiting.

My companion's name is not really Andreas, nor (for reasons that will become clear) will I share much about his appearance or background other than to say he is still young enough to find a midnight sprint through the empty countryside exhilarating. He works for BirdLife Cyprus, the local partner of BirdLife International, and like a small number of conservationists on this eastern Mediterranean island, he has embraced a fairly direct response to one of the most pressing issues facing migratory birds worldwide—their wholesale slaughter for the pot.

As we drove south, a dull orange crescent moon rose in the east. It mirrored the crescent and star of an enormous Turkish flag hundreds of meters long, made up of brightly shining lights on the slope of the Kyrenia Mountains just beyond Nicosia. It is a highly visible—from space, no less—symbol of the political mess that Cyprus represents. After generations of generally peaceful coexistence, living side by side across Cyprus, conflict between Greek and Turkish Cypriots erupted after an attempted coup in 1974 by forces hoping to unify the island with Greece. This was followed immediately by two waves of Turkish invasions and fighting that left the northern third of the island under Turkey's control. In the wake of the war, hundreds of thousands of refugees sorted themselves along ethnic and religious lines— Orthodox Greek Cypriots mostly moving south into the Republic of Cyprus, Muslim Turkish Cypriots flowing into the occupied "Turkish Republic of Northern Cyprus," an entity no country other than Turkey recognizes. Today, more than 40 years after the ceasefire, UN peacekeepers still patrol a barbed-wire-rimmed "Green Zone" buffer to keep the two sides apart, and that giant flag—white paint shimmering in the sun by day and lit up by night—stares down across the border at Nicosia.

Despite the political situation, these days the island of Cyprus is a popular tourist destination; the Republic is a member of the European Union, and draws holiday crowds from across the continent to its wide beaches and clear water, to trek in the Troödos Mountains, or sample a traditional Cypriot *meze* dinner. But lying as it does along the eastern toe of the Mediterranean, just south of Turkey and west of Lebanon, the island is also at the nexus of great migratory flyways connecting central Europe to Africa and the Middle East. Millions of birds—raptors, waterfowl, quail, doves, shorebirds, and passerines— pass through twice each year. Many of them do not leave.

At night, illegal bird trappers unfurl mist nets in olive groves or stands of exotic acacia trees, the latter carefully irrigated and tended to create a seemingly inviting oasis for weary, southbound migrants in this otherwise parched land. The trapper flips a switch on a digital recorder, and broadcasts from scratchy bullhorn speakers into the night sky the sound of a song thrush's or Eurasian blackcap's melody—the same approach my colleagues and I use when we're netting songbirds in Alaska to tag them with geolocators. But as the passing birds drop from the darkness in response to the call, building up in greater and greater numbers toward dawn in the thickets around the nets, they face a very different fate. As the dim twilight grows to dawn, the trapper and a few helpers begin tossing handfuls of pebbles into the trees, flushing the tired birds into the nets—where they are killed and heaped into bloody buckets. Other migrants, foraging through shrubs and low trees, will find themselves gummed fast to "lime sticks" coated with a fiendishly sticky natural glue boiled down from honey and a kind of local fruit, from which they must be ripped free, leaving skin and feathers behind. However they are caught, before the end of the day the birds will be cooked in hot oil, dusted in salt, and served—usually whole, the bare head still attached—clandestinely in homes and local restaurants. The dish is known in Cyprus as *ambelopoulia*; aficionados chomp down each small morsel in a few bone-and-guts-crunching bites.

Ambelopoulia and songbird trapping are old traditions in Cyprus, something passed down generationally and conducted mostly with lime sticks, which while gruesome are not as ruthlessly efficient at catching large numbers of birds as the mist nets that trappers increasingly favor. For an island less than two-thirds the size of Connecticut, the toll that *ambelopoulia* trapping exacts on the migrants of Cyprus is staggering. BirdLife estimated in 2016 that trappers were killing between 1.3 million and 3.2 million birds annually in Cyprus, making this small island one of the worst places in the Mediterranean for such slaughter—in fact, Cyprus is *the* worst on a per capita basis, given its relatively light human population. It also holds three of the dozen worst killing sites anywhere in the Mediterranean, which between them account for as many as 2.3 million dead songbirds each year. But Cyprus is far from the only death trap. Syrians illegally kill 3.9 million birds a year, while the annual carnage in Lebanon is 2.4 million, and Egypt another 5.4 million birds (though the BirdLife researchers admit the true number in all these places could be almost twice as high).

If that seems somehow unsurprising for such battle-scarred and troubled corners of the world as Syria and Egypt, consider that arguably the single most dangerous place to be a bird in Europe is actually in peacefully civilized Italy, where some 5.6 million passerines are killed each year, the ingredients for traditional dishes like *mumbulì*, spitted and grilled songbirds, or *polenta e osei*, which in its traditional form is cornmeal mush topped with whole grilled birds. The French gobble another half a million birds or so—thrushes, for example, lured with scarlet clumps of rowan berries and strangled by simple nooses of horsehair, a trapping specialty in the Ardennes region near the Belgian border, where some 100,000 thrushes die each year as French authorities turn a blind eye.

Most famously, though, the French have long relished the ortolan bunting, a handsome, six-inch-long bird with a peach-colored breast, pale yellow throat, and dark mustache marks; yellow eye-rings give it a slightly startled expression, but the French traditionally revered the ortolan for its meat, not its appearance. Trapped in August and September on their way to Africa and kept in the dark to scramble their natural rhythms (at one time, they were blinded to accomplish the same thing), the birds feed constantly until they are bulging with fat, then are drowned in Armagnac brandy, plucked and baked whole in a sizzling hot earthenware *cassole*. The diner—head and shoulders draped with a large white napkin, allegedly "to hide from the sight of God" but more practically to trap the aromas and block any splatter—severs the head with a snap of the front teeth, then chews the rest of the bird in a cascade of searing grease and juices, grinding up bones and all. It is considered the epitome of traditional French gastronomy. Former French president François Mitterrand, dying of cancer in 1996, ate two ortolans as part of his last meal, refusing all else thereafter until he died eight days later. The late Anthony Bourdain called ortolan "the grand slam of rare and forbidden meals," and recounted an illicit dinner with a number of fellow gourmands: "With every bite, as the thin bones and layers of fat, meat, skin and organs compact in on themselves, there are sublime dribbles of varied and wondrous ancient flavors: figs, Armagnac, dark flesh slightly infused with the salty taste of my own blood as my mouth is pricked by the sharp bones." Ortolan has, on paper at least, been protected in France since 1999, but the law has been largely ignored, especially in Landes, in southwest France along the Atlantic coast, where the cult of the ortolan is strongest. Trappers there use a rig known as a *matole*, which features a live ortolan in a central cage to act as a decoy, surrounded by up to 30 or more small wire drop-cages baited with grain. Until recently, as many as 300,000 ortolan buntings a year were being taken, each fetching as much as 150 euros (about $175) on the black market. Even though the ortolan kill has declined in recent years, the species' numbers have dropped like a stone. This is especially true of the small and increasingly fragmented population that breeds in western Europe and migrates through France, which has fallen more than 80 percent since 1980, faster than any other European songbird.

That is just one species out of scores that are targeted. In all, conservationists calculate, between 11 million and 36 million birds—songbirds, waders, ducks, quail, storks, raptors, basically anything with feathers—die each year during their migration through the Mediterranean basin. Only two countries in the region, Gibraltar and Israel, are relatively free from the problem. To this must be added the *legal* killing of migrants, not just traditional game species like waterfowl but enormous numbers of songbirds. Although the European Union's 1979 Birds Directive—the oldest conservation legislation in the EU—generally bans such killing, the union has issued what are known as "derogations," basically exemptions. Firm numbers are hard to come by, but BirdLife estimates that another 1.4 million birds are killed legally, including almost half a million finches and almost 300,000 thrushes. Cyprus, when entering the EU, negotiated the right to continue to hunt half a dozen species of thrushes with shotguns for traditional Christmas meals. Taken together, it's sometimes hard to imagine how any bird makes it out of Europe alive.

Conservationists—watching with growing alarm as the populations of almost all European migrants have cratered—are pushing back hard against the killing. In Cyprus, at least, they have made great strides in just the past few years. Previously unresponsive authorities, like British forces on two large UK-controlled military bases where trapping had been rampant, have poured manpower into aggressive patrols, using night-vision gear and sophisticated drones equipped with infrared cameras—in large part, it seems, because of well-organized letter-writing campaigns that deluged the Ministry of Defence with demands from conservationists reportedly including Prince Charles. BirdLife quietly conducts methodical on-the-ground surveys across hundreds of kilometers for mist-netting and lime-sticking, tracking activities and notifying Cypriot law enforcement. Another group, the Committee Against Bird Slaughter (CABS), takes this a step farther, mounting what are essentially guerilla operations against bird trappers and hunters in Cyprus and throughout the Mediterranean. Wildlife crime specialists from the Royal Society for the Protection of Birds (RSPB), clad in camouflaged ghillie suits and deploying hidden surveillance cameras, capture video footage that is finally resulting in serious fines and punishment for trappers.

Which is why I'd been tagging along with Andreas for several days, following him as we ducked through olive groves, sneaking up behind houses to check backyard trees for lime sticks, alert to barking dogs or the sudden, angry appearance of potentially armed trappers. When I arrived in the autumn of 2018, the tide seemed to be turning in the birds' favor—but Cyprus had been down this road before. In the 1990s, the annual toll from trapping on the island was estimated at a staggering 10 million songbirds a year, but with Cyprus under consideration for EU membership, the Republic's government cracked down hard in the early 2000s. Special antipoaching squads were fielded, and by 2005–06, the incidence of trapping, as measured by those standardized BirdLife surveys, had fallen by almost 80 percent. Conservationists cautiously congratulated themselves on turning a critical corner. But after EU membership was secured in 2004, a lot of the air went out of enforcement efforts, and some of the most aggressive trappers shifted to the British bases, where the Cypriot Republic authorities had no jurisdiction. That was especially true of the Dhekelia base along the south coast near Larnaca, where a peninsula named Cape Pyla juts into the Mediterranean, a geographic funnel concentrating southbound birds headed to Africa and the Middle East.

"Everyone was thinking this would be sorted out in a few years," Tassos Shialis had told me earlier that day. Tassos, slim and darkly bearded, once had Andreas's job, and conducted the same clandestine antipoaching surveys; that anonymity is long gone, and he's now the public face of BirdLife's antitrapping campaign, a regular presence on television and in the newspapers. "But instead it started picking up, especially at the bases. There was quite a big increase, and by 2014 we were back to where we'd been in 2002, about 2.5 million birds [trapped] a year. Compared to the 1990s, when the RSPB guys estimated it was 10 million birds, that's a big improvement, obviously, but there are a lot more threats today—tourist development, climate change, habitat loss, they're just adding to the problems of migration the birds face. It's not acceptable anymore, these numbers of birds, even though comparatively they look much lower than the 1990s."

Halting the bird-trapping on Cyprus had turned into a game of whack-a-mole, with the poachers changing tactics and locations to evade prosecution. When the Republic of Cyprus had first cracked down, the trappers moved to the bases; now that the bases were responding, the poachers were finding fresh areas in which to trap, both in the Republic and, if reports were accurate, increasingly in the Turkish North.

Andreas took another turn at speed, and I was grateful the adrenaline was keeping me awake as we headed for the British base at Dhekelia; it had been a 19-hour day already. The previous morning I'd hitched a ride with Roger Little, a British volunteer who has been coming to this country for years to help combat illegal netting. Tall and fit, his gray hair cut close, Roger is a retired finance expert given to wearing red socks when he feels the Manchester United soccer club needs a psychic boost from afar. But his real passion is nature—"Not just birds—nature, full stop," he told me, as we drove to the BirdLife office to meet Andreas. Roger had no issue with my using his full name, since he is only in Cyprus for five or six weeks a year. Andreas,

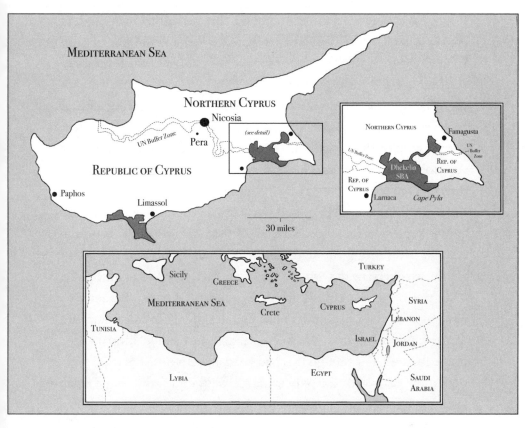

Cyprus and the eastern Mediterranean Sea.

on the other hand, is a native Cypriot, living in the communities he helps to police; like many islanders who work to curb bird-killing, he is sensitive about publicizing his work, both for security (many of the trappers are involved with organized crime, and grenades, homemade bombs, and arson are favorite weapons) and because a low profile makes his fieldwork easier and more effective.

Each autumn BirdLife searches for trapping activity in 60 randomly selected one-square-kilometer blocks scattered across a much larger 400-square-kilometer area of southern Cyprus where most of the netting occurs—a task that the RSBP started in 2002 and BirdLife Cyprus has managed since 2004. Our job that morning was to check a coastal block west of Larnaca—Andreas at the wheel and

Roger navigating via a tablet computer that contained maps and an extensive registry of all known trapping locations. On his lap also lay a clipboard with data sheets for a raptor survey, which serves as their cover story if they're stopped by locals or landowners, wondering what a couple of guys with binoculars are doing poking around back roads. "I kind of play it by ear," Andreas told me. "If the guy's charging up to the car and he's pissed off, I give him the raptor survey story. If he's just curious, I may tell him I'm lost, and ask for directions to somewhere nearby. People love to give you directions—suddenly they're the expert, and it usually defuses things."

We stopped at the edge of a plowed field within 100 yards of the sea, flat sun shining through high mare's tail clouds and twinkling off the Mediterranean with a sharp glare. Swifts crisscrossed the sky. Over the years, these BirdLife surveys and raids by CABS activists have allowed conservationists to assemble a highly detailed database of known trapping sites. We walked across the field, kicking up dust in the furrows, but though this had been an active mist-netting location a few years ago, the only evidence we found was a single rusty knee-high metal bar protruding from the ground at the edge of a streamside reedbed, once an anchor for a net pole. There was no sign that anyone had been there for a long time; the woody shrubs crowding up around the bar had several years' growth, Andreas judged. "That's what I like to see," he said, with satisfaction. "Abandoned trap sites." As we got back in the vehicle, he spotted a slender falcon passing by overhead. "There's a kestrel—Roger, write it down." A raptor survey form with no raptors on it wasn't going to fool anyone, if they needed their cover story.

The mid-morning sun was growing hot; it was early October, when daytime highs in the muggy lowlands are typically in the low 80s. There were few birds moving, and any trappers would have stopped hours earlier. That timing was intentional. Andreas and Roger actually didn't want to encounter any trappers; better, from their perspective, to identify an active location, slip away, and alert the Cypriot authorities so that the netters could be caught later and prosecuted.

To be spotted would mean, at a minimum, that the trappers would likely pack up and quickly move, escaping arrest. CABS takes a much more in-their-face approach, fielding teams at night or early morning when the netters are setting up and luring birds, sometimes tearing down or destroying nets and lime-stick arrays themselves and freeing trapped birds if they can. Not surprisingly, confrontations between trappers and activists can easily turn violent. In 2010, American author and avid birder Jonathan Franzen, working on a story for the *New Yorker* in Cyprus, escaped at a dead run while several Cypriot trappers beat the CABS activists he was with, smashing one in the head with the man's own video camera, leaving them battered and bloody.

The word *lime*, when referring to limesticks, has its root not in the familiar green citrus fruit, whose name is of Arabic or Persian origin, but in an ancient Indo-European word meaning "slimy" or "sticky." The trapper cooks up a gooey concoction; in some areas the main ingredient was boiled holly bark, in others the gummy fruit of mistletoe, though synthetic substitutes are now common in some regions of Europe, such as Spain. Cypriots still generally make their birdlime using Syrian plums, a cluster of which Andreas plucked from a tree and handed to me. When squeezed, the hard-skinned, marble-sized fruits exude a clear, decidedly snot-like gum that I realized, a little too late, was almost impossible to remove from my fingers without a lot of water and cursing. Boiled up with honey and several other additives, it creates an astonishingly strong natural glue, which the trappers use to coat long, straight wooden dowels (young shoots from a pomegranate tree are preferred for this purpose). Dried in the sun and gathered into bundles, the sticks are carried upright in specially woven tall baskets; as the trapper works, he or she (one of the most notorious lime-stickers on Cyprus is a woman) peels each stick from the gumbo mass and arranges it horizontally among trees and shrubs that have been carefully pruned to create open spaces within the canopy of leaves. Making and selling lime sticks is a cottage industry in parts of Cyprus, though many trappers make their own.

A songbird, exploring the tree for food, has only to brush the lime with the tip of a wing or tail feather, its beak, a toe, or some other body part to become stuck. Some hang miserably from wings or legs dislocated by their struggles, while others thrash and twist trying to escape until they are gummed fast at multiple points, glued and immobile. Sometimes, a single two-foot-long limestick may trap half a dozen or more birds. The technique is, of course, completely nonselective—BirdLife has documented 155 species caught in limesticks, from warblers and thrushes to hawks, falcons, bee-eaters, nightjars, shrikes, and owls. Even those few birds that manage to free themselves are usually so coated in lime that they're unable to fly or (if their bills are gummed shut) to eat or drink. The trapper can only remove them by ripping them loose, tearing away feathers and skin; activists raiding lime-sticking sites sometimes use their own saliva as a cleaning agent to free trapped birds, sucking at the clotted feathers to remove the worst of the gunk. To conservationists, this all seems barbaric. A local pro-trapping group, known as Friends of the Limestick, argue it's a tradition—to which one CABS leader responded that he was from Germany, where they once burned witches: not every tradition, he said, is worth preserving.

As the day wore on we turned and zigzagged on small roads, some paved, some dirt, driving slowly as we peered between long lines of citrus or olive trees for signs of netting. Carefully pruned trees with wide avenues between them might be the sign of conscientious husbandry—or it may mean the owner doesn't want the branches snagging his mist nets. Andreas and Roger were alert for metal poles and stakes, or tires filled with concrete into which a short pole had been set for a somewhat portable pole base. If they saw something suspicious we got out, searching for feathers, discarded dead birds, or their heads; there is a tradition that even nontarget birds caught in the nets are killed, if only because it's simpler and quicker to remove a dead bird than to fuss around with a live and struggling one. More than once, men in passing vehicles slowed to glower at us suspiciously, since these were not places where outsiders wander very often. We found no active net-rides, as they are known in Europe, though we did encounter a few old limesticks that appeared to have been missed when trappers abandoned one site; because the sticks posed a continuing hazard to birds, Roger lashed them back and forth across the ground, covering the goo with dead grass and dirt, then broke them into small, harmless pieces.

Many of the olive groves were fenced; one in particular caught their eyes, for it was topped with barbed wire and the fence was cloaked in green mesh netting that blocked any sight of what was inside. "Suspicious, that," Andreas said quietly. "Makes you wonder what they're so keen on nobody seeing from the road, doesn't it?" Despite our best efforts, though, we couldn't make out what might be inside, and Roger noted the location for a return visit. The only legal prohibition on trespass in the Republic of Cyprus is that fenced land like this is off-limits. But my companions' antennae were twitching. A bit farther, they noticed a tall metal pole rising 10 feet above the top of a lone carob tree beside a series of greenhouses. Just an antenna? Maybe, but there were no wires. "The trappers will put up a tall bamboo pole with lime sticks at an angle from the sides, to catch bee-eaters," Andreas said, but he couldn't decide if this was such a trap. As we drove off, a pair of bee-eaters swooped in to land on a neighboring tree, long, falconish wings flashing, the sun catching the russet and blue of their plumage, and their bell-like calls carrying across the warm air.

Down another lane, where a chukar scurried ahead of us, Andreas slowed. There was a small house on a knoll overlooking the olive grove; parking where the sight lines from the house were blocked, and leaving Roger to keep watch, he slipped out of the truck and into the cover of the trees, looking to either side as I followed him through the low olives, hung with blue-black fruit. We heard a car and he froze, gesturing for me to step into a dense copse; the car seemed to be passing on the paved road some distance away, but Andreas looped around to head back to the vehicle—and stopped. "A lime stick," he whispered, pointing up the small hill, where I now noticed a weirdly horizontal branch in a small carob tree just behind the house. "Give me your phone. I forgot mine and I need to get pictures," he said. He sprinted up a set of wooden steps in the hillside, peered into and around the low tree, grabbed a few images, and came back down, walking fast with me in tow. "There were seven lime sticks in that carob, and I expect there are more in all of those trees around the house. No birds caught, fortunately," he told Roger as we backed the truck around and made a quick exit. Once out of sight, he sent the photos to the police, with at best a cautious hope that they would follow up. "Seven's not a lot, but there could be many more in there."

"How's the cooperation from the police?" I asked, having heard disparaging remarks from Cypriot conservationists on that score.

" 'Mixed' is a great word," Roger replied drily. In the early 2000s, when Cyprus was pushing for EU membership, the Game and Fauna Service (the government ministry responsible for enforcing wildlife laws) had an active antipoaching squad and a reputation for being fairly hard-assed about trapping, but in recent years the agency has, in the view of BirdLife and other conservation groups, turned an increasingly disinterested eye to the practice. (The Game and Fauna Service disputes this, pointing to recent changes in regulations that increased penalties for mist-netting, if not lime-sticking.) Partly it's political pressure from the large and influential hunting lobby; partly a function of the fact that the police are from the local community, and the trappers are often their relatives or friends. Partly, some officials have privately admitted, it is because some of the poachers have ties to violent, organized criminals and the authorities are somewhat intimidated. Whatever the reason, poachers sometimes seem to operate with near-impunity. Andreas pulled out his tablet and zoomed in on one area with a marker pin—a netting site for song thrushes CABS had found the previous January, with a loud audiolure playing thrush songs through the night. "And that"—he pointed to a building just across some fields—"is the local police station, not 200 meters away. How could they not have heard that thing, blaring all night?"

Throughout the trapping season, the conservationists play a cat-

and-mouse game. "The trappers aren't stupid. Sometimes they'll set up a few lime sticks in a really obvious, visible place—I think the idea is, if the sticks disappear, then they know CABS is around," Andreas said. "Sometimes they'll string really fine, thin threads [at the trapping site], and if it's broken they know someone has been through."

After we finished for the day, I had a couple of hours before Andreas and I were supposed to drive south to rendezvous with the military base police for their night patrols, and my stomach was growling. I was staying in the village of Pera Orinis, in the hills of central Cyprus about 15 kilometers from Nicosia, in a lovely bed and breakfast run by Tassos and his Italian biologist girlfriend, Sara. The streets in the old center of Pera are narrow—so much so that my first evening, exhausted from travel and trying to navigate a warren of alleys just wide enough for my small rental car, I scraped the left side mirror against an open door while avoiding by an inch or two a rock wall on the right. But on foot, as I walked through the soft twilight, the narrow streets were inviting canyons walled in pale gray stone, the doors of houses made of dark, ancient wood with wrought-iron hasps and hinges. At one corner, four old men sat around a small table by the curb drinking frappe, iced coffee, the dry clack of counters as they played a Cypriot form of backgammon called *tavli*; one gent, hair and thick mustache white, looked up in unconcealed surprise as I walked by, and watched me out of sight. I turned down another lane, which opened onto the village square, the old Archangel Michael Church and neighboring boys' school forming one side. A father watched his two young daughters as they played. I was struck by how utterly quiet it was. My footsteps on the large, uneven cobblestones echoed.

I followed my nose, and the smell of wood smoke and meat, around the corner to a small tavern, a single room with a beer cooler, some old posters of Greek movie stars, and a shaded patio outside, run by Costas, a man in his mid-thirties with short dark hair and horn-rimmed glasses. I ordered souvlaki—chunks of grilled pork, fresh tomatoes, and cucumbers wrapped up in a whole pita and drizzled with lemon juice. "Thirty minutes, my friend," he said. We have been friends since my first evening in town, when Tassos had brought me in for a late meal. Tassos had errands to complete, so Costas sat and talked as I ate sizzling hot kebabs. Birds? Oh, yes, they taste very good, he told me after we'd chatted for an hour. "Hot oil. In hot oil only. And salt of sea. Not, not—" he grabbed a salt shaker and spilled some in his hand, then shook his head emphatically, throwing it away. "Only salt of sea. And *so* good."

Tonight, as Costas worked on my order, his brother invited me to join him as he, too, waited for his order and drank a beer at one of the small tables. He was more of a closed book than his younger brother, tapping the ash from his cigarette and observing me through hooded eyes. His English was basic, my Greek nonexistent, but we managed. Where do I live? What is the weather like? What is my job? He is a policeman. "Serious crimes only. I am like a detective," he said, rolling another slender cigarette. And why am I in Pera? "Bird-watching" was the easiest explanation, and—as it had with his brother the previous evening—it sparked an immediate reverie on *ambelopoulia*. "It is very good. Not so many birds to watch in Cyprus, eh? We eat them, many, so many." I ask: Is he worried that many, so many, are eaten, that the birds might disappear? A shrug and a smile. "So few, the price high—for 12 birds, 80 euros. Was 60, now 80." He shakes his head; the tragedy of it all. Supply dwindles, prices rise—but with the rising price, a powerful incentive for some trappers. But isn't *ambelopoulia* illegal, I ask? Another expressive shrug, but no smile this time. No doubt he knows I'm staying with that BirdLife trouble-maker Tassos. His meal is ready; he takes it out to the patio to eat, and this time he does not ask me to join him.

Killing migrant birds is hardly a Mediterranean problem; several million birds annually are killed illegally in northern Europe and (especially) the Caucasus region. Some, like waterfowl, are shot as much for sport as the plate, while others like thrushes snared by the neck in France are killed strictly for consumption. Raptors remain a target in much of Europe, like the 41 golden eagles in Scotland fitted with satellite transmitters between 2004 and 2016 that disappeared without a trace of either a carcass or the device, strongly suggesting they'd been killed and disposed of by those who resent eagles for preying on red grouse.

Asia, not surprisingly, is a death trap for many migrant birds. Netting and trapping, after all, were likely the main reason for the spoon-billed sandpiper's catastrophic collapse toward extinction. But while the spoonie has become an international conservation icon, another Asian bird, the yellow-breasted bunting, has fallen much further, much faster, yet with relatively little notice.

A very close relative of the ortolan bunting that is famously consumed in France, the yellow-breasted bunting occupied an immense breeding range stretching from Finland to the Russian Far East, some 6 million square miles (15.7 million square kilometers) in extent, and across which it nested in incalculable numbers; scientists used the term *superabundant* to convey the magnitude of its population. The males are marvelously patterned, with citrine breasts, black faces, and chestnut heads and backs, set off with bold white shoulders; the females and juveniles are, as is often the case, more subtle and complexly plumaged, streaked in brown and looking as though they'd been steeped in saffron tea. The great majority of all the yellow-breasted buntings from across this sweeping range funneled each year through coastal China, forming along the way tremendous flocks, and gathering to roost by the millions both in migration and on their far more restricted wintering grounds in southeast Asia.

The buntings' numbers, gregarious nature, and plump, tasty flesh made them a traditional target for subsistence bird-trappers, but in recent decades—as China forged a middle class with disposable income, and wild meat became a status symbol rather than a mere source of protein—the pace of slaughter moved to hyperspeed. Like *ambelopoulia*, bunting became a luxury item in China, with a single six-inch bird fetching the equivalent of up to 30 or 40 US dollars. In 2001, scientists estimated, as many as a million a year were being consumed in Guangdong Province alone. China's legal ban on killing wild birds for market, in place since 1997, is basically ignored in practice, and as early as 1980 declines in breeding populations of yellow-breasted buntings were being reported; since 2000, the collapse has been nearly universal in scope and terrifying in scale and rapidity. The bunting's range has contracted east by more than 3,000 miles. At current rates, what was once among the most common landbirds in the world will be extinct in a few years. "The magnitude and speed of the decline is unprecedented among birds with a comparable range size, with the exception of the Passenger Pigeon," an international team of scientists warned in 2015.

To North Americans, the idea of eating songbirds is shocking and abhorrent, but that just shows that we have short memories—after all, as the authors of the bunting article pointed out, we ate the passenger pigeon into limbo. The combination of railroads and telegraph lines allowed commercial trappers and dealers to find the nomadic pigeon colonies, decimate them, and easily send the birds to market, as when 1.8 million pigeons were shipped for sale from a single nesting site in Plattsburgh, New York. Histories of the American market hunting era in the nineteenth and early twentieth centuries usually focus on the immense carnage inflicted on more typical game species like ducks and geese, such as the lone hunter in North Dakota who, in a few hours of butchery, killed 700 pounds of waterfowl (including 46 Canada geese and 37 sandhill cranes) and kept up that pace for a month; or a gunner on Long Island Sound around 1870 who killed or crippled 127 scaup with a single discharge from a massive double-barreled four-gauge shotgun. Grouse and quail were shipped via railroad by the ton, including an estimated 30,000 prairie chickens and 15,000 bobwhite carted out of Nebraska in 1874. As the wild pigeons and grouse grew increasingly scarce in the late nineteenth century, gunners and dealers turned to shorebirds, most notably Eskimo curlews and American golden-plovers, but because the whole shorebird family tastes good and decoys well, all came under the gun. Small size was no protection; even least sandpipers, weighing barely half an ounce, were shot and marketed as "mud peeps."

But songbirds, too, were a frequent target. There may be no better window into that period than a remarkable book called *The Market Assistant*, published in 1867 by a New York butcher named Thomas F. De Voe, who delivers on the promise of his subtitle, *A Brief Description of Every Article of Human Food Sold in the Public Markets of New York, Boston, Philadelphia, and Brooklyn.* Several long chapters are taken up with wild game, from raccoon and skunk to caribou and bighorn sheep, as well as waterfowl, turkeys, grouse, and shorebirds—what was common in shops, what was rare, what tasted good and during which season. But De Voe also detailed dozens of songbirds, from woodpeckers to sparrows, that were regular fare in eastern markets. Some like "reed-birds," or bobolinks, were abundant in season, shipped into urban centers by the millions, while others were more sporadic, like the "gray shore-finch," as the seaside sparrow was known. "It is sometimes found in our markets in the summer months, but its flesh is quite indifferent eating, being somewhat fishy," De Voe warned.

Only a few birds merited such a warning. The northern flicker, which De Voe called a high-hole or golden-winged woodpecker, was "common in our markets in the fall months, when it is fat, and its flesh quite savory. . . . I have shot hundreds—sometimes as many as 20 or 30 from off one tree, in an afternoon." Meadowlarks were "almost as good as quail, but not so plump and large," and he recalled shooting many "in the neighborhood of the present Twenty-fourth Street, New York." Among the many other songbirds mentioned in *The Market Assistant* are horned larks and American pipits; snow buntings ("[w]hite snow-bird . . . much fatter and better in January or February, when its flesh is much admired by the epicure"); red-headed woodpeckers ("quite as good eating as the golden-winged woodpecker, but smaller"); cedar waxwings ("occasionally found in large numbers in our markets. Their flesh is but a morsel of delicate eating, and only fit to eat in the fall months"); gray catbirds ("very small, but the flesh is sweet and good"); eastern towhees, juncos, pine grosbeaks, yellow- and black-billed cuckoos ("their flesh is quite sweet, but rather a small body for so large-looking a bird"); and purple finches ("very delicate when in good condition"). The meat of the hermit thrush, he wrote, is sweet, "but not worth a charge of powder, even to a starving man."

In fact, it's surprising that amid page after page describing the gustatory virtues of American songbirds, De Voe often set aside his role as objective reporter to editorialize on the birds' behalf. Having noted the delicacy of waxwing flesh, he wrote, "I think they should never be killed, as they destroy more destructive worms than perhaps any bird in existence." With an ecological understanding rare for the day, he lamented that "[t]housands of birds of the small species are wantonly killed for the sport, or a few pence. These slaughtered birds, when alive, destroy millions of insects, flies, worms, slugs, etc." He pleaded with readers not to buy Baltimore orioles ("except for a collection") and to shun buying robins in the spring, when the birds were pairing up to mate. "I, however, think that these birds are more useful to man living than dead," he argued.

On the plate, the American robin was the standard against which most other small birds were measured—toothsome, common, relatively cheap. Blue jays, for example, were good but "not so well flavored as the robin," De Voe said, while flicker was "not so tender as the robin." Of the latter, he said, "[l]arge numbers of these well-known birds are found in our markets, and thousands are also shot by *all sorts* of sportsmen, in the months of September and October, when they are fat and delicate eating." In the South, robins were a particular target for the pot and the market. In central Tennessee, where in winter robins gathered by the hundreds of thousands in nighttime roosts, hunters at the beginning of the twentieth century worked in teams to kill them. While one man with a torch climbed a high tree, others with clubs and poles startled the sleeping birds into flight; drawn to the light, the robins were grabbed, decapitated, and stuffed into sacks. One team could often kill 300 or 400 robins a night, and "[m]any times, 100 or more hunters with torches and clubs would be at work," filling wagons with dead robins, one observer noted. So pervasive was robin-hunting throughout the South, where a number of states classified them as game birds, that the National Association of Audubon Societies (the forerunner of today's National Audubon Society) worked with teachers in the region to create grade-school clubs to teach "sympathy" for birds—an effort that became the astonishingly successful Junior Audubon program.

As Thomas De Voe had noted, the other mass-marketed song-bird was bobolinks. In 1912, Massachusetts ornithologist Edward Howe Forbush traveled to the Low Country of South Carolina to see the annual slaughter of "rice birds," as migrating bobolinks in their streaky, brown autumn plumage were sometimes known. Some 60,000 dozen—almost three-quarters of a million bobolinks—were shipped that year from the city of Georgetown alone, Forbush wrote, "killed for the twenty-five cents a dozen received by the shooters, and marketed for the seventy-five cents to one dollar per dozen that the marketmen received. In the height of their abundance the birds were shipped in quantity to the great markets in New York, Philadelphia, Paris, etc., and were eaten mostly by wealthy or spendthrift epicures."

A few states offered piecemeal protection. California imposed an eight-month closed season on robins in 1895, though such protection was absent for the many other species, from red-winged blackbirds to house finches, that were also prey for Golden State market gunners. Not until 1918, with the passage of the federal Migratory Bird Treaty Act, did robins, bobolinks, cedar waxwings, and most other wild birds receive legal protection, and the age of market hunting ended in North America. "New Game Bird Law Bars Pot Hunting," a headline in the *New York Times* announced on August 18 of that year. "Under Canada Treaty No Migratory Bird May Now be Sold or Offered for Sale." ("Migratory" is somewhat misleading; the act now protects all native North American birds, with the only exceptions being nonmigratory game species, which are protected under state law, and exotics like house sparrows and mute swans, though it initially excluded some raptors.)

Europe took a very different path. There were early agreements, like the International Convention for the Protection of Birds Useful to Agriculture, signed by 11 European powers in Paris in 1902, and considered the first multilateral agreement to protect wildlife—though it, too, singled out "noxious" birds like many raptors, herons, corvids, and loons that did not deserve protection. In England, a welter of sometimes conflicting laws passed in the first half of the twentieth century were replaced by the Protection of Birds Act, which in 1954 gave significant safeguards to a variety of wild birds (with the usual exceptions for "pest" species like crows and magpies). That law was largely superseded by the Wildlife and Countryside Act of 1981, which remains the primary bird-protection law in the UK. Across Europe, however, the controlling statute is the European Union's Birds Directive, first passed in 1979, amended in 2009, and the oldest of the EU's environmental legislation.

It was the power of the Birds Directive, and the need to begin to meet its mandate if Cyprus was going to join the EU, that drove the first downturn in songbird trapping there in the early 2000s. While there was never an overt quid pro quo—a public demand by the European Union that Cyprus deal with bird trapping or it wouldn't enter the EU—there reportedly was plenty of backdoor pressure on the Republic. "The commission never came out and said, 'Cyprus isn't acceding [into the EU] unless they deal with this problem,'" BirdLife Cyprus director Martin Hellicar told me. "We wish they had, but they didn't, and I understand that's how it works. But behind the scenes, we have it from a lot of sources there was a lot of pressure." Guy Shorrock, a senior investigations officer with the RSPB who began doing antipoaching work in Cyprus in 2000, recalled that initially he and his colleagues had very good cooperation from both the Republic's Game and Fauna Service, and from the British leadership on the bases. "After the enforcement work started and with Cyprus's accession to the EU, there was quite a significant drop in trapping, on the order of a 70 or 80 percent drop. We thought we were all going in the right direction."

With time, though—and perhaps with EU membership secured in 2004, an easing of pressure on authorities—the trapping resumed. The Republic's effective antipoaching squad was apparently disbanded, and more and more large-scale trappers were operating on the British bases, where they remade the very landscape to make their operations vastly more effective. At the same time that the first European conference on the illegal killing of birds was convened in Larnaca, Cyprus, in 2011—resulting in the so-called Larnaca Declaration, which called on Europeans to "pledge a zero tolerance approach to illegal killing, and a full and proactive role in fighting against this illegal activity"—the killing was on the rise again, especially on the Dhekelia base. Pressured by bird activists and swelling public opinion back home (expressed through mail bags heavy with petitions and angry letters delivered to the Ministry of Defence), but up against sometimes violent opposition from Cypriots to antipoaching efforts, British base commanders have found themselves dealing with what has, at times, been literally an explosive problem.

I ate my souvlaki back at my room, lay down for a while but tried not to fall asleep, and at two in the morning met Andreas at the BirdLife office for our trip to Dhekelia. The Brits have a long and tangled history on Cyprus. From 1878 to 1914 the island was a British protectorate; from 1914 to 1925 a militarily occupied territory; and from 1925 to 1960—a period that saw increasing violence against British forces—a Crown colony. Even after independence, the UK retained control of two large military bases, Dhekelia and Akrotiri, totaling about 100 square miles, with their own legal jurisdictions and law enforcement, known as the Sovereign Base Areas. These SBAs do not match the mental image of a typical military base. They are largely unfenced, crisscrossed with public roads and dotted with Greek Cypriot villages that are technically part of the Republic, but

whose residents often lease land for farming on the bases. This makes
for an unusually porous security situation, especially at night, which
is when the trappers begin their operations. What's more, the bases
remain a source of considerable anger and discontent on the part of
Cypriots, so for years the British military had sought to avoid aggra-
vating the locals by turning an intentionally blind eye to the ever-
growing trapping problem.

"The bases have always treaded this awkward line where they
know they're not welcome here, so they avoid, if at all possible, antag-
onizing the local residents. And for decades, poaching was an issue
they completely avoided," Martin Hellicar had told me a day or so
earlier, outlining the complex political situation in which lax enforcement on the bases provided an opening for a far more intensive form
of poaching to blossom there, one underpinned by organized crime
and operating at what conservationists have described as industrial
scale. If the stakes for bird-trapping might seem too petty for gang-
land action, consider that according to the SBA police, a successful trapper might be able to pull down as much as 70,000 euros
(roughly $78,000) a year, while the Game and Fauna Service esti-
mates that the *ambelopoulia* industry in Cyprus as a whole is worth
15 million euros ($16.8 million) annually. As gangland activity there
increased, Cape Pyla on the Dhekelia base had become (as Guy Shorrock told me) "the black hole of Cyprus, as Cyprus is the black hole
of the Mediterranean."

Conservation NGOs were for much of the past decade fairly
unanimous in their frustration with the perceived lack of interest
on the part of the SBA police in enforcing poaching laws, which
essentially gave the trappers free rein on base lands. But a change in
personnel, coupled with a lot of public pressure back in the UK from
bird conservation groups and changes in the laws under which the
base operates (which are set by the British but tend to reflect Cypriot
Republic rules), all had dramatically altered the picture two years
before my arrival. The SBA authorities have since poured men and
resources into the trapping problem. Base authorities can levy heavy

fines and require cash bail, as well as seize vehicles used by trappers. And because the base surrounds many small villages, whose residents lease base land for farming, such rental agreements can be negated if a lessee is caught trapping—another powerful deterrent.

So as Andreas and I hurtled through the Cypriot night toward that black hole, it was at a time of cautious but rising optimism on the part of bird activists. At four in the morning, Andreas parked under some palm trees in front of a long, low stone block building with an orange tile roof, surrounded by a high chain-link fence topped with double rows of barbed wire—the Dhekelia SBA police station. We were buzzed through an entrance gate and escorted back through hallways to the office of Inspector Nicos Alambritis, a burly Cypriot officer with close-cropped hair and a trimmed stubble beard, wearing—like all the members of his antipoaching team—a green tee-shirt, camouflaged slacks, and a black tactical vest with POLICE in large white letters. Yellow Taser guns in webbing holster were the only weapon the officers carried, even though the trappers are often armed with shotguns or rifles. We shook hands with Alambritis's sergeant and two police constables, all Cypriots and all, by prior request, to remain anonymous—like Andreas, they live and work in the communities they patrol, and as antipoaching enforcement has increased, so has violence from the trappers.

The entire wall behind Alambritis's desk was filled with a satellite photo of the Dhekelia SBA, over which were superimposed more than 100 green or red geometric shapes, most clustered near the coast on Cape Pyla. "These are the acacias," Alambritis said, waving his hand over the blocks—stands of a nonnative Australian tree sometimes known as golden-wreath wattle, which the trappers planted in this otherwise arid habitat to lure southbound migrants. The acacia plantings represent a huge artificial forest covering a couple of hundred acres and supplanting the native phrygana, a low, scrubby, and increasingly threatened Mediterranean coastal plant community rich in rare wildflowers. The acacia thickets were supported by countless kilometers of black plastic irrigation pipe, snaking across the Cape from illegal wells to water the trees. Anyone who doubts the economic heft of the *ambelopoulia* trade in Cyprus need only look at the herculean effort to which the trappers in Cape Pyla have gone to remake the landscape for commercial songbird netting.

And netting here is the key. Effective as lime-sticking can be at a backyard level, it doesn't lend itself to the scale of trapping undertaken at Cape Pyla, where the trappers use mist nets (illegal, but smuggled into Cyprus as "fishing" nets). The acacia plantations are scored with long lanes, known in Europe as net-rides, along which hundreds of meters of nets could be strung—the biggest, most concentrated trapping site in the Mediterranean, according to BirdLife. But with a dramatic increase in arrests and prosecutions, the authorities have seen a steep drop in trapping activity on Cape Pyla and on the bases as a whole. In 2017 the SBA police prosecuted more than 80 trapping cases; that had fallen to 35 in the spring of 2018, and by that autumn, when I was there, only five arrests had been made despite nightly patrols. But the rise in enforcement has been met with pushback, some of it violent. CABS has had some of their members shot at, and their vehicles rammed and shot up. ("No one was in them at the time—yet," Andreas told me.) In 2017 alone, a house and four cars belonging to a forestry department employee and his family were torched; a bomb damaged the home of a game warden; and someone on a motorcycle hurled a hand grenade over the fence into the courtyard of the SBA police headquarters at three in the morning, blowing out windows, spraying everything with shrapnel, and slightly injuring one officer who had, miraculously, just stepped inside. Lobbing explosives to settle scores is a hallmark of criminal gangs in Cyprus, with some gangsters bringing their past military skills to the task.

"The guy who threw the grenade knew what he was doing," a police constable I'll call Yiorgos said as we walked out of the SBA station and past the spot where the blast hit. "He held it and counted it down, so it would explode immediately when it landed." Andreas and I squeezed into the back of Yiorgos's unmarked patrol truck, with constable Nicholas (another pseudonym) riding shotgun on his

left, one of several teams operating that shift. So far that night, Yiorgos said, only one active trapping operation had been detected, and
we would steer well clear of it, because Guy Shorrock and a colleague
from the RSPB had installed covert surveillance cameras on the site,
hoping to record slam-dunk video evidence for the prosecution once
the SBA made arrests. Shorrock and his colleague had been creeping
about the base in camouflaged ghillie suits, placing hidden cameras
near known and suspected trapping sites, the video from which had
dramatically increased conviction rates. The previous year, Shorrock
told me, they had filmed 19 trappers at work, most of them established, major operators. Faced with video evidence, all 19 pleaded
guilty—and those who still trap have taken to wearing balaclavas
to mask their identities, and using metal detectors to find hidden
cameras. Because the bases operate under what is essentially UK law,
surveillance cameras—which are ubiquitous in Britain—pose no
legal issues there. Shorrock said the situation is much different in the
Republic, where covert video would be inadmissible in court. (Video
they obtained that autumn, showing a team of trappers working mist
nets in an orchard, extracting and killing songbirds and eventually
filling several plastic grocery bags with the dead bodies, resulted in
three of the men receiving suspended jail sentences, being banned
from hunting on the bases for 10 years, and paying hefty fines—
with the threat of immediate and lengthier jail time if they were
caught again.)

    We bounced and thumped along rutted back roads across the base,
Yiorgos behind the wheel—another driver young and cocky enough
to love killing the headlights while roaring down a winding road,
then braking hard to a stop on a high hill with the windows down,
listening. The trappers use a simple rig—an MP3 player, a car battery, long wires, and speakers—to blare the recorded songs of blackcaps, song thrushes, and other target species into the night sky. The
migrating birds, tired after a long night of flying and hearing these
calls, drop down to what must sound like a crowded and therefore
safe oasis. As the first light starts to dawn, the trappers will take

handfuls of smooth beach pebbles—often trucked in for just this purpose—and toss them into the trees and bushes, flushing the birds into the nets.

The same recorded calls that attract the birds also pinpoint the trappers' location for the police, but stop after stop, it was silent. "Not even a quail caller," Nicholas said. Hunting migratory quail, which like the songbirds fly from Europe to Africa each fall, is legal—but luring them with a caller is not. Yiorgos started the truck and turned on his lights, dropping down into a low valley. We were likely not the only ones with a wary eye and ear peeled for trouble. Careful trappers will watch the normally empty nighttime hills of the base, looking for distant lights or the sound of an engine; some caller rigs have remote controls, so the trapper can remain a safe distance from the nets and turn off the sound if there is any sign of the cops.

We were a long way from the lights of Larnaca, the stars bright overhead. Canis Major loomed high in the sky, Sirius a brilliant pinprick at its head. We flushed a barn owl from an old roadside post, and from time to time nightjars, grayish and ghostlike, fluttered through our headlight beams. We rendezvoused with another patrol, three officers who had encountered nothing suspicious all night. "We get bored—really bored," one of them told me. "Obviously that's good, but it can be really quiet."

Yiorgos agreed. "There are only a few trapping sites left, and at most of them we've placed cameras. Sometimes it's long nights with nothing. We're counting down the days until the end of the blackcap trapping season."

"But then it's song thrush season," Andreas reminded him. In late autumn the trappers shift their focus away from netting blackcaps and other small passerines, which by then have moved on to Africa, to killing song thrushes that breed in western Eurasia and winter along the Mediterranean and in the Middle East. Serving them for Christmas meals is another Cypriot tradition, and shooting song thrushes is legal—but netting them is not, and that season adds to the workload. "My wife wondered the other day when we'll finally be able to wake up at the same time in the same bed," one of the officers said.

We split up again, crossed a divide, and turned off a dirt track and onto a paved road, Yiorgos killing the lights as he did so—but just as his headlights, sweeping left, went dark, their dying light illuminated two vehicles, a car and a pickup truck sitting nose-to-tail beside the road about 100 yards ahead. Yiorgos immediately flashed his lights back on, passing them—I had a brief impression of faces inside the car, lit by the dim orange glow of cigarettes—and pulled a quick three-point turn, saying, "*That's* suspicious." Suspicious, indeed—the pickup had already peeled out and was speeding away, but Yiorgos blocked the car, and he and Nicholas got out and began questioning the occupants.

"What are you doing here?" Nicholas asked, Andreas translating quietly for me.

"We are here, um, because we are here," the driver answered—the kind of non sequitur response unlikely to deflect a cop's mistrust. They asked the three men to step out, Nicholas checking their papers while Yiorgos looked inside the vehicle, where he found an electronic caller and speakers. The trio claimed, somewhat improbably, that they were using the rig to listen to music. The possession of the devices is not, in itself, a crime, though, so there was nothing to do but let them go.

"Do you think the truck had the mist nets?" I asked Yiorgos as they drove off.

"Hard to say. They often hide the nets and the poles near the trapping site, so they might not have had anything with them. The guy driving the car, though, his name is familiar, but I don't recall if it's from trapping birds or drugs. I'll have to check."

By now there was a faint band of light on the eastern horizon. We bumped up a rutted dirt road that climbed along the rising edge of a steep cliff, the semicircular sweep of Lanarca Bay behind us to the west, the last lights of the city twinkling in the gloom. The sea was flat and inky; I tried not to think of it as "wine-dark"—such a

Mediterranean cliché—but the Homeric image fit. Boulders of sand-colored limestone, pocked and weathered, formed jumbled barricades, shining gently against the blacker shrubland as though they had borrowed some inner light from the coming dawn.

At the top of the hill, Inspector Alambritis and seven or eight other SBA cops smoked and talked, their nighttime patrols finished. Two of the men were operating a drone, which was invisible and inaudible hundreds of feet in the air until it was pointed out to me—but when I looked at the computer screen on the drone's controller I realized I was seeing our group in infrared, each tiny body glowing yellow-white as it moved against the dark blue of the chilly ground. Toggling a switch, the officer flying the drone zoomed in the camera farther and farther, until nothing but my own head and upper torso filled the screen. I glanced back up, but I'd lost the drone, invisible again against the half-lit sky. Like the RSPB with its surveillance cameras, the base police have been deploying drones to great effect; Inspector Alambritis had earlier shown us incredibly clear video (shot in visible light, not infrared) of an elderly man they had already arrested several times in the past for mist-netting, as he opened and cleared his single mist net.

"This is a stubborn one," Alambritis said of the old man. "You feel sorry for him, because this is the way he was raised—his grandfather taught him. That was survival in those days, but today it's a cheap excuse—you can get a chicken in the market. This time, he's going to have to go to jail." That risk, combined with greatly increased fines of up to 6,000 euros (about $6,700) for the biggest offenders, along with vehicle confiscations and the loss of leases, has finally begun to make trapping not worth the risk. Overall, BirdLife Cyprus later said, the pace of trapping on the island in 2019 had fallen to its lowest level in 17 years.

Nevertheless, it remains a politically fraught issue for the British. The base commanders did have some initial success removing the artificial acacia groves, but in October 2016 they sent 150 soldiers on a clandestine operation to rip out more on the cape. Someone

spotted them, and church bells in the nearby village of Xylofagou—
a pro-trapping hotspot—began ringing at three in the morning, rous-
ing the residents, hundreds of whom mobbed the infantrymen and
kept them penned up for six or seven hours during a tense standoff.
Today, the base leadership admits that physically removing the acacia
is a nonstarter. "What we cannot have is soldiers in confrontation
with locals, because that would be a public relations disaster," Deputy Chief Constable Jon Ward, the divisional commander of the base
police, told me. "There is a hard core of people who will do whatever
they can to oppose what we're doing." But if the trees themselves are
off-limits, their water supply is not, and in the previous year or so
(and officially to protect Cyprus's scarce water supply), the base had
begun removing many kilometers of illegal irrigation pipes, leaving
a lot of the acacias to wither and die. (The RSPB, doing its own
mapping, believes the base has removed less acacia than the 50 acres
officials suggest, and far less than the 100 kilometers of pipe the base
claims—"but the bottom line is, they've cleared a shitload of pipe-
work," Guy Shorrock said. "Quite a lot of acacia is dying back, so it's
less attractive to birds, and less attractive to trappers, which is great.")

Overall Shorrock, who has been tackling this problem for almost
two decades, thinks the picture is far brighter than just a few years
ago. "By 2017, it was immediately evident that we were having problems finding places to put our cameras—all seven sites where we'd
caught people the year before were inactive. It's a conservation success story, hundreds of thousands of birds off to Africa that wouldn't
be getting there otherwise. It's really a question now of trying to
maintain the pressure." But if the intensity of trapping on the
British-controlled bases has diminished, there is some indication it
has increased elsewhere in the Republic of Cyprus, like a balloon
squeezed in one place and bulging in another. Worse, the penalties in
the Republic for lime-sticking have been watered down by regulatory
changes that went into effect in 2017; now, a trapper caught during
the peak fall trapping season with up to 72 limesticks—what's seen as
a traditional threshold marking a small-scale operation—gets at most

a slap-on-the-wrist fine of 200 euros. (There are additional penalties if the trapper has birds in his or her possession, or has more than 72 sticks, but even in aggregate they don't pose much of a deterrence.)

"The situation in the Republic is incredibly depressing," Shorrock said. "They did have what seemed to be quite an efficient antipoaching squad that worked very well with CABS, catching lots of people, but there just seems to be no political will to drive over the problem."

Toward the end of my stay in Cyprus, I had a chance to meet with Pantelis Hadjiyerou, the director of the Republic's Game and Fauna Service. His office in Nicosia featured a lion skin on one wall, which he was at pains to stress he did not shoot, and the pelt of a European brown bear on the other, which he was pleased to say he had. I was surprised to see a stuffed wood duck behind his desk; turns out he did his graduate work in New Jersey, and spent weekends hunting in my old home state of Pennsylvania.

The biggest threat to Cyprus's birdlife, he told me, wasn't trapping but development—"for birds, for everything. And unfortunately, there is nothing you can do." He rejected any notion that the Game and Fauna Service had backed off on its antipoaching efforts. Just the previous day, Hadjiyerou said, his officers had caught a man with two mist nets and a calling device; should the offender be convicted, the fine would total 9,600 euros ($10,750), and if it was not paid in 30 days, he said, it would jump by half again as much. As for reductions in penalties, those for mist-netting had increased and that, he insisted—not limesticks, not "small scale, traditional trappers for their own consumption"—was the real threat to birds.

It was a civil but not especially cordial conversation, and after 45 minutes, I thanked him for his time and asked a final question: Did he ever see bird-trapping disappearing from Cyprus?

"Perhaps if there was a way to make it regulated," he replied, a little more animated than he had been. "If people only caught a quota of blackcaps to consume—then the illegal trapping would stop." Now, he said, 80 percent of the trapping had ended, but prices for *ambelopoulia* had gone up, and so there was actually a greater incentive for those trappers who remained to continue poaching. The blackcap is an increasing species, he said—and it's true, the International Union for the Conservation of Nature puts the global blackcap population at a minimum of 100 million and increasing.

"So there is no ecological reason not to take some," he continued. "The other birds [caught by trappers] are bycatch, so it would have to be some method that is selective, not netting or limesticks—maybe hunting with BB guns? But people will always want to consume *ambelopoulia*. If it's illegal there will always be some poaching, so then, have it be with guns. It's not something that's ever going to stop."

Maybe Ulysses the Blackcap can change that. The next morning I was in Athalassa Park in Nicosia, a lovely weekend day with bright skies and a light breeze, families crowding the walkways and pushing prams or riding bikes along the trails. BirdLife Cyprus had taken over a bunch of picnic tables in the shade of the woods for one of its periodic bird-appreciation outreach days; Tassos was helping to hang colorful banners, and other staff and volunteers, wearing snazzy matching cobalt-blue shirts with the organization's logo, an endemic Cyprus wheatear, were handing out free cold beverages, bird-themed coloring books, and informational fliers. Kids were coloring posters of larger-than-life flamingos or painting flowerpots to take home as bird houses, and adults were wearing enameled lapel pins of Ulysses the Blackcap, the organization's antipoaching cartoon mascot, who has been featured in a series of popular animated videos about bird migration and the dangers of poaching. Martin Hellicar and several of his young staff were manning spotting scopes along the lake, where 30 or 40 adults and kids were swarming the bird hide that perched out over the water. The lake was looking a bit sad in the dry autumn weather, with wide, muddy margins where the water level had dropped, exposing dead trees. In some respects the outreach had been a little too effective, since there were so many people around the lakeshore that the shier birds had moved away—but enough Eurasian coots, common moorhens, cormorants, and egrets remained that even those who needed a lot of help with their loaner binoculars could get a good look. A female marsh harrier, peat-brown but for her buff cap and shoulders, drifted over the edge of the trees, rocking back and forth on slender, uptilted wings.

If Cyprus is ever to shed its reputation as the black hole of poaching in the Mediterranean, it's as likely to come from changing demographics as from heightened law enforcement. "I don't think trapping will ever disappear completely," Hellicar told me as we watched kids stepping onto low stools to peer through spotting scopes at the egrets. "That would be too ambitious, probably unrealistic. Our hope would be that it's reduced to such a level that the impact isn't significant. And in the Republic, we're very encouraged—very encouraged—by what we see in the younger generation and the response we get to our awareness-raising efforts." Despite the black-market money, the 15 million tax-free euros that bird-trapping brings in, the allure of *ambelopoulia* may already be fading in some quarters, as Cypriots become more urban and less tied to rural traditions. For more and more of them, the idea of paying a premium price for a plate of small birds seems absurd and old-fashioned.

But what happens when killing migratory birds isn't just a dwindling tradition in an increasingly urban society, but an essential source of income for rural people who otherwise have very, very little? What happens when the demands of conservation collide with the needs of the poor, with the fate of an entire species in the balance? For that, I was going to have to make one last trip, to one of the most remote places I'd ever visited, to see if a story about slaughter and redemption—involving one of the most amazing migratory spectacles no one had ever heard of—was, as it seemed it had to be, too good to be true.

Ten

ENINUM

For hours, we fishtailed and jolted along a rutted, muddy single-track road through the low mountains, nervously watching the sun slide lower and lower. The forested Naga Hills looked lovely in the late buttery light—gently crumpled mountains through which flowed jumbled streams and fast-flowing rivers. But appearances aside, we'd been repeatedly warned to be off the road before dark, given the risk of bandits and armed insurgents in this remote and troubled corner of northeast India, not far from the Myanmar (Burmese) border.

We had no idea how much farther ahead lay our destination, a village known as Pangti, or whether we'd actually reach it before nightfall. The cratered morass of a road we were on didn't even show up on any of the maps of northeast India we had, and weeks earlier, staring at Google Earth satellite images of the area, I could only trace its path sporadically and with difficulty as it played peekaboo beneath the canopy of trees. Worse, the skies around us were largely empty of birds—which was more than a typical birding-trip disappointment. My colleagues and I had come here, to the state of Nagaland, in search of what's reputed to be the single greatest gathering of birds of prey on the planet. I wanted to learn more about how the recent discovery of one of the world's most astonishing migratory spectacles had also revealed a shocking conservation tragedy—and how that, in very short order, had reportedly become a stunning conservation success. Given the state of migratory birds around the world, I needed a little good news.

By all accounts, the skies should have been alive with lithe, sickle-winged Amur falcons, pausing here on an epic migration from eastern Asia to southern Africa that is perhaps the longest in the world for any bird of prey. Instead, hour after hour we'd seen little in the air except a few swallows. "I don't know. There should be many, many falcons," said Abidur Rahman, a young ornithologist who had visited the region the previous autumn. His forehead was creased with worry. "This should be just a highway of falcons." We were by now skirting the Doyang Reservoir, a hydroelectric impoundment along which the birds normally roost by the tens of thousands. We saw four.

There had been no mistaking the moment we entered Nagaland that morning. We'd already been driving for hours through Assam, a culturally Indian state that occupies the immense Brahmaputra River valley, framed to the north by the foothills of the Himalayas, whose white peaks occasionally nudged the horizon. Our drivers wove through an endlessly flowing torrent of noise and color, people and vehicles and livestock—cars and mopeds, bikes and motorcycles; cows, goats, dogs, donkeys, and bullock carts; three-wheeled jitneys and traditional rickshaws pulled by wiry men. We passed through throngs of schoolgirls, afoot or on bikes, in brightly pigmented saris or uniform dresses, and schoolboys in crisp matching shirts and ties, color-coded by grade—a phalanx of teenaged girls in turquoise, another batch in lemon-yellow, then mobs of small boys sporting white shirts and maroon ties. Behind a sea of young girls in green plaid came a convoy of older boys on bicycles riding two abreast, their freshly pressed, cobalt blue shirts identical in the cool morning air.

The border between Assam and Nagaland is stark enough to show up from space, a sinuous line as precise as a topographic map's tracing the contours of the first rise in elevation between the flat Assam lowlands, almost entirely cleared for rice paddies, and the forested hills of Nagaland. Not that we needed to look at the terrain; the road was evidence enough. As we passed out of the Assamese town of Merapani, clearing a police checkpoint and crossing a small river, what had been a perfectly respectable paved road on one side of the small bridge became a tortured, potholed mess on the other. We skirted huge washouts where monsoon floods had almost entirely chewed away the roadbed, slabs of optimistic macadam perched precariously over the void with just barely enough room for our tires, and before long came to a massive landslide, which had scraped off hundreds of feet of mountainside, leaving an orange-brown scar of bare earth and taking the road with it. Undaunted, the locals had simply bulldozed another across the unstable slide zone, which was now a quagmire of churned mud in the middle of which was stuck a large lorry, listing badly to port. When it was at last yanked free by the same bulldozer, a flurry of lesser vehicles—ours included—slewed and slipped frantically through the newly opened gap in single file, like water from an uncorked jug, none of us willing to linger in the slide a fraction of a second longer than necessary, lest the whole thing let loose again. Peering down out the side window, I could see nothing but a sheer drop a couple hundred feet below, and tried not to think about how close to the gooey edge our tires must be.

The road was not the only thing that changed at the border. Gone were signs in Hindi and the ubiquitous Hindu religious imagery we'd seen everywhere in Assam; everything now was in English, the official language of Nagaland, one of many ways in which that state, which has for decades tried to break away on its own, is a defiantly un-Indian part of India. Similarly, the people we passed were starkly different in appearance and dress from those just a few kilometers back, on the Assamese side of the border. The Naga are a Tibeto-Burmese ethnicity; there were no saris, no Muslim men with long beards and white skullcaps. Many of the older Naga women wore long traditional skirts known as *mekhalas*, with white blouses and colorful shawls (often done up as head-wraps) whose patterns vary by village, tribe, and social station. Most arresting, though, were the men. Many that we passed—walking along the roadsides, riding three at a time on motorcycles, or clinging to the bumpers of trucks—were armed, with small-caliber rifles or double-barreled shotguns slung across their backs.

We finally pulled into Pangti, exhausted but relieved, just as the sun dropped below the horizon. The village of perhaps 500 households sat at the peak of a broad, defensible ridge—typical of the Naga, whose tribes were headhunters for generations, warring constantly with their neighbors. Today, in another un-Indian twist, they are overwhelmingly Baptists. Nzam Tsopoe, our host and the village's assistant schoolteacher, greeted each of us in turn, enfolding one of our hands in both of his and making a slight bow; he and his wife would be sharing their small three-room house with us for the next week. Mrs. Tsopoe brought the evening meal from their dirt-floored kitchen—delicious pork that had seasoned for weeks in the smoke above the open cooking hearth, pots of sticky rice and *dahl*, long beans, and steaming boiled squash. As we ate, we tried to ease the kinks from our long-suffering bones.

There was a more immediate issue, though. Mr. Tsopoe introduced us to two young men, whom he said would be our guides in the morning. Are the falcons here? we asked. How many?

"Um, one, two thousand," one of the young men replied. Surely we'd misunderstood him, but no. Far from the sky-darkening multitudes we'd expected, he said there were hardly any birds on the roosts at all. The monsoons, which usually end in September, had continued for week after rainy, flooding week through October, their southwesterly winds holding back the migrant falcons coming from the northeast. After two years of planning, and days of wearying travel, it seemed the journey might all have been in vain.

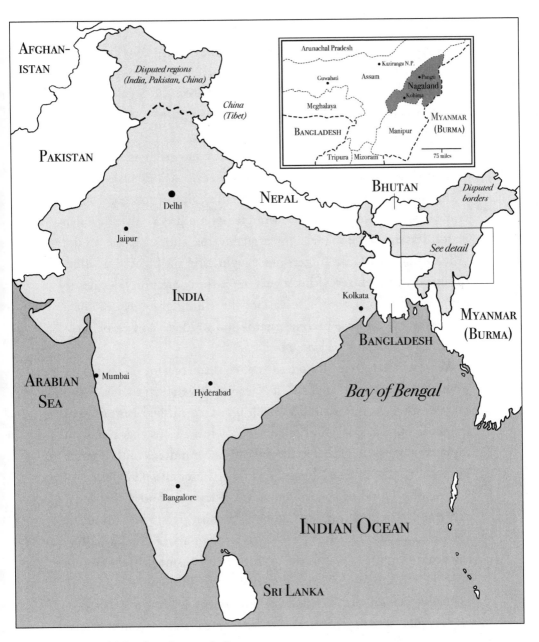

Nagaland and northeastern India.

I slept poorly—partly because the Naga don't use mattresses, and my wooden-plank bed had just a thin cotton blanket for a cushion, but mostly because the whole effort to come to Pangti now seemed likely to have been an enormous waste.

I'd been lured here with a few friends by stories about Amur falcons in Nagaland, coming out of this secluded and little-known region over the past few years, that seemed almost too good to be true anyway: Conservationists stumble upon a previously unknown concentration of raptors that is arguably the largest in the world, only to find that local hunters are slaughtering the birds at a wildly unsustainable rate. Yet within a year or so, the community decides to embrace protection and preservation; the killing grounds become a sanctuary, the trappers become guards and wardens, and residents of the village prepare to welcome birders.

As we would learn in the days to come, the bare bones of that story were basically correct. In 2012, a Naga conservationist named Bano Haralu, along with two Indian colleagues, confirmed rumors that immense numbers of Amur falcons had—for reasons that remain a mystery—begun to gather each night by the hundreds of thousands in densely packed roosts along the Doyang Reservoir, perhaps a million or more total in this single valley. They also found that local fishermen, stringing their nets among the roost trees, were killing an estimated 140,000 falcons in little more than a week and a half at the peak of migration—plucking the carcasses, smoking them over open fires to preserve them, then selling the birds in larger towns for badly needed cash. The disturbing videos Haralu and her colleagues shot, showing trappers ripping tangled falcons from the nets, and small boys bent beneath the weight of hundreds of dead and dying birds, went viral among outraged conservationists worldwide. Quickly, leading bird-protection groups within India and abroad, like the Bombay Natural History Society and BirdLife International, decried the killing, as online petitions battered the government for action, and viewers around the world reacted with horror to the images. "I witnessed a massive swarm of these little falcons flying into the South African town of Cradock to roost for the night," one commenter wrote on YouTube. "There were tens of thousands and I stood in awe. I cannot believe that they are slaughtered like this in India. There must be a special hell reserved for these bastards."

Of course, the reality was a bit more complicated—and less morally simplistic. In fact, Pangti and its neighboring villages, where most of the trappers lived, agreed to abandon the hunt in surprisingly short order. In barely more than a year, the villages made a hard transition with serious economic consequences, giving up the income that falcon meat represented—partly because it was the right thing to do; partly because the authorities made clear they would no longer turn a blind eye to what was already illegal killing; and partly because they'd been told by conservationists that tourism could make up the loss.

But as we'd found out, just getting to Pangti was not for the faint of heart, and not surprisingly, tourists have been thin on the land. Lying on my wooden bed in the damp and chilly night, I wondered in the dark: What happens when poor people make a wrenching decision, expecting an outcome that may take years to materialize, if ever?

Hot water, instant coffee, and tea were waiting for us at three in the morning when we climbed stiffly from bed. Along with Abidur and our drivers for the trek was my friend Kevin Loughlin, owner of Wildside Nature Tours, who was exploring the feasibility of bringing American tourists to Pangti to see the falcons. For that, Kevin needed willing guinea pigs—me; Catherine Hamilton, a California bird artist whose participation was underwritten by Zeiss Sports Optics; and birders Peter Trueblood of California and his cousin Bruce Evans of Maryland, who was delayed by a visa snafu but would be joining us the following day.

The ride to the main roost site down by the reservoir took 45 minutes, and given the hour and the mood, no one had much energy or inclination to talk. Once or twice we were startled by the explosive warning "bark" of the small forest deer known as muntjacs off in the blackness. We covered the last half-kilometer on foot, still walking in silence, passing beneath tall elephant grass and arching bamboo. It was cool, with a light breeze and no stars, but soon I could see the silhouette of a 40-foot wooden watchtower, newly built for visiting birders, that rose against the slightly lighter sky as we emerged along the edge of the lake. We climbed to a roofed platform just barely large enough for us, and waited.

Except for the chirping of frogs and the hushed voices of our guides below, there was no sound except for a dry rustle that I took to be the breeze in bamboo. But when Catherine raised her binoculars to peer through the murky twilight, she gasped.

"Oh, my God. Look. *Look!*"

Binoculars revealed what our eyes alone could not yet see—that the dimly lit air was filled with tens of thousands of falcons, rising in the gloom like a dense insect swarm from their roost a few hundred meters away and spreading out overhead. As the light grew, so did the number of birds, the whisper of their wings rising now to an omnipresent susurration, like fast-flowing water. No one spoke; this time not from disappointment, but awe.

"So . . . way more than a thousand," I said at last, when I was finally able to say anything. "Maybe—what? Fifty thousand? And that's only what's in the air."

"Maybe twice that," Catherine said in a hoarse whisper. Kevin was glued to the viewfinder of his camera, making the most of the growing light; Peter just stared, wide-eyed. For the next hour, the falcons would rise from the roost in great tides, enveloping us in a chaos of wings and movement, then settling back down again until the air was empty. Then something—on one occasion, a jungle crow dive-bombing the trees—would set them off once more, and they would erupt in fresh waves tens of thousands strong, layer upon layer of slender birds on long, narrow wings, swirling in counterclockwise gyres. The movement was hypnotic, disorienting, and I frequently found myself leaning, slowly and imperceptibly, in the same direction of the flow, caught in some sympathetic current.

I have seen some of the world's greatest gatherings of raptors, most notably the planet's largest migration of birds of prey, which passes each autumn through the narrow coastal plain of the eastern Mexican state of Veracruz. As I'd mentioned earlier, it's not unheard of for trained counters there to tally half a million passing overhead in a single day—but those birds, numerous as they are, often appear as little more than flyspecks in the sky, lofted high into the hazy tropical air by powerful thermals and traveling at the limit of binocular-aided vision. Nor do they linger—some, scientists believe, do not even pause to feed while they push south as rapidly as possible through Mexico and Central America. Only in Nagaland do such overwhelming numbers of any raptor gather for weeks in one area, creating the kind of all-encompassing spectacle we were witnessing.

Across the landscape, like smoke pushed by a light breeze, gauzy columns of thousands of falcons rose from other roosts and bent with the wind as they caught the morning's first thermals. This went on for hours, each new departing rush of birds seemingly certain to finally empty the roost—yet when we'd peer through our spotting scope, the trees would appear as heavily laden with perching falcons as before.

The Amur falcon is a slim, dove-sized little raptor, slightly bigger than an American kestrel. The males are gray—darker above, paler below, with elegantly contrasting white wing linings and a splash of bright rufous on the thighs and undertail coverts. The females and juveniles are very different, their white undersides barred with black and lightly washed with buff on the chest, the face distinctly "mustached" after the fashion of most falcons. All ages and sexes have bright carmine legs and feet, and the Amur was long lumped with the very similar red-footed falcon of western and central Eurasia. Amur falcons, however, breed in woody margins and the edges of savannahs from eastern China and North Korea to parts of Siberia and Mongolia—an area one-third the size of the contiguous United States, and from which they make one of the longest migrations of any raptor in the world, some 8,000 miles one way to southern Africa.

Some of the other migrants passing through eastern India take the most direct path, obstacles be damned. As we've seen, bar-headed geese fly at nosebleed altitudes of more than four and a half miles through the Himalayas en route to southern India, and ruddy shelducks taking the same path—the highest migratory route in the world—have been tracked at only slightly lower altitudes. The falcons avoid this frigid, oxygen-starved approach by swinging east and south, skirting the edge of the Tibetan Plateau through the lower hills of southeastern China, northern Vietnam, and Laos, then turning northwest through Myanmar and into this finger of India. But having avoided one record-setting migratory trial, they face another even more daunting, for after leaving India they make the greatest over-water crossing of any bird of prey, traversing as much as 2,400 miles across the Indian Ocean to Africa. The hot-air thermals and deflection currents that assist migrating raptors over land, allowing them to soar for hours and save energy, are largely absent over the ocean, meaning that the falcons must beat their wings continuously on their transoceanic trip, which may take them four or five days. If they're to survive, it's absolutely critical that they top off their tanks before they leave land.

And so in late October and early November, the migrant falcons pause for some weeks in Nagaland. At this same time of year, just after the monsoon, there is a great stirring underground as countless

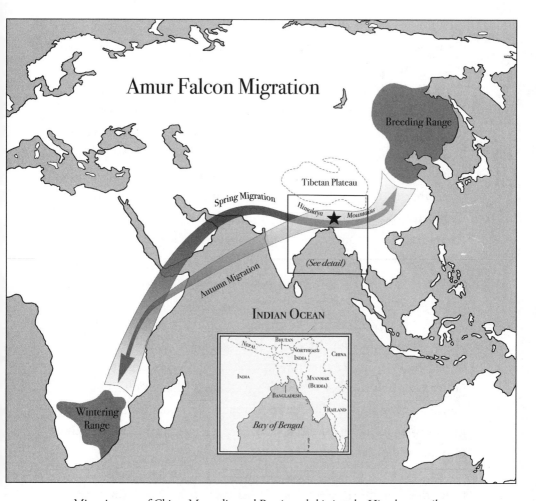

Migrating out of China, Mongolia, and Russia and skirting the Himalayas, millions of Amur falcons gather each autumn in northeastern India to feed before making the longest over-water migration of any raptor on their way to southern Africa.

subterranean termite colonies prepare for the mating season. These termites make no above-ground mounds, like those in other parts of Asia or Africa, but live out of sight most of the year. In autumn, though, worker termites chew tunnels to the surface, out of which emerge trillions of winged, inch-long fertile adults known as alates, which rise into the sky in vast mating flights, the air twinkling as sunlight flashes off their transparent wings. The fat-rich termites are the perfect food for an insectivorous falcon about to risk an ocean crossing, and the Amurs gorge.*

It seems Amur falcons have always stopped off in northeastern India during migration to feed on termites, but the completion of the Doyang Reservoir in 2000 dramatically altered the situation, both for the falcons and for local villagers. Although the Naga live in hilltop communities, their terraced fields, orchards, and rice paddies lay primarily in the valleys—in the case of Pangti, along the narrow floodplain of the Doyang River. Although the 6,500-acre impoundment brought welcomed electricity to the region, it also flooded many of those farms, including more than 2,000 acres of land cultivated by the people of Pangti. New fields on the mountain slopes were less productive, and wild elephants often trample the crops. (Some villages have even resorted to explosives to deter or kill elephants.) Several hundred men shifted to fishing, despite the many uncut trees below the lake surface that tore up their nets. But the fishermen also noticed something they'd never seen before—that the falcons were now gathering in incredible numbers each night in autumn in small groves of trees along the impoundment, then fanning out to hunt termites and other insects during the day.

No one knows why the falcons began to form such dense nighttime roosts along the reservoir—whether it was a function of some local microclimate, proximity to drinking water, or rich hunting areas. Unlike most raptors, Amur falcons are highly social in all but the breeding season. They travel in tremendous flocks, often with large numbers of lesser kestrels, and on their wintering grounds in southern Africa they gather by the hundreds or thousands each night

---

* Recent research suggests that millions of dragonflies also migrate each autumn from India to Africa, and thus the falcons (along with several other India-Africa migrants like Eurasian rollers, several bee-eaters, cuckoos, hobbies, and kestrels) may be able to feed on the swarms along the way. The dragonflies, in turn, may be preying on trillions of tinier insects carried along by the prevailing winds—an aerial food chain borne through the sky to the shores of Africa.

in communal roosts. The numbers gathering at Doyang, however, were unlike anything else in the world—or anything the Naga had seen before, despite their long familiarity with the species. And the Naga, with their torn nets and flooded fields (and being the good Baptists that most of them are), couldn't help but see this as manna from heaven, simply and literally. By 2003, fishermen were stringing their monofilament nets near the roost trees and across the lakeside gorges, and returning in the morning to retrieve hundreds of falcons.

"It was 2010 that I first came to this area with some birder friends, in the month of April. That's when I first heard about the massacre," Bano Haralu recalled as she poured us some wine (illicit, since Nagaland is officially dry). She was helping the state government produce a book, the *Birds of Nagaland*. "But people said you've come at the wrong season. They said there's this *enormous* harvest of birds, just sack-loads of birds. I said please, don't bluff so much—there can't be so many! But my birder friends said no, they must be Amur falcons."

It was after dark on our first day, and we were still having a hard time processing the spectacle we'd seen that morning. Bano had planned to meet us on our arrival, but had been delayed; her brother in Kolkata had just been diagnosed with cancer, and she'd only just come herself from her home in Dimapur, Nagaland's largest city many hours to the east. We met her in the small wooden house in the village that serves as the headquarters for the Nagaland Wildlife and Biodiversity Conservation Trust, a nonprofit that she founded. We walked in out of the chilly night to a warm room filled with savory but unfamiliar food smells coming from hot plates and bubbling crockpots in the little adjoining kitchen.

A Naga herself, Bano had activism more or less baked into her genome. Her father, Thepofoorya Haralu, received one of India's highest civilian honors for his service as a government official during the border war with China in 1962. Her mother, Lhusi, was a founder of the Nagaland branch of the Indian Red Cross, an ardent social worker, and a noted peace activist who directed the Nagaland Peace Centre until her death in 2015. Bano was educated in a convent

school ("It was full of Irish nuns," she told us), and then did graduate work in New Delhi before spending two decades as a respected
broadcast journalist and the editor of the English-language *Eastern
Mirror*. Now in her mid-fifties, she left television news in 2009 and
has focused heavily on conservation since then.

Those comments about a falcon massacre stuck with her, although
it was two more years, in October 2012, before she came back to
investigate with several colleagues—Ramki Sreenivasan, a Bangalore-
based nature photographer who had cofounded Conservation India
just a few months earlier; Shashank Dalvi, a birder and researcher
specializing in India's northeast; and Rokohebi Kuotsu, a young
Naga naturalist from a village about 50 miles south of Pangti. Their
first morning along the Doyang, they were thunderstruck to find
thousands of falcons jammed shoulder to shoulder on sagging electri-
cal wires near the reservoir. As they were filming the birds, they met
two Naga women carrying what Bano thought at first were dressed
chickens—but which, when she looked more closely, proved to be 60
or so plucked raptors. The women said they were taking the birds to
the complex at the base of the dam housing hydroelectric workers.
"And while we were there, watching the birds on the wires, they came
back with a second load," Bano said.

Now badly worried, the team drove up to Pangti. "We saw birds
in almost every home. It was overwhelming." Hundreds of plucked
falcons, skewered through the heads, hung smoking over fires; hundreds more were being kept alive in zippered mosquito nets that
functioned as holding cages until they, too, could be killed. Trapping
and selling falcons, Bano and her friends quickly discovered, was a
universal cottage industry in Pangti.

"You don't know where to start, what to do, what to say, how not
to offend people, not create a scene," Bano told us. They made calls
to the district commissioner, prompting an official order reinforcing formal protection for the birds; the forest department deployed
guards to enforce it, and a few arrests were made. Several European
scientists, tracking one of the first Amur falcons to be fitted with

satellite transmitters, were frantic; the latest satellite fixes showed
that the bird was in the Doyang area. "And we were also in a panic,
because we didn't want this bird to go down here—definitely not.
[We did] whatever we could do, pulling down the nets and all that.
But then the bird signaled that it had left the killing fields, and we
thought great, this is something we can share now with the people in
this village, that something wonderful has happened. This one bird
that has been saved. This can go out to the world that, because you
stopped hunting this season, this bird has got its freedom."

In the months that followed, conservationists met with village
leaders to describe the global migration of the falcons. In addition
to the local wildlife trust and Conservation India, support came
from a variety of organizations, including BirdLife International, the
Wildlife Conservation Society, the venerable Bombay Natural History Society, the Royal Society for the Protection of Birds, and several others. Together, they started eco-clubs for children in Pangti
and surrounding communities, gave "Amur Ambassador Passports"
to those who pledged to protect the birds, and created a local-pride
campaign of the sort that has been successful in protecting threatened wildlife in other regions—organizing falcon-celebration festivals, bringing in governmental dignitaries to issue "Falcon Capital
of the World" proclamations while choruses of schoolkids sang pro-
falcon songs they had written themselves, and handing out "Friends
of the Amur Falcon" buttons to villagers. Baptist ministers were per-
suaded to preach pro-falcon sermons and conduct special church services, taking for their text Leviticus 11:13–19: "And these are they
which ye shall have in abomination among the fowls; they shall not
be eaten. . . ." The former trappers and hunters formed AFRAU,
the Amur Falcon Roost Area Union, which posted guards, certified
guides, and worked with the landowners of the roosts to build view-
ing towers like the one we'd visited.

In 2013, an international group of scientists from India and Hun-
gary, the latter already working with the closely related red-footed falcon, cooperated with former hunters to catch three falcons near the

reservoir—an adult male they named Naga, and two adult females they dubbed Wohka and Pangti. Fitted with tiny satellite transmitters, the raptors were tracked to southern Africa, where Wohka's signal was lost, but the next spring Pangti and Naga were followed back to different regions of northern China, along the margins of Inner Mongolia, where they settled down to nest. The tagged falcons made their extraordinary round-trip journey three more times before their signals were also lost—not only a dramatic insight into one of the world's greatest migrations, but another boost to the sense of local pride and ownership as Nagaland newspapers breathlessly reported on the birds' latest whereabouts.

None of this obscured the fact that Pangti and neighboring villages like Ashaa and Sungro had taken a financial hit when they suspended the hunt. Villagers had been able to sell four of the falcons for 100 rupees, Bano said, the equivalent of a little more than a $1.50 in US funds and about half a day's wages in the region. Assuming that about 140,000 falcons were being killed annually (very much a lowball estimate, since it only encompassed the peak 10 days of the migration), then the end of falcon-trapping meant forgoing about 3.5 million rupees annually—roughly $56,000, an enormous sum in such a remote, cash-strapped area, especially because many people used that money to pay their children's school fees. "That's a huge loss of money," Bano admitted. Some, however, saw the potential for tourism. Several families in Pangti invested in improvements so their houses could serve as home-stays for visitors. The Tsopoes, with whom we were staying, built a two-stall Western-style bathroom in their side yard, with flush toilets (which needed a little help from handy buckets of water, but were a big step up from the squat pits typical in the region) and a dirt-floored washroom with a sink. The Wildlife Trust of India had just built a small guesthouse for tourists in Pangti, but it wasn't yet furnished or ready by the time of our visit.

One of those to whom the hunting ban was a hard change was Nchumo Odyuo, a slender, soft-spoken man who is a neighbor of the Tsopoes, a former trapper now active in the protective union, who became our primary guide during our time in Pangti. Losing the money from selling falcons was difficult, he told me one morning as we watched hundreds of Amurs flying in to perch around the edges of a small teak plantation, some miles from the main roost. Buoyant, they spun and pivoted in the air, wheeling against a clear blue sky, preening in the sun and occasionally dropping to the ground to snag large mantises or grasshoppers.

Nchumo and his wife have several children at home, and two older kids at boarding school in the city, the only choice for more than a grade-school education in a rural village like Pangti. The village had lost much of its best farmland when the reservoir was constructed, so when the falcons began to gather, trapping seemed like a godsend. Some of the men and boys had already been shooting falcons with slingshots and guns, but nets proved to be a vastly more efficient way of killing the birds. The trappers would stay in small fishermen's shanties that dot the edge of the reservoir, and in late afternoon Nchumo and the others would shinny up slender trees near the roosts, tossing rocks tied to ropes over the higher limbs, and hoist their fishing nets into the branches. Other nets were strung across narrow gorges leading down to the water, along which the falcons flew. The next morning, each 50-foot net would be sagging under the weight of 150 or more wriggling, writhing falcons. Tens of thousands a week would be taken back to the villages, plucked and smoked, and the majority sent on to larger towns like Wohka for sale. It was technically illegal, but so is most of the hunting in Nagaland—and that stops almost no one; in our time there, I rarely saw a man in the forest without a gun, or a boy without a slingshot.

But the falcons were different. All the attention, the spotlight of publicity, the sudden increase in formerly lax law enforcement—little wonder the killing stopped so abruptly. "At first [the residents] were angry, because the government has not compensated us," Nchumo said. "But slowly, we have understood."

"Do you miss trapping falcons?" I asked.

"We call them *eninum*, 'two-love,' because they like to sit—" he pantomimed with his hands, tightly together. "Like them," he said, pointing to two juvenile Amurs perched side-by-side. "I am glad the falcons are protected, but . . ." He trailed off into silence, and I raised an eyebrow. Nchumo laughed nervously. "I do wish I could eat one! They taste *very* good."

While the falcon trapping benefited almost everyone in the local communities, the new tourism-based paradigm helps a narrower segment, Deven Mehta told me the next day, as we watched the morning flight at the lakeside tower. Deven was a junior research fellow at the Wildlife Institute of India who is studying the diet of the Amur falcons; he showed me dozens of alcohol-filled vials in which he'd collected the regurgitated pellets from roosting falcons. Later, he would sift through the chitinous bits that make up the crumbly pellets, examining them under a dissecting microscope, hoping to discover more about what the falcons eat besides termites—though he admitted he still had a lot to learn about identifying insect body parts.

Where once many people made some money from the falcons, Deven said, the primary beneficiaries now are guides like Nchumo; landowners like Nchumo's uncle Zanimo, who owns the property where the watchtowers sit, and had come out to greet us that morning with fresh bananas; and families like the Tsopoes who had enough extra cash to invest in creating home-stay operations. Pangti may be the "Falcon Capital," as now-weathered signs proclaimed, but a lack of community-wide equity is a major threat to the future of their protection, Deven believes. And even with incentives, there is no guarantee that people will make the best long-term decisions. Since the previous falcon season, Deven said, Zanimo had cut down trees on the edge of the roost grove to plant teak seedlings—a common agroforestry practice, but one that obviously could threaten the whole local operation were the falcons to abandon the Pangti roost for a less disturbed site.

Although Pangti is far from the tourist track, we were pleasantly surprised to find we weren't the only visitors. During our stay a few small groups of Indians from outside the region—in twos and threes, and most (to judge from their lack of equipment) not birders—showed up at the watchtowers. An Indian documentary filmmaker and his friends spent several days, as did a large, enthusiastic bunch from BNGBirds, a huge online birding group from Bangalore in southern India. I struck up a conversation with one of them, Ulhas Anand, and discovered we had several mutual birding friends from his time living in Philadelphia. "The birding in Bangalore is incredible—we have some very bird-rich areas. But *nothing* like this," he said, gesturing to the multitudes of falcons emerging from their roosts. Then he was gone—someone had spied a rare Philippine shrike along the edge of the lake and hollered for him.

In most wild parts of the world, conservationists bemoan roads, but Bano Haralu and others see the awful condition of Nagaland's roads—often cited as the worst in India—as a major hurdle to conservation and the tourism that could support it. "You see the state of our roads," she said, "the worst roads on the entire planet. You can see the roads in Nagaland from the moon, I think, they are so bad." The peak of the falcon migration neatly coincides with the seasonal, postmonsoon opening of Kaziranga National Park in Assam state to the north, a UNESCO World Heritage site that attracts visitors from around the globe. And for good reason: a week later, looking out on Kaziranga's grassy floodplain, I counted 59 endangered Indian one-horned rhinos in a single, wide sweep of my binoculars, as wild elephants, buffalo, wild hogs, and swamp deer grazed, and Pallas's fish-eagles soared overhead. Combining the two sites would be an ecotourism no-brainer—if the travel time between them were a couple of easy hours on well-paved roads, instead of the bone-grinding, eight- or nine-hour marathon that travelers now face.

For those willing to make the journey, though, the rewards go beyond one of the world's great raptor spectacles. As a battleground in Asia's longest-running guerilla war, Nagaland was officially closed to outsiders (including other Indians) for decades. It is less restricted now, but with its rugged terrain and awful roads it still very much carries a sense of being the back side of beyond. Tourism officials have pushed tribal culture as a marketing point, with a number of big festivals throughout the year in major towns, featuring traditional dress and dance, and culminating in the annual, 10-day Hornbill Festival every December in Kisama. Foodies have also discovered Nagaland and its cuisine, which varies among the Naga tribes like the Lotha who inhabit Pangti. The *bhut jolokia*, or ghost pepper, with its off-the-chart Scoville heat-unit rating, is a regional specialty, although Bano dismissed the notion that Naga cooking is routinely fiery. Certainly, what the Tsopoes prepared was only modestly spicy and uniformly delicious.

Although Nagaland is no longer sealed off from the rest of India, visitors must still register with the police, and unfamiliar faces of any sort, much less a bunch of Americans, are still a profound novelty—as we discovered again and again. After crossing from Assam on our first day, we stopped (as required by law) at the police station in the first sizable community along the way. It was a three-story building on a high hilltop; off-duty officers clustered at the windows of the barracks on the upper floors, peering down with obvious bewilderment while Abidur disappeared inside with our passports. Nor did he reappear for a very long while, as we nibbled snacks and scanned the high, cloudy sky for passing raptors. Eventually we heard a motorbike approaching at speed, bouncing up the rutted road. A compact, muscular man with cropped hair and wearing a tee-shirt parked it and sprinted into the station; the commander had been called in on his day off to deal with this new and perplexing situation, foreigners coming through his district. Another long passage of time, and eventually Abidur returned with our passports, trailing the commander—who wanted selfies with us to show his family.

Still, navigating the tense political landscape was not always so easy. A few days into our visit, having again risen at three o'clock, we found a heavy, overcast sky and the world wrapped in mist. Catherine had come down with a cold, and opted to stay back at the village— a wise decision, since the morning flight was a bust as the mist became drizzle, then light rain. Worse, we discovered that what we assumed

were four-wheel-drive vehicles supplied by our Indian contacts were
in fact rear-wheel-drive look-alikes, and not up to the long, steep,
and now slickly muddy road leading up from the reservoir. Getting
out was a process of pushing them forward a few feet as the tires
sprayed us with mud, jamming large rocks behind the wheels to keep
the vehicle from sliding back, and repeating this exhausting process
again and again and again for more than an hour, slowly clawing our
way up the hill.

Bedraggled and tired, we finally reached the AFRAU welcome station at the top, along the main road to Pangti, only to find two dozen
camouflage-clad Indian soldiers and an enormous Taka truck idling
nearby, a .50-caliber machine gun mounted in its bed. The soldiers'
surprise at seeing us soon morphed into another selfie free-for-all, as
the grinning officers handed their phones to our drivers for group
shots. But the looks we were getting from the AFRAU guides, grimly
watching us, were at odds with their cheerful, matching red tee-shirts;
the Assam Rifles, a paramilitary police unit, have been accused of
countless human rights abuses in Nagaland over the years, including
massacres and torture, backed up by a 1958 law giving them carte
blanche to arrest, detain, or shoot almost anyone under almost any
circumstance in such so-called protected areas. The AFRAU guides
were openly glowering at the soldiers—and us.

"Maybe we shouldn't act so chummy, guys," I said quietly out
of the corner of my mouth, as the seemingly endless photo session
continued.

"Is there a problem?" Peter asked.

"Think, British Army in northern Ireland," I said under my breath,
as his grin crumpled. There is plenty of tension on both sides. Despite
a decades-long peace process, the Rifles continue to battle Maoist
insurgents that filter across the porous Myanmar border. Just a week
or two before our arrival, the Indian military claimed to have killed
40 militants who had crossed the border less than 50 kilometers from
the area we were visiting.

We finally extricated ourselves, anxious for a chance to clean up and eat a very late breakfast. But when we got back to the Tsopoes at midday, we found waiting on the porch of the house a very slender man in his early thirties, hair neatly combed and dressed in a sharply creased gray suit and black shoes that were immaculate, despite the sea of orange mud that now surrounded the house. A young woman in a white dress and scarlet lipstick, her hair perfectly coiffed and looking shyly down, sat nearby beside Catherine, who had a decidedly nervous look on her face.

It was an almost absurd scene—the rain was by now pouring, and we stood in the deluge, momentarily dumbfounded. The man leapt to his feet, extended his hand toward me and asked, "Hello, who are you?" Weary, short on patience and a little taken aback, I shook it once, pushed past him out of the rain and said, tersely, "Scott, and who are *you*?" As Catherine's look shaded to panic, the man informed us with a thin smile that he was with a special intelligence unit in Kohima—police or military, we never learned—sent to investigate a report of highly suspect foreigners who had failed to follow protocol. While we had registered with the police on our arrival we had not, it seems, registered with the *correct* police; instead of stopping in the first town after entering Nagaland, we should have gone many hours out of our way to the district capital of Wohka. Clearly not believing Catherine's explanations for why we were there—Birds? Really?—he had already spent much of the morning interrogating her, a weird mix of overly friendly joshing and direct threats: Why are you truly here? Perhaps I should arrest you for failing to register. Ha ha, not really, that's just a joke. Except no, perhaps I really will take you to Kohima and lock you up now. Ha ha.

There have been cases of westerners coming to Nagaland to join the insurgents, so his suspicion may not have been entirely unwarranted. He asked for our passports—no, no, not the real thing, he said as we pulled out our documents. Photocopies. Only photocopies, please. We all had multiple paper copies (a sensible precaution against theft or loss)—all save for Peter, whose backup was an electronic scan of his passport on his laptop. There was no copier in the village, nor a way to print the scan, and Peter understandably balked when the officer suggested he accompany them all the way to Kohima, at least a day's travel in each direction. By this point, any sense of jokiness in our visitor had disappeared, and an impasse was setting in. For some reason, he showed no interest in Peter's actual passport, focusing single-mindedly on having a copy. I was emptying a flash drive so he could take the electronic file when Peter, opening his own laptop, showed the officer the scan. For reasons known only to him and the Gods of Bureaucracy, this suddenly satisfied the man. Oily smiles reappeared, more handshakes, and then he and his wife—who had uttered not a word nor made eye contact through the whole visit—abruptly popped open a large, black umbrella and departed into the rain.

"That was freaking *weird*," said Kevin, watching them drive away.

"Weird and scary," Catherine said. "For a while there, I really thought you'd come back to find that I'd been arrested. You have no idea how happy I was to see you guys."

Given such difficulties, it's not surprising how rare outsiders are in the Naga hinterlands. Though not cut off from the world—there are plenty of cell phones, and some of the homes in Pangti sprout satellite TV dishes—villagers rarely see anyone from outside the region. One afternoon, in breezy shirt-sleeve weather after the damp chill of daybreak, Mr. Tsopoe led us on a walk through Pangti, eliciting everything from shy glances to spontaneous embraces from the Lotha we met. The streets and alleys twisted and turned unpredictably, doubling back on themselves between the close-hemmed houses, high drystone walls, and masses of flowers—orchids, beds of marigolds, begonias, coreopsis, and many others that made turning each corner a fresh palette of color. Blankets and quilts lay airing on the metal roofs, while herbs, beans, peppers, and ground rice dried in the sun. From the walls of many houses hung the skulls of muntjacs, good luck totems for hunting, and Mr. Tsopoe pointed out the heavy, short-horned skull of a mithun, one of the huge, semi-wild cattle that still roam the Naga Hills. "That was feast, New Year feast for village last year!" he told us proudly.

The homes ran the gamut from a gleaming four-story edifice being constructed by a well-heeled government official to traditional Naga houses with woven bamboo-mat walls and dirt floors. We were invited into one of the latter by the housewife and her mother—who, like everyone we met, made a particular fuss over Catherine, who was told again and again that she was the first Western woman anyone had ever seen in Pangti. A kitten followed us inside, meowing by the smoldering fire, above which were bamboo lattice shelves holding meat and vegetables, curing in the ever-present smoke; spotless pots and pans lined one wall, while on the other side of the room was a sleeping platform, also woven of bamboo, and the family's possessions. Outside, laundry fluttered in the wind above a bamboo deck that bowed under our weight, as we drank in a stunning view over the deep, convoluted valleys and peaks surrounding Pangti.

The enormous steepled Baptist church, white and green, occupied the top of the hill, while the Catholic and Assembly of God churches were considerably more modest. The Naga were originally an animist culture, and pushed back ferociously against British control starting in the 1830s. The next half-century was especially violent; one scholar has called the British-Naga conflict among the bloodiest of Britain's brutal history in India. Pangti did not escape; after a survey party was attacked in 1875 and its commander killed, the British military burned the village to the ground in reprisal. But evangelizing by American Baptist missionaries that started in the mid-nineteenth century made slow progress and accelerated rapidly in the twentieth. Ironically, the decision by the newly independent Indian government to expel foreign missionaries caused the conversion rate to explode— in direct reaction, some scholars have argued, by the Naga to the widely despised government's attacks on churches and ministers. The wholesale conversion of the Naga was, in one expert's view, "the most massive movement to Christianity in all of Asia, second only to that of the Philippines." Such is the underlying weirdness of Nagaland that today, even the Maoist militants coming across from Myanmar are also mostly Baptists.

Earlier in the day, a group of women had gathered at the Tsopoes' house to do a "sing," wearing matching scarlet shawls knotted around their waists or heads, and navy blue *mekhalas* into which silver and crimson bands had been woven. Rice is a staple here, as in much of Asia, and must be pounded to remove the hulls and break up the grains into flour. Six women lined up facing each other, three to a side, across a low, six-foot-long wooden mortar made from a carved log; along the flat upper surface were three deep cavities, each the size of a coffee can. Into these holes one of the other women poured rice, then the six lifted heavy wooden shafts taller than themselves, and began singing a rhythmic song in Lotha as they slammed the butts of these pestles into the mortar holes, alternating sides with unflinching aim and perfect timing. At the end of each song, they paused as pounded rice was scooped out into a large, open-ended triangular basket, which one of the women would hold against her waist and begin shaking rhythmically with her back to the breeze, winnowing the hulls and dust onto a bamboo mat as she, too, kept time with a song, and chickens skittered around her feet for the lost grains.

Everyone was a little stiff and formal at the start of the "sing"—the women were dressed up a bit more than would normally be the case, this being largely for our benefit—but as the work progressed they loosened up, cracking jokes between songs, sipping water from throwaway cups made from strips of expertly folded banana leaves. One woman was a bit of a showboat, and once the rice had been ground she called for a battered guitar, which she strummed (tunelessly but enthusiastically) while leading the group in Lotha-language hymns.

Still, signs of the Naga's more traditional past were literally in many backyards: tombstones. As we walked through the village, we paused to read one for Chonchio Lotha, who died July 13, 1947. Into the slab were engraved commemorations of a life of dangerous hunting, the simple shapes of five tigers, two leopards, one elephant—as well as six human heads, commemorating the deceased's skill at war. Although headhunting was banned by the British in the 1940s the practice continued, at least rarely, through the 1960s and '70s before completely dying out. But the reputation remains, and even today, you're liable to get a wide-eyed stare of disbelief from an Indian elsewhere in the country when they learn you're going to the land of "those backwards headhunters."

At length we came to the home of Mr. Tsopoe's parents—his mother, who at 98 is now blind, and his frail but alert 102-year-old father, attended by a younger relative. Both were sitting in plastic chairs, enjoying the sun by the woven-bamboo walls of their home, wearing shawls in the village's crimson, black-striped pattern over their shoulders. "My father, he was a *great* hunter," Mr. Tsopoe said several times for emphasis, showing us the skulls and antlers of a number of deer his father had killed in years past. During the Second World War, when the Japanese invaded India, the elder Tsopoe carried food for British and Indian troops, who fought a bloody siege in Kohima, about 40 miles to the southwest. (Some Naga took a much more direct role in the battle, and I couldn't help but wonder if some of the carved heads we'd seen on that tombstone earlier had been Japanese—or perhaps even British, since some Naga sided with Imperial Japan, seeing it as a path to independence.) From a corner of the house, Mr. Tsopoe emerged with his father's handmade spear, eight feet long with a polished (and still wickedly sharp) leaf-shaped point. Such spears were used by the Naga in war, along with the square-tipped, machete-like swords known as *daos*, as well as for hunting. The old man, who had been smiling through the introductions, suddenly became very serious, gripping the spear with both hands as he showed us how he used it—and as his son pointed out the white scars on his father's right hand from fighting off a tiger that had jumped him in the forest one day many years ago.

Such an encounter would be unlikely today. Tigers are essentially gone from Nagaland, although every so often a lone animal, dispersing from more secure populations in Assam to the north or Myanmar to the southeast, wanders in. That happened in 2016, when a tiger killed a cow and two pigs in the hills near the Assamese border, only to be hunted down and shot by local residents. The evening that we met with Bano at the conservation trust headquarters, we were joined by a group of young biologists from the Indian branch of the Wildlife Conservation Society, who were conducting mammal surveys in Nagaland. The results thus far had been fairly grim, they said; hunting had greatly reduced much of the game, and while muntjacs and a few elephants remained, residents around Pangti told the crew they hadn't seen signs of tigers for more than 15 years. But times are changing, Bano told us; a seasonal six-month closure of the hunting season had been put in place, as well as a ban on air guns—though we saw many children with rubber slingshots, and could not help but notice the scarcity of songbirds, especially close to the villages.

Even the lushly forested nature of Nagaland itself is somewhat deceptive. True, compared with much of India Nagaland is overwhelmingly wooded, with little large-scale agricultural clearing, like the endless rice fields we encountered in Assam. But we soon realized that essentially all the forest we saw was young, scrubby, early-successional stuff—there was a complete absence of old, mature stands, and there are few even theoretically protected reserves in the state, compared with immense parks and preserves in nearby states like Assam and Arunachal Pradesh. Instead, the forests showed the effects of generations of small-scale, constantly shifting slash-and-burn agriculture, an approach known locally as *jhum*, as well as countless teak plantations that look like forest from a distance, but are wooded monocultures, harvested when the teak trees are only a few decades old. While Bano believes *jhum* is a sustainable approach when practiced in a traditional manner, with long periods between clearing and cultivation, those fallow intervals keep shrinking, and between *jhum* and the constant turnover of teak plantings, there seemed to be almost no mature, naturally functioning forest left. Add to that the Naga mania for shooting almost everything that moves, and it makes the abundance of the falcons and the local success at protecting them all the more astonishing.

Although the greatest spectacle in Pangti was the morning liftoff, on our last evening we returned to the roost area at dusk, hoping to see the falcons come in for the night. The sky was a pale, hazy blue, the horizons rimmed with puffy cumulus clouds, as we hiked down from the hills and onto what remained of the unflooded valley. Stonechats—plump songbirds down from the Himalayas for the winter, the males in mottled brown nonbreeding plumage—perched high in the elephant grass. Other birds flitted about in the failing light—white-browed scimitar-babblers, all lanky and brown like thrashers; pairs of red-vented bulbuls, charcoal-black with a splash of crimson under their tails; and flocks of hyperactive yellow-bellied fairy-fantails, round-headed and long-tailed, with black masks across their lemon-colored faces. We passed trails pushed through the dense vegetation, the paths of wild elephants, perhaps the same ones we'd heard trumpeting across the lake that morning.

We climbed up to a different observation tower this time, one set farther back from the lake but with a wide, open view over the elephant grass and shrubs that covered most of the plain. Off behind us were small, thatched fishermen's huts, where the falcon trappers once stayed during the migration, but where now AFRAU guards camp. About 45 minutes after sunset, falcons began streaming in—first hundreds of birds a minute, then thousands, a sheet of movement against the shrinking band of orange and purple light on the western horizon. We were near the convergence of a great inrushing of wings and movement, coming from all points on the compass, like a black hole drawing everything toward itself. The falcons flew with smooth, languid wingbeats, mostly gliding toward the roost trees, the morning's shivering noise of tens of thousands of rising wings replaced now with an almost eerie silence. Soon, the silhouettes of the distant ridges were lost in the darkness, replaced by the lights of faraway villages spangling the crests of mountains, and still the numbers of falcons continued to rise.

For now, the falcons are safe—not only in Pangti, but across Nagaland, where conservationists were unaware of even a single bird being trapped during the past several migration seasons. As interest in the Amur falcon spreads throughout northeast India, reports have emerged of other major roost sites in neighboring states like Assam and Manipur—and local efforts have emerged in those places as well, to end hunting and to protect and celebrate the raptors. In Nagaland, other villages have jostled with Pangti for the title of "Falcon Capital," claiming concentrations as impressive. Tourism is building slowly, as we saw, but Kevin planned to be back the next autumn with a full group of Americans (and with mattresses for their beds) to buttress the nascent industry.

The intensity of the flight only increased with darkness, as a nearly full moon hung overhead. The white disk flickered and trembled with streams of black silhouettes as falcons beyond count came home to roost, their bellies full and their instincts already pulling them toward the next stop on a global journey—one that at least here, at least now, at least for this one precious species—is not as dangerous as it was just a few years before.

EPILOGUE

One o'clock in the morning is early by anyone's standards, but during the summer in central Alaska, at least it doesn't look like the middle of the night. The sky at that hour is a dim, gentle gray, light enough to easily read by, as dark as it will get here at this time of year. That helped, as my friends and I stumbled out of our rooms in the Toklat dorm, halfway back the 90-mile gravel road that bisects Denali National Park—rubbing sleep from our eyes, pouring coffee, fixing breakfast, making up sandwiches for the day's work. We were eight days into a two-week field season, and we had a long drive to reach that day's site.

It had been five years since we launched this study of Denali's migratory birds—five years since that grizzly charge that terrified my colleague Iain Stenhouse and the rest of us. Over the years, our crew had encountered a lot of Denali's wildlife, from moose and caribou to many more bears, but more importantly, we'd begun to gain a far better understanding of how the migratory connections that radiate out from this part of Alaska link the park, through the travels of its birds, with the rest of the world—fox sparrows migrating to Georgia, blackpoll warblers to the Amazon, Swainson's thrushes to Bolivia, Wilson's warblers to Central America.

This summer felt strange, though. We'd been snakebit at every turn. For more than a week we'd been trying to recapture Arctic warblers, slender, brownish-green songbirds with pale eyebrows and a staccato, metallic trill that chatters like a machine gun, *seet-seet-seet-*

seet-seet-seet-seet-seet! It is a species with Old World roots that breeds only in central and western Alaska and winters somewhere in southeast Asia, presumably in Borneo or the Philippines; no one had ever tracked them to their nonbreeding grounds, or recovered a banded Alaskan bird there. The very first morning, right after we started netting, we caught one of the 15 birds we'd tagged with geolocators the previous summer—but then for the next seven days, as we marched methodically across an entire valley, setting up nets and using our audiolures every 100 meters for miles, blanketing the willow habitat with a precise gridwork of trapping locations, we caught bupkus. That is to say, we caught dozens and dozens of Arctic warblers, just not ones that were tagged. The warblers were scattered, moving all over the place, not showing the kind of site fidelity we normally expect from songbirds.

Maybe it was the weather. Alaska was broiling under the most extreme heat wave in its history; Anchorage had hit 90 degrees for the first time ever, and kept on hitting it, day after sweltering day. The heat capped a year—several years—of bizarre twists and turns in Alaska's climate. The previous winter, the Bering and Chukchi Seas were as much as 20°F warmer than normal, melting the sea ice months early and leading one climate expert to describe the oceans around the state as "baking." Whether or not he meant to make a bad pun, it was fitting. Even up in the Alaska Range, working as we were at elevations of 4,000 feet, the heat was oppressive, made worse by dense smoke from wildfires burning all over the state—including one outside of Fairbanks that was threatening the home of our project leader, Carol McIntyre. For two weeks she and her husband had been under a level-two evacuation notice, meaning they had to be ready to flee their house at a moment's notice. Instead of working with us in the park, she and Ray had been moving their belongings out of their home, finding temporary shelter for their sled dogs, clearing fire lanes and removing brush, and watching anxiously as fire crews conducted back-burn operations and aerial water-bombing to keep the 10,000-acre blaze away from their neighborhood.

Because of the heat, we hadn't been seeing the numbers of wildlife we'd come to expect. The snowshoe hare cycle was on the rise, with dozens of hares per mile along the road at daybreak, nibbling on the calcium chloride that the park service uses to try to hold down the dust from passing tourist buses—and because the hare numbers were climbing, so too were those of their main predator, the lynx, several of which we encountered. Otherwise, though, we'd seen little; very few bears, an occasional flock of Dall sheep high on the mountainsides. The heat had driven a small herd of bull caribou into the camp at Toklat, where among the campus of maintenance sheds, cabins, and other housing for NPS staff they found some relief from the heat. We always looked both ways when we stepped off the dorm porch, because the caribou liked to lounge in the shade of the buildings, and more than once we'd startled one or two of them into hoof-clattering flight. In that part of the world, the sudden appearance of a large, brown, fast-moving animal really gets your blood pumping.

The first day we were in the field—a cool, rainy morning just before the heat wave started—we'd come across an immense bull moose browsing on willows, down a steep bank some yards below us. He was right where we'd been heading, so we were fortunate to spot him first—moose injure more people every year in Alaska than do bears. Instead, we had a good angle at a safe distance from which to watch one of the biggest moose I've seen in decades of working in Alaska, his velvet-covered, only partially grown antlers already more than six feet across. Nothing but the great animal's hump, neck, and head rose above the gray-green shrubs, which thrashed and tossed as he ripped away mouthfuls of willow leaves, flashing their white, fuzzy undersides; the effect was very much as though the moose were breasting rushing waves, green ocean combers that broke and fell across his back. Then he took a few more steps and was gone, swallowed by the thickets as completely as if he'd submerged. "How can something that big just, just *disappear*?" my friend George Gress whispered.

It was a reminder of how little we could actually see when we're in the brush. "I hate this shit," Iain said quietly, as he and I detoured around the bull moose and, splitting off from George and Tucker, headed off in our own direction to start a new transect line of netting locations. The tangled willows rose above our heads, their twisted, interlocking trunks barking our shins and snagging our ankles at every step. Every few yards we shouted to let moose or bears or anything else bigger and more dangerous know we were coming. It was all we could do, other than make sure that our bear spray, riding at our hips in quick-draw holsters, was free of our rain gear.

Iain had never really gotten over that close call five years earlier with the charging grizzly and her cub, the first summer we were tagging birds in Denali. He felt, at times, as though the bears had it out for him, though the evidence for this was shaky. The next summer, for example, working in that same general area as the charge, we'd encountered what was most likely the same female bear and her by-now nearly full-grown cub. Iain, Carol, and I were kneeling around a folded tarp on the soggy tundra, banding a blackpoll warbler we'd just captured, when Carol's head snapped around. "Bears!" she said, and to this day I have no idea how she'd sensed them; more than 30 years of working in grizzly country hones your instincts, I guess. Because there they were, 200 yards off but moving fast in our direction, heads down, with that peculiar rolling gait a big bear employs to cover ground. The bears of Denali are known as Toklat grizzlies, their coats a distinctly straw-blonde color, and these two glowed in the morning sunlight like gold, their faces and lower legs chocolate brown. I tossed the warbler, we scrambled into our packs, and at Carol's direction we simply lifted the corners of the tarp together like Santa's gift bag and half-lifted, half-dragged it and all our gear within it toward the truck, a few hundred yards away across the tundra and at right angles from the bears' approach.

"Don't run, don't run," Carol said a few times as we hustled through the brush.

"I'm not running," I said testily, nerves making me edgy.

"I was talking to myself," Carol replied. The bears were only 100 yards off now, reaching a sidehill at which they turned, angling down and—dammit—again on an intercept course with us, though they seemed not to have noticed us at all. Our paths were rapidly converging, and we had no choice but to hurry to the truck, reaching it shortly before the bears crossed 15 or 20 yards below us. The cub stopped in the middle of the gravel road, looking back at us once, but the sow simply ignored us.

"Did you see that?" Iain asked in his Glaswegian accent. "She was giving me the hairy eyeball."

We'd had no such near brushes during this smoky, hot season. This particular morning driving east we did finally see a few bears, a sow and two big cubs just a few yards off the road in Sable Pass, but we were safely inside our vehicle and able to enjoy their presence without worry. The morning was cooler than any had been the past week, 39°F on the van's dashboard thermometer. It was three o'clock, just about sunrise, when we got to our study site, a little stream called Hogan Creek that runs through a narrow draw of willows and spruce. A bit of fog hung in the damp, chill air, and my fingers ached with cold as we assembled the metal net poles and opened four or five mist nets along a couple hundred meters of the scrubby woods. Then we turned on the recordings—the almost-dog-whistle-high trill of a blackpoll warbler at some of the nets, the buzzy flute notes of a gray-cheeked thrush at the others. The previous year, our team had tagged several birds of both species here; had they survived the long migration, and had they come back to the same territories, we would catch them and learn the secrets of their migration.

It didn't take long. Within half an hour we'd caught three blackpoll warblers and one thrush—all of them unbanded, unfortunately. Most songbirds show a fair degree of fidelity to their nesting sites from year to year, but it's not absolute, and territories can shift. Our sour luck seemed destined to continue. An hour later, we pulled up stakes and moved the nets farther downstream—and hit pay dirt. Iain's whoop of excitement wasn't for the intricate beauty of the male blackpoll he'd just caught: white with elegant, overlapping black chevrons that created long, streaky patterns on its flanks and back, with an inky cap—the black "poll" of its name—and black mustache marks. His shout was entirely due to the small, black geolocator riding on the bird's back. Soon we'd caught two more tagged warblers in the same net at the same time as they chased each other in, and then a thrush with a GPS logger, which proved to be a particular treasure—a male we'd first tagged in 2015, the Year of the Grizzly, whose geolocator we recovered the following year, and which we again recaptured on territory in 2017. The next summer he was back in Hogan Creek for the fourth season, so we tagged him with one of the new GPS loggers—and now, after his fifth encounter with us, we would get another and far more precise picture of his travels.

After a week of empty nets and dashed hopes, we were pumped—and all the more so that night, when Emily connected the gray-cheeked thrush's GPS tag to her computer, and the long tale of the bird's last 10 months spooled out across Google Earth's globe on the screen, a luminous green line across the planet. The thrush had remained in Denali through early September the year before, but by the middle of that month it was on the wing, passing Whitehorse in the Yukon. It flew through the Cassair Mountains in British Columbia, skirted the northern edge of the Canadian prairies, and on October 5 was near Akley, Minnesota, along the evocatively named Tenth Crow Wing Lake. Ten days later the thrush was resting in woodlots along the Ohio River in the western toe of Kentucky; three days after that it was in the swampy bottomlands along the Big Black River in Yazoo County, Mississippi. A week more, having crossed the Gulf of Mexico and the western Caribbean, it was in rain forest in Veraguas Province in Panama. Turning due east, it followed the isthmus to South America and turned southeast, finally reaching its wintering grounds in Venezuela's remote Serrania De La Neblina National Park by November 30. Having flown 6,500 miles, however, the thrush appears to have spent its next four-and-a-half months at rest, in a patch of rain forest just 90 acres in extent, before heading north again in mid-April.

It's hard for me to say what emotion was strongest just then; excitement to see this long-hidden glimpse of a hemispheric journey revealed in such extraordinary detail; gratitude to an individual bird that, again and again for five years, had provided a window into what its species has been doing for eons; simple awe, knowing that so small and seemingly fragile an animal could link, by so many miles and across so many years, a vast tundra wilderness at the planet's far north with the humid rain forest in an equally remote corner of the tropics, and all the lands and wide seas in between.

Or maybe it was . . . reverence? Yes, that was it; reverence for a creature that, despite every obstacle we as a species have placed in its path, continues to hold faith with the wind and the far horizon, with its genes and with the seasons. Reverence for an endurance and tenacity I cannot match nor fully comprehend, but which leaves me breathless when I am confronted with it. Reverence for this extraordinary bird and the billions more like it, which by obeying their ancient rhythms knit up the scattered and beleaguered wild places of the world into a seamless whole through the simple act of flight. May it always be so.

ACKNOWLEDGMENTS

While this book took many years to research and write, its roots go back to some of my earliest memories, and the lifelong hold that birds and bird migration have had on me—an instinct that, if they did not fully understand, my parents at least encouraged.

I am, as always, deeply grateful to Hawk Mountain Sanctuary in Pennsylvania for helping spark my interest in migration as a wide-eyed kid, and allowing me later to enter the world of migration research three decades ago as a volunteer raptor-bander. I've tried to pay back that debt in some small way ever since. The staff there remain not only good friends, but unfailing resources and the patient answerers of countless questions. I am especially grateful to Kutztown University, through its partnership with Hawk Mountain, for online access to scientific journals, an otherwise serious challenge for a writer without an institutional affiliation.

Dr. Peter Marra, formerly of the Smithsonian Migratory Bird Center and now director of the Georgetown Environment Initiative at Georgetown University, has long been a valued resource for insights into many facets of migratory bird ecology and research. I was especially grateful to him, and to Dr. Ken Rosenberg of the Cornell Lab of Ornithology, for an advance look at their research into continental bird population declines, and for Pete's initial suggestion that the carry-over work the Smithsonian was conducting with Kirtland's warblers would be especially interesting—as it was. Dr. Nathan Cooper and his crew, both in the Bahamas and Michigan, were wonderfully hospitable despite being really, really tired most of the time.

In China, particular thanks to Professor Zhengwang Zhang and Wei-pan Lei of Beijing Normal University for their assistance in arranging my Yellow Sea visit. Theunis Piersma and his Global Flyway Network colleagues (Chris Hassell, Matthew Slaymaker, Adrian Boyle, and Katherine Leung) were wonderful hosts in Nanpu. Thanks to Dr. Jianbin Shi, Rose Nui, and Kathy Wang of the Paulson Institute for logistical assistance in Jiangsu, where Jing Li, Ziyou Yang, Zhang Lin, Chen Tengyi, and Dongming Li were wonderful field companions. Thanks to Terry Townsend of Birding Beijing for his insights into conservation in China. Wendy and Hank Paulson were very generous with their time and long experience working on conservation issues in China, and I especially appreciated their making time to join us in the field.

For initial assistance in arranging my travels in Nagaland, thanks to Dr. Keith Bildstein, now retired from Hawk Mountain Sanctuary, and to noted raptor expert Bill Clark. Dr. Asad Rahmani, former director of the Bombay Natural History Society, was a crucial and much appreciated early contact. Thanks to my good friend Kevin Loughlin of Wildside Nature Tours, for embracing the idea of an Amur falcon adventure and seeing it through a host of hurdles. We were all grateful to Zeiss Sports Optics for underwriting Catherine Hamilton's participation and for supplying optics for the team to use. Many thanks to Amit Sankhala of Encounters India for handling logistics for much of the trip, including vehicles and drivers, and to our guide, Abidur Rahman of Jungle Travels India, whose good nature never flagged. Particular thanks to the Tsopoe family and the people of Pangti, notably Nchumo Odyuo and his family, for their warm hospitality and window into Naga culture. And most of all, thank you to Bano Haralu for her help in the two years it took to arrange the visit, and making time to come meet us in Pangti at a time when she was dealing with a serious family medical crisis.

It was serendipitous that I ran into Chris Vennum of Colorado

State University at a Raptor Research Foundation conference and learned about the recent developments with Swainson's hawks in the Butte Valley. Thanks to Chris Briggs of Hamilton College, the legendary Pete Bloom, and Melissa Hunt for their help and hospitality while there—and to Brian Woodbridge for saying yes more than 20 years ago when I asked to join his field work in Argentina during the original pesticide crisis.

My good friend Ben Olewine IV was happy to make connections for me with BirdLife International, where Jim Lawrence (UK) was helpful with both the subject of illegal bird killing and with Amur falcons in Nagaland. Barend Van Gemerden and Willem Van den Bossche, with BirdLife in the Netherlands, were very generous with their time in helping me understand the appalling scope of the subject throughout Europe, the Mediterranean, and Middle East, and put me in touch with folks in the field, especially Tassos Shialis and Martin Hellicar at BirdLife Cyprus. I am particularly grateful to "Andreas" with BirdLife Cyprus for allowing me to join him and Roger Little on their patrols, and to Deputy Chief Constable Jon Ward of the Sovereign Base Area police force for allowing me to accompany his officers on patrol.

I remain indebted to my longtime agent Peter Matson of Sterling Lord Literistic. It has been a pleasure to work with John Glusman, vice president and editor in chief, and Helen Thomaides, assistant editor, at W. W. Norton on this project—and I am thankful for their patience with a longer gestation period than is typical.

Why on earth my wife Amy puts up with me, I cannot say. But she does, and I am more grateful for that than anything else.

Versions of some of the stories here have appeared earlier in a variety of publications. Elements of the Preface first appeared in *Bird Watcher's Digest*, while aspects of chapters 1, 3, 4, 7, and 10 appeared in Cornell's *Living Bird*. A much-condensed version of chapter 5 appeared in *Audubon*. I am grateful to Dawn Hewitt and my late good friend Bill Thompson III at *BWD*, Gus Axelson at *Living Bird*, and the editorial staff at *Audubon* for their help and support.

REFERENCES

Works directly quoted or cited in the text for each chapter are documented with end-notes; works referenced while writing the chapter are listed in a brief bibliography.

One: SPOONIES

31 **"stunned joy":** Nicola Crockford, quoted in Benjamin Graham, "A Boon for Birds: Once Overlooked, China's Mudflats Gain Protections," Mongabay.com, May 11, 2018, https://news.mongabay.com/2018/05/a-boon-for-birds-once-overlooked-chinas-mudflats-gain-protections/.

32 **"although this is thought":** BirdLife International, "*Calidris pygmaea* (amended version of 2017 assessment)," The IUCN Red List of Threatened Species 2017: e.T22693452A117520594. http://dx.doi.org/10.2305/IUCN.UK.2017–3.RLTS .T22693452A117520594.en.

61 **"The spoon-billed sandpiper is one of the best":** Debbie Pain, Baz Hughes, Evgeny Syroechkovskiy, Christoph Zöckler, Sayam Chowdhury, Guy Anderson, and Nigel Clark, "Saving the Spoon-billed Sandpiper: A Conservation Update," *British Birds* 111 (June 2018): 333.

Battley, Phil F., Theunis Piersma, Maurine W. Dietz, Sixian Tang, Anne Dekinga, and Kees Hulsman. "Empirical Evidence for Differential Organ Reductions During Trans-oceanic Bird Flight." *Proceedings of the Royal Society of London B: Biological Sciences* 267, no. 1439 (2000): 191–195.

Bijleveld, Allert I., Robert B. MacCurdy, Ying-Chi Chan, Emma Penning, Rich M. Gabrielson, John Cluderay, Eric L. Spaulding, et al. "Understanding Spatial Distributions: Negative Density-dependence in Prey Causes Predators to Trade-off Prey Quantity with Quality." *Proceedings of the Royal Society of London B: Biological Sciences* 1828 (2016): 20151557.

Brown, Stephen, Cheri Gratto-Trevor, Ron Porter, Emily L. Weiser, David Mizrahi, Rebecca Bentzen, Megan Boldenow, et al. "Migratory Connectivity of Semipalmated Sandpipers and Implications for Conservation." *Condor* 119, no. 2 (2017): 207–224.

Gill, Robert E., T. Lee Tibbitts, David C. Douglas, Colleen M. Handel, Daniel M. Mulcahy, Jon C. Gottschalck, Nils Warnock, Brian J. McCaffery, Philip F. Battley, and Theunis Piersma. "Extreme Endurance Flights by Landbirds Crossing the Pacific Ocean: Ecological Corridor Rather Than Barrier?" *Proceedings of the Royal Society of London B: Biological Sciences* 276, no. 1656 (2009): 447–457.

Gupta, Alok. "China Land Reclamation Ban Revives Migratory Birds' Habitat." Feb. 2, 2018. China Global Television Network. https://news.cgtn.com/news/3049544f 30677a6333566d54/share_p.html.

International Union for the Conservation of Nature. *IUCN World Heritage Evaluations 2019.* Gland, Switzerland: IUCN, 2019.

McKinnon, John, Yvonne I. Yerkuil, and Nicholas Murray. "IUCN Situation Analysis on East and Southeast Asian Intertidal Habitats, with Particular Reference to the Yellow Sea (including the Bohai Sea)." Gland, Switzerland: International Union for the Conservation of Nature, 2012.

Melville, David S., Ying Chen, and Zhijun Ma. "Shorebirds Along the Yellow Sea Coast of China Face an Uncertain Future—A Review of Threats." *Emu-Austral Ornithology* 116, no. 2 (2016): 100–110.

Murray, Nicholas J., Robert S. Clemens, Stuart R. Phinn, Hugh P. Possingham, and Richard A. Fuller. "Tracking the Rapid Loss of Tidal Wetlands in the Yellow Sea." *Frontiers in Ecology and the Environment* 12, no. 5 (2014): 267–272.

Piersma, Theunis. "Why Marathon Migrants Get Away with High Metabolic Ceilings: Towards an Ecology of Physiological Restraint." *Journal of Experimental Biology* 214, no. 2 (2011): 295–302.

Stroud, D. A., A. Baker, D. E. Blanco, N. C. Davidson, S. Delany, B. Ganter, R. Gill, P. González, L. Haanstra, R. I. G. Morrison, T. Piersma, D. A. Scott, O. Thorup, R. West, J. Wilson, and C. Zöckler. "The Conservation and Population Status of the World's Waders at the Turn of the Millennium." In *Waterbirds Around the World,* ed. G. C. Boere, C. A. Galbraith, and D. A. Stroud, 643–648. Edinburgh, UK: The Stationery Office, 2007.

Zoeckler, Christoph, Alison E. Beresford, Gillian Bunting, Sayam U. Chowdhury, Nigel A. Clark, Vivian Wing Kan Fu, Tony Htin Hla, et al. "The Winter Distribution of the Spoon-billed Sandpiper *Calidris pygmaeus.*" *Bird Conservation International* 26, no. 4 (2016): 476–489.

Zoeckler, Christoph, Evgeny E. Syroechkovskiy, and Philip W. Atkinson. "Rapid and Continued Population Decline in the Spoon-billed Sandpiper *Eurynorhynchus pygmeus* Indicates Imminent Extinction Unless Conservation Action is Taken." *Bird Conservation International* 20, no. 2 (2010): 95–111.

Two: QUANTUM LEAP

67   **"The metaphor of marathon running":** Christopher G. Guglielmo, "Move that Fatty Acid: Fuel Selection and Transport in Migratory Birds and Bats," *Integrated and Comparative Biology* 50 (2010): 336.

73   **"By human standards":** Paul Bartell and Ashli Moore, "Avian Migration: The Ultimate Red-eye Flight," *New Scientist* 101 (2013): 52.

85 **"I got the paper back":** Klaus Schulten, quoted in Ed Yong, "How Birds See Magnetic Fields: An Interview with Klaus Schulten," Nov. 24, 2010, https://www.nationalgeographic.com/science/phenomena/2010/11/24/how-birds-see-magnetic-fields-an-interview-with-klaus-schulten.html.

87 **"something one gets 'for free'":** P. J. Hore and Henrik Mouritsen, "The Radical-pair Mechanism of Magnetoreception," *Annual Review of Biophysics* 45 (2016): 332.

88 **"a second magnetic sense":** Dmitry Kishkinev, Nikita Chernetsov, Dominik Heyers, and Henrik Mouritsen, "Migratory Reed Warblers Need Intact Trigeminal Nerves to Correct for a 1,000 km Eastward Displacement," *PLoS One* 8, no. 6 (2013): e65847, 1.

90 **"not significantly different":** Tyson L. Hedrick, Cécile Pichot, and Emmanuel De Margerie, "Gliding for a Free Lunch: Biomechanics of Foraging Flight in Common Swifts (*Apus apus*)," *Journal of Experimental Biology* 221, no. 22 (2018): jeb186270, 1.

Bairlein, Franz. "How to Get Fat: Nutritional Mechanisms of Seasonal Fat Accumulation in Migratory Songbirds." *Naturwissenschaften* 89, no. 1 (2002): 1–10.

Barkan, Shay, Yoram Yom-Tov, and Anat Barnea. "Exploring the Relationship Between Brain Plasticity, Migratory Lifestyle, and Social Structure in Birds." *Frontiers in Neuroscience* 11 (2017): 139.

———. "A Possible Relation Between New Neuronal Recruitment and Migratory Behavior in *Acrocephalus* Warblers." *Developmental Neurobiology* 74, no. 12 (2014): 1194–1209.

Biebach, H. "Is Water or Energy Crucial for Trans-Sahara Migrants?" In *Proceedings International Ornithological Congress*, 19 (1990): 773–779.

Chernetsov, Nikita, Alexander Pakhomov, Dmitry Kobylkov, Dmitry Kishkinev, Richard A. Holland, and Henrik Mouritsen. "Migratory Eurasian Reed Warblers Can Use Magnetic Declination to Solve the Longitude Problem." *Current Biology* 27, no. 17 (2017): 2647–2651.

Edelman, Nathaniel B., Tanja Fritz, Simon Nimpf, Paul Pichler, Mattias Lauwers, Robert W. Hickman, Artemis Papadaki-Anastasopoulou, et al. "No Evidence for Intracellular Magnetite in Putative Vertebrate Magnetoreceptors Identified by Magnetic Screening." *Proceedings of the National Academy of Sciences* 112, no. 1 (2015): 262–267.

Einfeldt, Anthony L., and Jason A. Addison. "Anthropocene Invasion of an Ecosystem Engineer: Resolving the History of *Corophium volutator* (Amphipoda: Corophiidae) in the North Atlantic." *Biological Journal of the Linnean Society* 115, no. 2 (2015): 288–304.

Elbein, Asher. "Some Birds Are Better Off with Weak Immune Systems." *New York Times*, June 26, 2018, D6.

Fuchs, T., A. Haney, T. J. Jechura, Frank R. Moore, and V. P. Bingman. "Daytime Naps in Night-migrating Birds: Behavioural Adaptation to Seasonal Sleep Deprivation in the Swainson's thrush, *Catharus ustulatus*." *Animal Behaviour* 72, no. 4 (2006): 951–958.

Gerson, Alexander R. "Avian Osmoregulation in Flight: Unique Metabolic Adaptations Present Novel Challenges." *The FASEB Journal* 30, no. 1 supplement (2016): 976.1.

―――. "Environmental Physiology of Flight in Migratory Birds." PhD diss., University of Western Ontario, 2012.

―――. and Christopher Guglielmo. "Flight at Low Ambient Humidity Increases Protein Catabolism in Migratory Birds." *Science* 333, no. 6048 (2011): 1434–1436.

Gill, Robert E., Jr., Theunis Piersma, Gary Hufford, Rene Servranckx, and Adrian Riegen. "Crossing the Ultimate Ecological Barrier: Evidence for an 11,000-km-long Nonstop Flight from Alaska to New Zealand and Eastern Australia by Bar-tailed Godwits." *The Condor* 107, no. 1 (2005): 1–20.

Guglielmo, Christopher G. "Obese Super Athletes: Fat-fueled Migration in Birds and Bats." *Journal of Experimental Biology* 221, Suppl. 1 (2018): jeb165753.

Hawkes, Lucy A., Sivananinthaperumal Balachandran, Nyambayar Batbayar, Patrick J. Butler, Peter B. Frappell, William K. Milsom, Natsagdorj Tseveenmyadag, et al. "The trans-Himalayan Flights of Bar-headed Geese (*Anser indicus*)." *Proceedings of the National Academy of Sciences* 108, no. 23 (2011): 9516–9519.

Hawkes, Lucy A., Beverley Chua, David C. Douglas, Peter B. Frappell, et al. "The Paradox of Extreme High-altitude Migration in Bar-headed Geese *Anser indicus*." *Proc. Royal Society-B* 280 (2013): 20122114. http://dx.doi.org/10.1098/rspb.2012.2114.

Hedenström, Anders, Gabriel Norevik, Kajsa Warfvinge, Arne Andersson, Johan Bäckman, and Susanne Åkesson. "Annual 10-month Aerial Life Phase in the Common Swift *Apus apus*." *Current Biology* 26, no. 22 (2016): 3066–3070.

Hua, Ning, Theunis Piersma, and Zhijun Ma. "Three-phase Fuel Deposition in a Long-distance Migrant, the Red Knot (*Calidris canutus piersmai*), Before the Flight to High Arctic Breeding Grounds." *PLoS One* 8, no. 4 (2013): e62551.

Jones, Stephanie G., Elliott M. Paletz, William H. Obermeyer, Ciaran T. Hannan, and Ruth M. Benca. "Seasonal Influences on Sleep and Executive Function in the Migratory White-crowned Sparrow (*Zonotrichia leucophrys gambelii*)." *BMC Neuroscience* 11 (2010).

Landys, Meta M., Theunis Piersma, G. Henk Visser, Joop Jukema, and Arnold Wijker. "Water Balance During Real and Simulated Long-distance Migratory Flight in the Bar-tailed Godwit." *The Condor* 102, no. 3 (2000): 645–652.

Lesku, John A., Niels C. Rattenborg, Mihai Valcu, Alexei L. Vyssotski, Sylvia Kuhn, Franz Kuemmeth, Wolfgang Heidrich, and Bart Kempenaers. "Adaptive Sleep Loss in Polygynous Pectoral Sandpipers." *Science* 337, no. 6102 (2012): 1654–1658.

Liechti, Felix, Willem Witvliet, Roger Weber, and Erich Bächler. "First Evidence of a 200-day Non-stop Flight in a Bird." *Nature Communications* 4 (2013): 2554.

Lockley, Ronald M. "Non-stop Flight and Migration in the Common Swift *Apus apus*." *Ostrich* 40, no. S1 (1969): 265–269.

Maillet, Dominique, and Jean-Michel Weber. "Relationship Between n-3 PUFA Content and Energy Metabolism in the Flight Muscles of a Migrating Shorebird: Evidence for Natural Doping." *Journal of Experimental Biology* 210, no. 3 (2007): 413–420.

McWilliams, Scott R., Christopher Guglielmo, Barbara Pierce, and Marcel Klaassen. "Flying, Fasting, and Feeding in Birds During Migration: A Nutritional and Physiological Ecology Perspective." *Journal of Avian Biology* 35, no. 5 (2004): 377–393.

Nießner, Christine, Susanne Denzau, Katrin Stapput, Margaret Ahmad, Leo Peichl, Wolfgang Wiltschko, and Roswitha Wiltschko. "Magnetoreception: Activated Cryptochrome 1a Concurs with Magnetic Orientation in Birds." *Journal of The Royal Society Interface* 10, no. 88 (2013): 20130638.

O'Connor, Emily A., Charlie K. Cornwallis, Dennis Hasselquist, Jan-Åke Nilsson, and Helena Westerdahl. "The Evolution of Immunity in Relation to Colonization and Migration." *Nature Ecology & Evolution* 2, no. 5 (2018): 841.

Piersma, Theunis. "Phenotypic Flexibility During Migration: Optimization of Organ Size Contingent on the Risks and Rewards of Fueling and Flight?" *Journal of Avian Biology* (1998): 511–520.

——— and Robert E. Gill, Jr. "Guts Don't Fly: Small Digestive Organs in Obese Bar-tailed Godwits." *The Auk* (1998): 196–203.

———, Gudmundur A. Gudmundsson, and Kristján Lilliendahl. "Rapid Changes in the Size of Different Functional Organ and Muscle Groups During Refueling in a Long-distance Migrating Shorebird." *Physiological and Biochemical Zoology* 72, no. 4 (1999): 405–415.

———, Renée van Aelst, Karin Kurk, Herman Berkhoudt, and Leo R. M. Maas. "A New Pressure Sensory Mechanism for Prey Detection in Birds: The Use of Principles of Seabed Dynamics?" *Proceedings of the Royal Society of London B: Biological Sciences* 265 (1998): 1377–1383.

Rattenborg, Niels C. "Sleeping on the Wing." *Interface Focus* 7, no. 1 (2017): 20160082.

———, Bryson Voirin, Sebastian M. Cruz, Ryan Tisdale, Giacomo Dell'Omo, Hans-Peter Lipp, Martin Wikelski, and Alexei L. Vyssotski. "Evidence That Birds Sleep in Mid-flight." *Nature Communications* 7 (2016): 12468.

Ritz, Thorsten, Salih Adem, and Klaus Schulten. "A Model for Photoreceptor-based Magnetoreception in Birds." *Biophysical Journal* 78 (2000): 707–718.

Schulten, Klaus, Charles E. Swenberg, and Albert Weller. "A Biomagnetic Sensory Mechanism Based on Magnetic Field Modulated Coherent Electron Spin Motion." *Zeitschrift für Physikalische Chemie* 111, no. 1 (1978): 1–5.

Scott, Graham R., Lucy A. Hawkes, Peter B. Frappell, Patrick J. Butler, Charles M. Bishop, and William K. Milsom. "How Bar-headed Geese Fly Over the Himalayas." *Physiology* 30, no. 2 (2015): 107–115.

Tamaki, Masako, Ji Won Bang, Takeo Watanabe, and Yuka Sasaki. "Night Watch in One Brain Hemisphere in Sleep Associated with the First-Night Effect in Humans." *Current Biology* 26 (2016): 1190–1194.

Treiber, Christoph Daniel, Marion Claudia Salzer, Johannes Riegler, Nathaniel Edelman, Cristina Sugar, Martin Breuss, Paul Pichler, et al. "Clusters of Iron-rich Cells in the Upper Beak of Pigeons are Macrophages Not Magnetosensitive Neurons." *Nature* 484, no. 7394 (2012): 367.

Viegas, Ivan, Pedro M. Araújo, Afonso D. Rocha, Auxiliadora Villegas, John G. Jones, Jaime A. Ramos, José A. Masero, and José A. Alves. "Metabolic Plasticity for Subcutaneous Fat Accumulation in a Long-distance Migratory Bird Traced by ^2H_2O." *Journal of Experimental Biology* 220, no. 6 (2017): 1072–1078.

Wallraff, Hans G., and Meinrat O. Andreae. "Spatial Gradients in Ratios of Atmospheric Trace Gases: A Study Stimulated by Experiments on Bird Navigation." *Tellus B: Chemical and Physical Meteorology* 52, no. 4 (2000): 1138–1157.

Weber, Jean-Michel. "The Physiology of Long-distance Migration: Extending the Limits of Endurance Metabolism." *Journal of Experimental Biology* 212, no. 5 (2009): 593–597.

Weimerskirch, Henri, Charles Bishop, Tiphaine Jeanniard-du-Dot, Aurélien Prudor, and Gottfried Sachs. "Frigate Birds Track Atmospheric Conditions Over Months-long Transoceanic Flights." *Science* 353, no. 6294 (2016): 74–78.

Wiltschko, Wolfgang, and Roswitha Wiltschko. "Magnetic Orientation in Birds." *Journal of Experimental Biology* 199, no. 1 (1996): 29–38.

Winger, Benjamin M., F. Keith Barker, and Richard H. Ree. "Temperate Origins of Long-distance Seasonal Migration in New World Songbirds." *Proceedings of the National Academy of Sciences* 111, no. 33 (2014): 12115–12120.

Zink, Robert M., and Aubrey S. Gardner. "Glaciation as a Migratory Switch." *Science Advances* 3, no. 9 (2017): e1603133.

Three: WE USED TO THINK

91 **"Perhaps other observers":** Ronald M. Lockley, "Non-stop Flight and Migration in the Common Swift *Apus apus*," *Ostrich* 40, no. S1 (1969): 265.

124 **"butter dripping from the trees":** Christopher M. Tonra, quoted in Ben Guarino, "Songbird Migration Finds a Tiny, Vulnerable Winter Range," *Washington Post*, June 21, 2019, https://www.washingtonpost.com/science/songbird -migration-study-finds-a-tiny-vulnerable-winter-range/2019/06/20/1bffa6fe -92cb-11e9-b570–6416efdc0803_story.html.

Anders, Angela D., John Faaborg, and Frank R. Thompson III. "Postfledging Dispersal, Habitat Use, and Home-range Size of Juvenile Wood Thrushes." *The Auk* 115, no. 2 (1998): 349–358.

Delmore, Kira E., James W. Fox, and Darren E. Irwin. "Dramatic Intraspecific Differences in Migratory Routes, Stopover Sites, and Wintering Areas, Revealed Using Light-level Geolocators." *Proceedings of the Royal Society B: Biological Sciences* 279, no. 1747 (2012): 4582–4589.

Delmore, Kira E., and Darren E. Irwin. "Hybrid Songbirds Employ Intermediate Routes in a Migratory Divide." *Ecology Letters* 17, no. 10 (2014): 1211–1218.

DeLuca, William V., Bradley K. Woodworth, Stuart A. Mackenzie, Amy E. M. Newman, Hilary A. Cooke, Laura M. Phillips, Nikole E. Freeman, et al. "A Boreal Songbird's 20,000 km Migration Across North America and the Atlantic Ocean." *Ecology* (2019): e02651.

Finch, Tom, Philip Saunders, Jesús Miguel Avilés, Ana Bermejo, Inês Catry, Javier de la Puente, Tamara Emmenegger, et al. "A Pan-European, Multipopulation Assessment of Migratory Connectivity in a Near-threatened Migrant Bird." *Diversity and Distributions* 21, no. 9 (2015): 1051–1062.

Haddad, Nick M., Lars A. Brudvig, Jean Clobert, Kendi F. Davies, Andrew Gonzalez, Robert D. Holt, Thomas E. Lovejoy, et al. "Habitat Fragmentation and its Lasting Impact on Earth's Ecosystems." *Science Advances* 1, no. 2 (2015): e1500052.

Hahn, Steffen, Valentin Amrhein, Pavel Zehtindijev, and Felix Liechti. "Strong Migratory Connectivity and Seasonally Shifting Isotopic Niches in Geographically Separated Populations of a Long-distance Migrating Songbird." *Oecologia* 173, no. 4 (2013): 1217–1225.

Hallworth, Michael T., and Peter P. Marra. "Miniaturized GPS Tags Identify Nonbreeding Territories of a Small Breeding Migratory Songbird." *Scientific Reports* 5 (2015): 11069.

Hallworth, Michael T., T. Scott Sillett, Steven L. Van Wilgenburg, Keith A. Hobson, and Peter P. Marra. "Migratory Connectivity of a Neotropical Migratory Songbird Revealed by Archival Light-level Geolocators." *Ecological Applications* 25, no. 2 (2015): 336–347.

Koleček, Jaroslav, Petr Procházka, Naglaa El-Arabany, Maja Tarka, Mihaela Ilieva, Steffen Hahn, Marcel Honza, et al. "Cross-continental Migratory Connectivity and Spatiotemporal Migratory Patterns in the Great Reed Warbler." *Journal of Avian Biology* 47, no. 6 (2016): 756–767.

Lemke, Hilger W., Maja Tarka, Raymond H. G. Klaassen, Mikael Åkesson, Staffan Bensch, Dennis Hasselquist, and Bengt Hansson. "Annual Cycle and Migration Strategies of a Trans-Saharan Migratory Songbird: A Geolocator Study in the Great Reed Warbler." *PLoS One* 8, no. 10 (2013): e79209.

Pagen, Rich W., Frank R. Thompson III, and Dirk E. Burhans. "Breeding and Postbreeding Habitat Use by Forest Migrant Songbirds in the Missouri Ozarks." *The Condor* 102, no. 4 (2000): 738–747.

Priestley, Kent. "Virginia's Wild Coast." *Nature Conservancy*, Dec. 2014/Jan. 2015. https://www.nature.org/magazine/archives/virginias-wild-coast-1.xml.

Rivera, J. H. Vega, J. H. Rappole, W. J. McShea, and C. A. Haas. "Wood Thrush Postfledging Movements and Habitat Use in Northern Virginia." *The Condor* 100, no. 1 (1998): 69–78.

Rohwer, Sievert, Luke K. Butler, and D. R. Froehlich. "Ecology and Demography of East–West Differences in Molt Scheduling of Neotropical Migrant Passerines." In *Birds of Two Worlds: The Ecology and Evolution of Migration*, edited by Russell Greenberg and Peter P. Marra, 87–105. Baltimore: Johns Hopkins University Press, 2005.

Rohwer, Sievert, Keith A. Hobson, and Vanya G. Rohwer. "Migratory Double Breeding in Neotropical Migrant Birds." *Proceedings of the National Academy of Sciences* 106, no. 45 (2009): 19050–19055.

Stanley, Calandra Q., Emily A. McKinnon, Kevin C. Fraser, Maggie P. Macpherson, Garth Casbourn, Lyle Friesen, Peter P. Marra, et al. "Connectivity of Wood Thrush Breeding, Wintering, and Migration Sites Based on Range-wide Tracking." *Conservation Biology* 29, no. 1 (2015): 164–174.

Tonra, Christopher M., Michael T. Hallworth, Than J. Boves, Jessie Reese, Lesley

P. Bulluck, Matthew Johnson, Cathy Viverette, et al. "Concentration of a Widespread Breeding Population in a Few Critically Important Nonbreeding Areas: Migratory Connectivity in the Prothonotary Warbler." *Condor* (2019). https://doi.org/10.1093/condor/duz019.

Vitz, Andrew C., and Amanda D. Rodewald. "Can Regenerating Clearcuts Benefit Mature-forest Songbirds? An Examination of Post-breeding Ecology." *Biological Conservation* 127, no. 4 (2006): 477–486.

Watts, Bryan D., Fletcher M. Smith, and Barry R. Truitt. "Leaving Patterns of Whimbrels Using a Terminal Spring Staging Area." *Wader Study* 124 (2017): 141–146.

Watts, Bryan D., and Barry R. Truitt. "Decline of Whimbrels Within a Mid-Atlantic Staging Area (1994–2009)." *Waterbirds* 34, no. 3 (2011): 347–351.

Four: BIG DATA, BIG TROUBLE

146 **"take all practical steps":** Federal Communications Commission Memorandum FCC-18–161 (Nov. 15, 2018), 13.

147 **"consistent, steep population loss":** Kennth V. Rosenberg, Adriaan M. Dokter, Peter J. Blancher, John R. Sauer, Adam C. Smith, Paul A. Smith, Jessica C. Stanton, Arvind Panjabi, Laura Helft, Michael Parr, and Peter P. Marra, "Decline of the North American Avifauna," *Science* 366, no. 6461 (2019): 120–124.

148 **"a poignant reminder":** Ibid.

149 **"The situation is catastrophic":** Benoit Fontaine, quoted in " 'Catastrophe' as France's Bird Population Collapses Due to Pesticides," *The Guardian*, March 20, 2018, https://www.theguardian.com/world/2018/mar/21/catastrophe-as-frances-bird-population-collapses-due-to-pesticides.

Cabrera-Cruz, Sergio A., Jaclyn A. Smolinsky, and Jeffrey J. Buler. "Light Pollution is Greatest Within Migration Passage Areas for Nocturnally-migrating Birds Around the World." *Scientific Reports* 8, no. 1 (2018): 3261.

Cohen, Emily B., Clark R. Rushing, Frank R. Moore, Michael T. Hallworth, Jeffrey A. Hostetler, Mariamar Gutierrez Ramirez, and Peter P. Marra. "The Strength of Migratory Connectivity for Birds En Route to Breeding Through the Gulf of Mexico." *Ecography* 42, no. 4 (2019): 658–669.

Golet, Gregory H., Candace Low, Simon Avery, Katie Andrews, Christopher J. McColl, Rheyna Laney, and Mark D. Reynolds. "Using Ricelands to Provide Temporary Shorebird Habitat During Migration." *Ecological Applications* 28, no. 2 (2018): 409–426.

Hausheer, Justine E. "Bumper-Crop Birds: Pop-Up Wetlands Are a Success in California." *Cool Green Science*, Jan. 29, 2018. https://blog.nature.org/science/2018/01/29/bumper-crop-birds-pop-up-wetlands-are-a-success-in-california/.

Horton, Kyle G., Cecilia Nilsson, Benjamin M. Van Doren, Frank A. La Sorte, Adriaan M. Dokter, and Andrew Farnsworth. "Bright Lights in the Big Cities: Migratory Birds' Exposure to Artificial Light." *Frontiers in Ecology and the Environment* 17, no. 4 (2019): 209–214.

Horton, Kyle G., Benjamin M. Van Doren, Frank A. La Sorte, Emily B. Cohen, Hannah L. Clipp, Jeffrey J. Buler, Daniel Fink, Jeffrey F. Kelly, and Andrew Farnsworth. "Holding Steady: Little Change in Intensity or Timing of Bird Migration Over the Gulf of Mexico." *Global Change Biology* 25, no. 3 (2019): 1106–1118.

Inger, Richard, Richard Gregory, James P. Duffy, Iain Stott, Petr Voříšek, and Kevin J. Gaston. "Common European Birds are Declining Rapidly While Less Abundant Species' Numbers are Rising." *Ecology Letters* 18, no. 1 (2015): 28–36.

La Sorte, Frank A., Daniel Fink, Jeffrey J. Buler, Andrew Farnsworth, and Sergio A. Cabrera-Cruz. "Seasonal Associations with Urban Light Pollution for Nocturnally Migrating Bird Populations." *Global Change Biology* 23, no. 11 (2017): 4609–4619.

Lin, Tsung-Yu, Kevin Winner, Garrett Bernstein, Abhay Mittal, Adriaan M. Dokter, Kyle G. Horton, Cecilia Nilsson, Benjamin M. Van Doren, Andrew Farnsworth, Frank A. La Sorte, et al. "MistNet: Measuring Historical Bird Migration in the U.S. Using Archived Weather Radar Data and Convolutional Neural Networks." *Methods in Ecology and Evolution* (2019): 1–15. https://doi.org/10.1111/2041–210X.13280.

McLaren, James D., Jeffrey J. Buler, Tim Schreckengost, Jaclyn A. Smolinsky, Matthew Boone, E. Emiel van Loon, Deanna K. Dawson, and Eric L. Walters. "Artificial Light at Night Confounds Broad-scale Habitat Use by Migrating Birds." *Ecology Letters* 21, no. 3 (2018): 356–364.

Powell, Hugh. "eBird and a Hundred Million Points of Light." *Living Bird* no. 1 (2015). https://www.allaboutbirds.org/a-hundred-million-points-of-light/.

Reif, Jiří, and Zdeněk Vermouzek. "Collapse of Farmland Bird Populations in an Eastern European Country Following its EU Accession." *Conservation Letters* 12, no. 1 (2019): e12585.

Reynolds, Mark D., Brian L. Sullivan, Eric Hallstein, Sandra Matsumoto, Steve Kelling, Matthew Merrifield, Daniel Fink, et al. "Dynamic Conservation for Migratory Species." *Science Advances* 3, no. 8 (2017): e1700707.

Sullivan, Brian L., Jocelyn L. Aycrigg, Jessie H. Barry, Rick E. Bonney, Nicholas Bruns, Caren B. Cooper, Theo Damoulas, et al. "The eBird Enterprise: An Integrated Approach to Development and Application of Citizen Science." *Biological Conservation* 169 (2014): 31–40.

Sullivan, Brian L., Christopher L. Wood, Marshall J. Iliff, Rick E. Bonney, Daniel Fink, and Steve Kelling. "eBird: A Citizen-based Bird Observation Network in the Biological Sciences." *Biological Conservation* 142, no. 10 (2009): 2282–2292.

Van Doren, Benjamin M., Kyle G. Horton, Adriaan M. Dokter, Holger Klinck, Susan B. Elbin, and Andrew Farnsworth. "High-intensity Urban Light Installation Dramatically Alters Nocturnal Bird Migration." *Proceedings of the National Academy of Sciences* 114, no. 42 (2017): 11175–11180.

Watson, Matthew J., David R. Wilson, and Daniel J. Mennill. "Anthropogenic Light is Associated with Increased Vocal Activity by Nocturnally Migrating Birds." *Condor* 118, no. 2 (2016): 338–344.

Zuckerberg, Benjamin, Daniel Fink, Frank A. La Sorte, Wesley M. Hochachka, and Steve Kelling. "Novel Seasonal Land Cover Associations for Eastern North American Forest Birds Identified Through Dynamic Species Distribution Modelling." *Diversity and Distributions* 22, no. 6 (2016): 717–730.

Five: HANGOVER

Angelier, Frédéric, Christopher M. Tonra, Rebecca L. Holberton, and Peter P. Marra. "Short-term Changes in Body Condition in Relation to Habitat and Rainfall Abundance in American Redstarts *Setophaga ruticilla* During the Non-breeding Season." *Journal of Avian Biology* 42, no. 4 (2011): 335–341.

Bearhop, Stuart, Geoff M. Hilton, Stephen C. Votier, and Susan Waldron. "Stable Isotope Ratios Indicate That Body Condition in Migrating Passerines is Influenced by Winter Habitat." *Proceedings of the Royal Society of London B: Biological Sciences* 271, no. Suppl 4 (2004): S215–S218.

Conklin, Jesse R., and Phil F. Battley. "Carry-over Effects and Compensation: Late Arrival on Non-breeding Grounds Affects Wing Moult But Not Plumage or Schedules of Departing Bar-tailed Godwits *Limosa lapponica baueri*." *Journal of Avian Biology* 43, no. 3 (2012): 252–263.

Cooper, Nathan W., Michael T. Hallworth, and Peter P. Marra. "Light-level Geolocation Reveals Wintering Distribution, Migration Routes, and Primary Stopover Locations of an Endangered Long-distance Migratory Songbird." *Journal of Avian Biology* 48, no. 2 (2017): 209–219.

Cooper, Nathan W., Thomas W. Sherry, and Peter P. Marra. "Experimental Reduction of Winter Food Decreases Body Condition and Delays Migration in a Long-distance Migratory Bird." *Ecology* 96, no. 7 (2015): 1933–1942.

Finch, Tom, James W. Pearce-Higgins, D. I. Leech, and Karl L. Evans. "Carry-over Effects from Passage Regions are More Important Than Breeding Climate in Determining the Breeding Phenology and Performance of Three Avian Migrants of Conservation Concern." *Biodiversity and Conservation* 23, no. 10 (2014): 2427–2444.

Gamble, Douglas W., and Scott Curtis. "Caribbean Precipitation: Review, Model and Prospect." *Progress in Physical Geography* 32, no. 3 (2008): 265–276.

Gunnarsson, Tomas Grétar, Jennifer A. Gill, Jason Newton, Peter M. Potts, and William J. Sutherland. "Seasonal Matching of Habitat Quality and Fitness in a Migratory Bird." *Proceedings of the Royal Society of London B: Biological Sciences* 272 (2005): 2319–2323.

Marra, Peter P., Keith A. Hobson, and Richard T. Holmes. "Linking Winter and Summer Events in a Migratory Bird by Using Stable-carbon Isotopes." *Science* 282, no. 5395 (1998): 1884–1886.

Marra, Peter P., and Richard T. Holmes. "Consequences of Dominance-mediated Habitat Segregation in American Redstarts During the Nonbreeding Season." *The Auk* 118, no. 1 (2001): 92–104.

Norris, D. Ryan, Peter P. Marra, T. Kurt Kyser, Thomas W. Sherry, and Laurene M. Ratcliffe. "Tropical Winter Habitat Limits Reproductive Success on the Temperate Breeding Grounds in a Migratory Bird." *Proceedings of the Royal Society of London B: Biological Sciences* 271, no. 1534 (2004): 59–64.

Ockendon, Nancy, Dave Leech, and James W. Pearce-Higgins. "Climatic Effects on Breeding Grounds are More Important Drivers of Breeding Phenology in Migrant Birds than Carry-over Effects from Wintering Grounds." *Biology Letters* 9, no. 6 (2013): 20130669.

Rhiney, Kevon. "Geographies of Caribbean Vulnerability in a Changing Climate: Issues and Trends." *Geography Compass* 9, no. 3 (2015): 97–114.

Rockwell, Sarah M., Joseph M. Wunderle, T. Scott Sillett, Carol I. Bocetti, David N. Ewert, Dave Currie, Jennifer D. White, and Peter P. Marra. "Seasonal Survival Estimation for a Long-distance Migratory Bird and the Influence of Winter Precipitation." *Oecologia* 183, no. 3 (2017): 715–726.

Schamber, Jason L., James S. Sedinger, and David H. Ward. "Carry-over Effects of Winter Location Contribute to Variation in Timing of Nest Initiation and Clutch Size in Black Brant (*Branta bernicla nigricans*)." *The Auk* 129, no. 2 (2012): 205–210.

Senner, Nathan R., Wesley M. Hochachka, James W. Fox, and Vsevolod Afanasyev. "An Exception to the Rule: Carry-over Effects Do Not Accumulate in a Long-distance Migratory Bird." *PLoS One* 9, no. 2 (2014): e86588.

Sorensen, Marjorie C., J. Mark Hipfner, T. Kurt Kyser, and D. Ryan Norris. "Carry-over Effects in a Pacific Seabird: Stable Isotope Evidence that Pre-breeding Diet Quality Influences Reproductive Success." *Journal of Animal Ecology* 78, no. 2 (2009): 460–467.

Studds, Colin E., and Peter P. Marra. "Nonbreeding Habitat Occupancy and Population Processes: An Upgrade Experiment with a Migratory Bird." *Ecology* 86, no. 9 (2005): 2380–2385.

Wunderle, Joseph M., Jr., and Wayne J. Arendt. "The Plight of Migrant Birds Wintering in the Caribbean: Rainfall Effects in the Annual Cycle." *Forests* 8, no. 4 (2017): 115.

Wunderle, Joseph M., Jr., Dave Currie, Eileen H. Helmer, David N. Ewert, Jennifer D. White, Thomas S. Ruzycki, Bernard Parresol, and Charles Kwit. "Kirtland's Warblers in Anthropogenically Disturbed Early-successional Habitats on Eleuthera, the Bahamas." *Condor* 112, no. 1 (2010): 123–137.

Wunderle, Joseph M., Jr., Patricia K. Lebow, Jennifer D. White, Dave Currie, and David N. Ewert. "Sex and Age Differences in Site Fidelity, Food Resource Tracking, and Body Condition of Wintering Kirtland's Warblers (*Setophaga kirtlandii*) in the Bahamas." *Ornithological Monographs* 80, no. 2014 (2014): 1–62.

Zwarts, Leo, Rob G. Bijlsma, Jan van der Kamp, and Eddy Wymenga. *Living on the Edge: Wetlands and Birds in a Changing Sahel.* Zeist, the Netherlands: KNNV Publishing, 2009.

Six: TEARING UP THE CALENDAR

194 **"Even [the] loss"**: Susan M. Haig, Sean P. Murphy, John H. Matthews, Ivan Arismendi, and Mohammad Safeeq, "Climate-Altered Wetlands Challenge Waterbird Use and Migratory Connectivity in Arid Landscapes," *Scientific Reports* 9, no. 1 (2019): 6.

198 **"Owing to":** Christiaan Both and Marcel E. Visser, "Adjustment to Climate Change is Constrained by Arrival Date in a Long-distance Migrant Bird," *Nature* 411, no. 6835 (2001): 296.

Andres, Brad A., Cheri Gratto-Trevor, Peter Hicklin, David Mizrahi, RI Guy Morrison, and Paul A. Smith. "Status of the Semipalmated Sandpiper." *Waterbirds* 35, no. 1 (2012): 146–149.

Bearhop, Stuart, Wolfgang Fiedler, Robert W. Furness, Stephen C. Votier, Susan Waldron, Jason Newton, Gabriel J. Bowen, Peter Berthold, and Keith Farnsworth. "Assortative Mating as a Mechanism for Rapid Evolution of a Migratory Divide." *Science* 310, no. 5747 (2005): 502–504.

Bilodeau, Frédéric, Gilles Gauthier, and Dominique Berteaux. "The Effect of Snow Cover on Lemming Population Cycles in the Canadian High Arctic." *Oecologia* 172, no. 4 (2013): 1007–1016.

Chambers, Lynda E., Res Altwegg, Christophe Barbraud, Phoebe Barnard, Linda J. Beaumont, Robert J. M. Crawford, Joel M. Durant, et al. "Phenological Changes in the Southern Hemisphere." *PloS one* 8, no. 10 (2013): e75514.

Chambers, Lynda E., Linda J. Beaumont, and Irene L. Hudson. "Continental Scale Analysis of Bird Migration Timing: Influences of Climate and Life History Traits—a Generalized Mixture Model Clustering and Discriminant Approach." *International Journal of Biometeorology* 58, no. 6 (2014): 1147–1162.

Corkery, C. Anne, Erica Nol, and Laura Mckinnon. "No Effects of Asynchrony Between Hatching and Peak Food Availability on Chick Growth in Semipalmated Plovers (*Charadrius semipalmatus*) near Churchill, Manitoba." *Polar Biology* 42, no. 3 (2019): 593–601.

Cornulier, Thomas, Nigel G. Yoccoz, Vincent Bretagnolle, Jon E. Brommer, Alain Butet, Frauke Ecke, David A. Elston et al. "Europe-wide Dampening of Population Cycles in Keystone Herbivores." *Science* 340, no. 6128 (2013): 63–66.

Eggleston, Jack, and Jason Pope. *Land Subsidence and Relative Sea-level Rise in the Southern Chesapeake Bay Region.* US Geological Survey Circular 1392. Reston, VA: US Geological Survey, 2013. http://dx.doi.org/10.3133/cir1392.

Fischer, Hubertus, Katrin J. Meissner, Alan C. Mix, Nerilie J. Abram, Jacqueline Austermann, Victor Brovkin, Emilie Capron, et al. "Palaeoclimate Constraints on the Impact of 2 C Anthropogenic Warming and Beyond." *Nature Geoscience* 11, no. 7 (2018): 474.

Ge, Quansheng, Huanjiong Wang, This Rutishauser, and Junhu Dai. "Phenological Response to Climate Change in China: A Meta-analysis." *Global Change Biology* 21, no. 1 (2015): 265–274.

Helm, Barbara, Benjamin M. Van Doren, Dieter Hoffmann, and Ute Hoffmann. "Evolutionary Response to Climate Change in Migratory Pied Flycatchers." *Current Biology* (2019). https://doi.org/10.1016/j.cub.2019.08.072.

Hiemer, Dieter, Volker Salewski, Wolfgang Fiedler, Steffen Hahn, and Simeon Lisovski. "First Tracks of Individual Blackcaps Suggest a Complex Migration Pattern." *Journal of Ornithology* 159, no. 1 (2018): 205–210.

Ims, Rolf A., John-Andre Henden, and Siw T. Killengreen. "Collapsing Population Cycles." *Trends in Ecology and Evolution* 23, no. 2 (2008): 79–86.

Iverson, Samuel A., H. Grant Gilchrist, Paul A. Smith, Anthony J. Gaston, and Mark R. Forbes. "Longer Ice-free Seasons Increase the Risk of Nest Depredation by Polar Bears for Colonial Breeding Birds in the Canadian Arctic." *Proceedings of the Royal Society B: Biological Sciences* 281, no. 1779 (2014): 20133128.

Kobori, Hiromi, Takuya Kamamoto, Hayashi Nomura, Kohei Oka, and Richard Primack. "The Effects of Climate Change on the Phenology of Winter Birds in Yokohama, Japan." *Ecological Research* 27, no. 1 (2012): 173–180.

Kwon, Eunbi, Emily L. Weiser, Richard B. Lanctot, Stephen C. Brown, H. River Gates, H. Grant Gilchrist, Steve J. Kendall, et al. "Geographic Variation in the Intensity of Warming and Phenological Mismatch Between Arctic Shorebirds and Invertebrates." *Ecological Monographs* (2019): e01383.

Lameris, Thomas K., Henk P. van der Jeugd, Götz Eichhorn, Adriaan M. Dokter, Willem Bouten, Michiel P. Boom, Konstantin E. Litvin, Bruno J. Ens, and Bart A. Nolet. "Arctic Geese Tune Migration to a Warming Climate But Still Suffer From a Phenological Mismatch." *Current Biology* 28, no. 15 (2018): 2467–2473.

Langham, Gary M., Justin G. Schuetz, Trisha Distler, Candan U. Soykan, and Chad Wilsey. "Conservation Status of North American Birds in the Face of Future Climate Change." *PloS One* 10, no. 9 (2015): e0135350.

La Sorte, Frank A., and Daniel Fink. "Projected Changes in Prevailing Winds for Transatlantic Migratory Birds Under Global Warming." *Journal of Animal Ecology* 86, no. 2 (2017): 273–284.

La Sorte, Frank A., Daniel Fink, Wesley M. Hochachka, Andrew Farnsworth, Amanda D. Rodewald, Kenneth V. Rosenberg, Brian L. Sullivan, David W. Winkler, Chris Wood, and Steve Kelling. "The Role of Atmospheric Conditions in the Seasonal Dynamics of North American Migration Flyways." *Journal of Biogeography* 41, no. 9 (2014): 1685–1696.

La Sorte, Frank A., Daniel Fink, and Alison Johnston. "Time of Emergence of Novel Climates for North American Migratory Bird Populations." *Ecography* (2019).

La Sorte, Frank A., Wesley M. Hochachka, Andrew Farnsworth, André A. Dhondt, and Daniel Sheldon. "The Implications of Mid-latitude Climate Extremes for North American Migratory Bird Populations." *Ecosphere* 7, no. 3 (2016): e01261.

La Sorte, Frank A., Kyle G. Horton, Cecilia Nilsson, and Adriaan M. Dokter. "Projected Changes in Wind Assistance Under Climate Change for Nocturnally Migrating Bird Populations." *Global change biology* 25, no. 2 (2019): 589–601.

Layton-Matthews, Kate, Brage Bremset Hansen, Vidar Grøtan, Eva Fuglei, and Maarten J. J. E. Loonen. "Contrasting Consequences of Climate Change for Migratory Geese: Predation, Density Dependence and Carryover Effects Offset Benefits of High-Arctic Warming." *Global Change Biology* (2019).

Lehikoinen, Esa, and Tim H. Sparks. "Changes in Migration." In *Effects of Climate Change on Birds*, edited by Anders Pape Møller, Wolfgang Fiedler, and Peter Berthold, 89–112. Oxford and New York: Oxford University Press, 2010.

Lewis, Kristy, and Carlo Buontempo. "Climate Impacts in the Sahel and West Africa:

The Role of Climate Science in Policy Making." *West African Papers* no. 2. Paris: OECD Publishing, 2016. http://dx.doi.org/10.1787/5jlsmktwjcd0-en.

Marra, Peter P., Charles M. Francis, Robert S. Mulvihill, and Frank R. Moore. "The Influence of Climate on the Timing and Rate of Spring Bird Migration." *Oecologia* 142, no. 2 (2005): 307–315.

Mettler, Raeann, H. Martin Schaefer, Nikita Chernetsov, Wolfgang Fiedler, Keith A. Hobson, Mihaela Ilieva, Elisabeth Imhof, et al. "Contrasting Patterns of Genetic Differentiation Among Blackcaps (*Sylvia atricapilla*) with Divergent Migratory Orientations in Europe." *PLoS One* 8, no. 11 (2013): e81365.

Møller, Anders Pape, Diego Rubolini, and Esa Lehikoinen. "Populations of Migratory Bird Species That Did Not Show a Phenological Response to Climate Change are Declining." *Proceedings of the National Academy of Sciences* 105, no. 42 (2008): 16195–16200.

Monerie, Paul-Arthur, Michela Biasutti, and Pascal Roucou. "On the Projected Increase of Sahel Rainfall During the Late Rainy Season." *International Journal of Climatology* 36, no. 13 (2016): 4373–4383.

Newson, Stuart E., Nick J. Moran, Andy J. Musgrove, James W. Pearce-Higgins, Simon Gillings, Philip W. Atkinson, Ryan Miller, Mark J. Grantham, and Stephen R. Baillie. "Long-term Changes in the Migration Phenology of U.K. Breeding Birds Detected by Large-scale Citizen Science Recording Schemes." *Ibis* 158, no. 3 (2016): 481–495.

Prop, Jouke, Jon Aars, Bård-Jørgen Bårdsen, Sveinn A. Hanssen, Claus Bech, Sophie Bourgeon, Jimmy de Fouw, et al. "Climate Change and the Increasing Impact of Polar Bears on Bird Populations." *Frontiers in Ecology and Evolution* 3 (2015): 33.

Samplonius, Jelmer M., and Christiaan Both. "Climate Change May Affect Fatal Competition Between Two Bird Species." *Current Biology* 29, no. 2 (2019): 327–331.

Senner, Nathan R. "One Species But Two Patterns: Populations of the Hudsonian Godwit (*Limosa haemastica*) Differ in Spring Migration Timing." *The Auk* 129, no. 4 (2012): 670–682.

———, Maria Stager, and Brett K. Sandercock. "Ecological Mismatches Are Moderated by Local Conditions for Two Populations of a Long-distance Migratory Bird." *Oikos* 126, no. 1 (2017): 61–72.

———, Mo A. Verhoeven, José M. Abad-Gómez, José A. Alves, Jos CEW Hooijmeijer, Ruth A. Howison, Rosemarie Kentie et al. "High Migratory Survival and Highly Variable Migratory Behavior in Black-Tailed Godwits." *Frontiers in Ecology and Evolution* 7 (2019): 96.

———, Mo A. Verhoeven, José M. Abad-Gómez, Jorge S. Gutiérrez, Jos CEW Hooijmeijer, Rosemarie Kentie, José A. Masero, T. Lee Tibbitts, and Theunis Piersma. "When Siberia Came to the Netherlands: The Response of Continental Black-tailed Godwits to a Rare Spring Weather Event." *Journal of Animal Ecology* 84, no. 5 (2015): 1164–1176.

Stange, Erik E., Matthew P. Ayres, and James A. Bess. "Concordant Population Dynamics of Lepidoptera Herbivores in a Forest Ecosystem." *Ecography* 34, no. 5 (2011): 772–779.

Tarka, Maja, Bengt Hansson, and Dennis Hasselquist. "Selection and Evolutionary Potential of Spring Arrival Phenology in Males and Females of a Migratory Songbird." *Journal of Evolutionary Biology* 28, no. 5 (2015): 1024–1038.

van Gils, Jan A., Simeon Lisovski, Tamar Lok, Włodzimierz Meissner, Agnieszka Ożarowska, Jimmy de Fouw, Eldar Rakhimberdiev, Mikhail Y. Soloviev, Theunis Piersma, and Marcel Klaassen. "Body Shrinkage Due to Arctic Warming Reduces Red Knot Fitness in Tropical Wintering Range." *Science* 352, no. 6287 (2016): 819–821.

Weeks, Brian C., David E. Willard, Aspen A. Ellis, Max L. Witynski, Mary Hennen, and Benjamin M. Winger. "Shared Morphological Consequences of Global Warming in North American Migratory Birds." *Ecology Letters* (2019). https://doi .org/10.1111/ele.13434.

Seven: *AGUILUCHOS* REDUX

Anderson, Dick, Roxie Anderson, Mike Bradbury, Calvin Chun, Julie Dinsdale, Jim Estep, Kristio Fien, and Ron Schlorff. *California Swainson's Hawk Inventory: 2005–2006, 2005 Progress Report.* Sacramento: California Department of Fish and Game, 2005.

Battistone, Carrie, Jenny Marr, Todd Gardner, and Dan Gifford. *Status Review: Swainson's Hawk (*Buteo swainsoni*) in California.* Sacramento: California Department of Fish and Wildlife, 2016.

Bechard, M. J., C. S. Houston, J. H. Saransola, and A. S. England. "Swainson's Hawk (*Buteo swainsoni*), version 2.0." In *The Birds of North America,* edited by A. F. Poole. Cornell Lab of Ornithology, Ithaca, NY, 2010. https://doi.org/10 .2173/bna.265.

Bedsworth, Louise, Dan Cayan, Guido Franco, Leah Fisher, and Sonya Ziaja. *Statewide Summary Report, California's Fourth Climate Change Assessment.* Sacramento: California Governor's Office of Planning and Research, Scripps Institution of Oceanography, California Energy Commission, and California Public Utilities Commission, 2018. Publication number SUM- CCCA4–2018–013.

Bloom, Peter H. *The Status of the Swainson's Hawk in California, 1979.* Federal Aid in Wildlife Restoration, Project W-54-R-12, Nongame Wildlife. Investment Job Final Report 11–8.0. Sacramento: California Department of Fish and Game, 1980.

Huning, Laurie S., and Amir AghaKouchak. "Mountain Snowpack Response to Different Levels of Warming." *Proceedings of the National Academy of Sciences* 115.43 (2018): 10932–10937.

Snyder, Robin E., and Stephen P. Ellner. "Pluck or Luck: Does Trait Variation or Chance Drive Variation in Lifetime Reproductive Success?" *American Naturalist* 191, no. 4 (2018): E90–E107.

Whisson, D. A., S. B. Orloff, and D. L. Lancaster. "Alfalfa Yield Loss from Belding's Ground Squirrels." *Wildlife Society Bulletin* 27 (1999): 178–183.

Eight: OFF THE SHELF

261 **"used the entire Pacific Ocean":** Scott A. Shaffer, Yann Tremblay, Henri Weimerskirch, Darren Scott, David R. Thompson, Paul M. Sagar, Henrik Moller, Graeme A. Taylor, David G. Foley, Barbara A. Block, and Daniel P. Costa, "Migratory Shearwaters Integrate Oceanic Resources Across the Pacific Ocean in an Endless Summer," *PNAS* 103, no. 34 (2006): 12799–12802, p. 12799.

Bolton, Mark, Andrea L. Smith, Elena Gómez-Díaz, Vicki L. Friesen, Renata Medeiros, Joël Bried, Jose L. Roscales, and Robert W. Furness. "Monteiro's Storm-petrel *Oceanodroma monteiroi*: A New Species from the Azores." *Ibis* 150, no. 4 (2008): 717–727.

Brown, S., C. Duncan, J. Chardine, and M. Howe. "Red-necked Phalarope Research, Monitoring, and Conservation Plan for the Northeastern U.S. and Maritimes Canada." Manomet Center for Conservation Sciences, Manomet, MA. Version 1 (2005). https://whsrn.org/wp-content/uploads/2019/02/conservationplan_rnph_v1.1_2010.pdf.

Caravaggi, Anthony, Richard J. Cuthbert, Peter G. Ryan, John Cooper, and Alexander L. Bond. "The Impacts of Introduced House Mice on the Breeding Success of Nesting Seabirds on Gough Island." *Ibis* 161, no. 3 (2019): 648–661.

Dias, Maria P., José P. Granadeiro, and Paulo Catry. "Do Seabirds Differ from Other Migrants in Their Travel Arrangements? On Route Strategies of Cory's Shearwater During its Trans-equatorial Journey." *PLoS One* 7, no. 11 (2012): e49376.

Dilley, Ben J., Delia Davies, Alexander L. Bond, and Peter G. Ryan. "Effects of Mouse Predation on Burrowing Petrel Chicks at Gough Island." *Antarctic Science* 27, no. 6 (2015): 543–553.

Duncan, Charles D. "The Migration of Red-necked Phalaropes: Ecological Mysteries and Conservation Concerns." *Bird Observer* 23, no. 4 (1996): 200–207.

Ebersole, Rene. "How Intrepid Biologists Brought Balance Back to the Aleutian Islands." Atlas Obscura, Aug. 6, 2019. https://www.atlasobscura.com/articles/fox-extermination-aleutian-islands-alaska.

Friesen, V. L., A. L. Smith, E. Gomez-Diaz, M. Bolton, R. W. Furness, J. González-Solís, and L. R. Monteiro. "Sympatric Speciation by Allochrony in a Seabird." *Proceedings of the National Academy of Sciences* 104, no. 47 (2007): 18589–18594.

Getz, J. E., J. H. Norris, and J. A. Wheeler. Conservation Action Plan for the Black-capped Petrel (*Pterodroma hasitata*). International Black-capped Petrel Conservation Group, 2011. https://www.fws.gov/migratorybirds/pdf/management/focal-species/Black-cappedpetrel.pdf.

Hedd, April, William A. Montevecchi, Helen Otley, Richard A. Phillips, and David A. Fifield. "Trans-equatorial Migration and Habitat Use by Sooty Shearwaters *Puffinus griseus* from the South Atlantic During the Nonbreeding Season." *Marine Ecology Progress Series* 449 (2012): 277–290.

Holmes, Nick D., Dena R. Spatz, Steffen Oppel, Bernie Tershy, Donald A. Croll, Brad Keitt, Piero Genovesi, et al. "Globally Important Islands Where Eradicating

Invasive Mammals Will Benefit Highly Threatened Vertebrates." *PloS One* 14, no. 3 (2019): e0212128.

Howell, Steve N. G. *Petrels, Albatrosses and Storm-Petrels of North America*. Princeton, NJ, and Oxford: Princeton University Press, 2012.

———, Ian Lewington, and Will Russell. *Rare Birds of North America*. Princeton, NJ, and Oxford: Princeton University Press, 2014.

Hunnewell, Robin W., Antony W. Diamond, and Stephen C. Brown. "Estimating the Migratory Stopover Abundance of Phalaropes in the Outer Bay of Fundy, Canada." *Avian Conservation and Ecology* 11, no. 2 (2016): 11.

Marris, Emma. "Large Island Declared Rat-free in Biggest Removal Success." National Geographic Online, May 9, 2018. https://news.nationalgeographic.com/2018/05/south-georgia-island-rat-free-animals-spd/.

Newman, Jamie, Darren Scott, Corey Bragg, Sam McKechnie, Henrik Moller, and David Fletcher. "Estimating Regional Population Size and Annual Harvest Intensity of the Sooty Shearwater in New Zealand." *New Zealand Journal of Zoology* 36, no. 3 (2009): 307–323.

Nisbet, Ian C. T., and Richard R. Veit. "An Explanation for the Population Crash of Red-necked Phalaropes *Phalaropus lobatus* Staging in the Bay of Fundy in the 1980s." *Marine Ornithology* 43 (2015): 119–121.

Pollet, Ingrid L., April Hedd, Philip D. Taylor, William A. Montevecchi, and Dave Shutler. "Migratory Movements and Wintering Areas of Leach's Storm-Petrels Tracked Using Geolocators." *Journal of Field Ornithology* 85, no. 3 (2014): 321–328.

Reynolds, John D. "Mating System and Nesting Biology of the Red-necked Phalarope *Phalaropus lobatus*: What Constrains Polyandry?" *Ibis* 129 (1987): 225–242.

Rubega, M. A., D. Schamel, and D. M. Tracy. Red-necked Phalarope (*Phalaropus lobatus*), ver. 2.0. In *The Birds of North America*, edited by A. F. Poole and F. B. Gill. Cornell Lab of Ornithology, Ithaca, NY, 2000. https://doi.org/10.2173/bna.538.

Ryan, Peter G., Karen Bourgeois, Sylvain Dromzée, and Ben J. Dilley. "The Occurrence of Two Bill Morphs of Prions *Pachyptila vittata* on Gough Island." *Polar Biology* 37, no. 5 (2014): 727–735.

Silva, Mauro F., Andrea L. Smith, Vicki L. Friesen, Joël Bried, Osamu Hasegawa, M. Manuela Coelho, and Mónica C. Silva. "Mechanisms of Global Diversification in the Marine Species Madeiran Storm-petrel *Oceanodroma castro* and Monteiro's Storm-petrel *O. monteiroi*: Insights From a Multi-locus Approach." *Molecular Phylogenetics and Evolution* 98 (2016): 314–323.

Silva, Mónica C., Rafael Matias, Vânia Ferreira, Paulo Catry, and José P. Granadeiro. "Searching for a Breeding Population of Swinhoe's Storm-petrel at Selvagem Grande, NE Atlantic, with a Molecular Characterization of Occurring Birds and Relationships within the Hydrobatinae." *Journal of Ornithology* 157, no. 1 (2016): 117–123.

Smith, Malcolm, Mark Bolton, David J. Okill, Ron W. Summers, Pete Ellis, Felix Liechti, and Jeremy D. Wilson. "Geolocator Tagging Reveals Pacific Migration of Red-necked Phalarope *Phalaropus lobatus* Breeding in Scotland." *Ibis* 156, no. 4 (2014): 870–873.

Weimerskirch, Henri, Karine Delord, Audrey Guitteaud, Richard A. Phillips, and
    Patrick Pinet. "Extreme Variation in Migration Strategies Between and Within
    Wandering Albatross Populations During their Sabbatical Year, and Their Fitness
    Consequences." *Scientific Reports* 5 (2015): 8853.
Wong, Sarah N. P., Robert A. Ronconi, and Carina Gjerdrum. "Autumn At-sea Dis-
    tribution and Abundance of Phalaropes *Phalaropus* and Other Seabirds in the
    Lower Bay of Fundy, Canada." *Marine Ornithology* 46 (2018): 1–10.

Nine: TO HIDE FROM GOD

281 **"the grand slam":** Anthony Bourdain, *Medium Raw* (New York: HarperCol-
     lins, 2010), xiii.
281 **"With every bite":** Ibid., xv.
295 **"It is sometimes found":** Thomas F. De Voe, *The Market Assistant* (New York:
     Hurd and Houghton, 1867), 168.
295 **"common in our markets":** Ibid., 175–176.
295 **"almost as good as quail":** Ibid., 175.
295 **"in the neighborhood":** Ibid.
295 **"[w]hite snow-bird":** Ibid., 176.
295 **"quite as good eating":** Ibid.
295 **"occasionally found":** Ibid., 178.
296 **"very small":** Ibid.
296 **"very delicate":** Ibid.
296 **"I think they should never be killed":** Ibid., 178.
296 **"[t]housands of birds":** Ibid., 146.
296 **"except for a collection":** Ibid., 176.
296 **"I, however, think":** Ibid., 175.
296 **"not so well flavored":** Ibid., 177.
296 **"not so tender":** Ibid., 175.
296 **"[l]arge numbers":** Ibid.
297 **"[m]any times":** P. P. Claxton, quoted in T. Gilbert Pearson, "The Robin,"
     Bird-Lore 11, no. 5 (Oct. 1, 1910): 208.
297 **"killed for the twenty-five cents":** Edward Howe Forbush, *Birds of Massa-
     chusetts and Other New England States*, vol. 2 (Norwood, MA: Norwood Press,
     1927), 417.
297 **"New Game Bird Law":** *New York Times*, Aug. 18, 1918, 14.
299 **"pledge a zero tolerance approach":** Larnaca Declaration, July 7, 2011, http://
     www.moi.gov.cy/moi/wildlife/wildlife_new.nsf/web22_gr/F5BC37B27C945E
     BCC22578410043F43F/$file/Larnaca%20%20Declaration.pdf.

Andreou, Eva. "Cypriot and Bases Authorities Slammed by Anti-poaching NGOs."
    Cyprus Mail, July 7, 2017. https://cyprus-mail.com/2017/07/20/cypriot-bases
    -authorities-slammed-anti-poaching-ngos/?hilite=%27poaching%27.
Anon. "Explosion Outside Dhekelia Police Station." *Cyprus Mail*, June 13, 2017.

https://cyprus-mail.com/2017/06/13/explosion-outside-dhekelia-police-station/#disqus_thread.

———. "Illegal Bird Trapping Begins to Fall. *KNEWS*, March 6, 2018. https://knews.kathimerini.com.cy/en/news/illegal-bird-trapping-begins-to-fall.

———. "The New Protection of Birds Act." *British Birds* no. 12 (Dec. 1954): 409–413.

Bicha, Karel D. "Spring Shooting: An Issue in the Mississippi Flyway, 1887–1913." *Journal of Sport History* 5 (Summer 1978): 65–74.

BirdLife International. *A Best Practice Guide for Monitoring Illegal Killing and Taking of Birds*. Cambridge, UK: BirdLife International, 2015.

Brochet, Anne-Laure, Willem Van den Bossche, Sharif Jbour, P. Kariuki Ndang'ang'a, Victoria R. Jones, Wed Abdel Latif Ibrahim Abdou, Abdel Razzaq Al-Hmoud, et al. "Preliminary Assessment of the Scope and Scale of Illegal Killing and Taking of Birds in the Mediterranean." *Bird Conservation International* 26, no. 1 (2016): 1–28.

Brochet, Anne-Laure, Willem Van Den Bossche, Victoria R. Jones, Holmfridur Arnardottir, Dorin Damoc, Miroslav Demko, Gerald Driessens, et al. "Illegal Killing and Taking of Birds in Europe Outside the Mediterranean: Assessing the Scope and Scale of a Complex Issue." *Bird Conservation International* (2017): 1–31.

Day, Albert M. *North American Waterfowl*. New York and Harrisburg, PA: Stackpole and Heck, 1949.

Eason, Perri, Basem Rabia, and Omar Attum. "Hunting of Migratory Birds in North Sinai, Egypt." *Bird Conservation International* 26, no. 1 (2016): 39–51.

European Union. "Directive 2009/147/EC Of the European Parliament and of the Council of 30 November 2009 on the Conservation of Wild Birds." *Official Journal L* 20 26.1.2010 (2010): 7–25.

Franzen, Jonathan. "Emptying the Skies." *The New Yorker*, July 26, 2010. https://www.newyorker.com/magazine/2010/07/26/emptying-the-skies.

Greenberg, Joel. *A Feathered River Across the Sky*. New York: Bloomsbury, 2014.

Grinnell, Joseph, Harold Child Bryant, and Tracy Irwin Storer. *The Game Birds of California*. Berkeley: University of California Press, 1918.

Hajiloizis, Mario. "Up to 300 British Soldiers 'Trapped' by Xylofagou Residents." *SigmaLive*, Oct. 20, 2016. https://www.sigmalive.com/en/news/local/149580/up-to-300-british-soldiers-trapped-by-xylofagou-residents.

Jenkins, Heather M., Christos Mammides, and Aidan Keane. "Exploring Differences in Stakeholders' Perceptions of Illegal Bird Trapping in Cyprus." *Journal of Ethnobiology and Ethnomedicine* 13, no. 1 (2017): 67–77.

Jiguet, Frédéric, Alexandre Robert, Romain Lorrillière, Keith A. Hobson, Kevin J. Kardynal, Raphaël Arlettaz, Franz Bairlein, et al. "Unravelling Migration Connectivity Reveals Unsustainable Hunting of the Declining Ortolan Bunting." *Science Advances* 5, no. 5 (2019): eaau2642.

Kamp, Johannes, Steffen Oppel, Alexandr A. Ananin, Yurii A. Durnev, Sergey N. Gashev, Norbert Hölzel, Alexandr L. Mishchenko, et al. "Global Population Collapse in a Superabundant Migratory Bird and Illegal Trapping in China." *Conservation Biology* 29, no. 6 (2015): 1684–1694.

Mark, Philip. "Xylofagou Residents Stop British Soldiers from Cutting Trees." *Cyprus Mail*, Oct. 20, 2016. https://cyprus-mail.com/old/2016/10/20/stop-soldiers-from-cutting-trees/.

McLaughlin, Kelly. "Police Officer Injured in Explosion at British Military Base in Cyprus as Police Open Criminal Investigation." *Daily Mail*, June 13, 2017. https://www.dailymail.co.uk/news/article-4598670/Small-blast-British-station-Cyprus-criminal-motive-seen.html.

Paterniti, Michael. "The Last Meal." *Esquire* 129, May 1998, 112–117.

Psyllides, George. "Cyprus a Bird 'Trapper's Treasure Island,' According to Survey." *Cyprus Mail*, Aug. 21, 2015. https://cyprus-mail.com/old/2015/08/21/cyprus-a-bird-trappers-treasure-island-according-to-survey/.

Shialis, Tassos. "Update on Illegal Bird Trapping Activity in Cyprus." BirdLife Cyprus (March 2018). https://www.impel-esix.eu/wp-content/uploads/sites/2/2018/07/BirdLife-Cyprus_Spring-2017-trapping-report_Final_for-public-use.pdf.

United States Entomological Commission, Alpheus Spring Packard, Charles Valentine Riley, and Cyrus Thomas. *First Annual Report of the United States Entomological Commission for the Year 1877: Relating to the Rocky Mountain Locust and the Best Methods of Preventing Its Injuries and of Guarding Against Its Invasions, in Pursuance of an Appropriation Made by Congress for this Purpose*. Washington, DC: US Government Printing Office, 1878.

Ten: *ENINUM*

334   **"the most massive movement":** Richard M. Eaton, "Comparative History as World History: Religious Conversion in Modern India," *Journal of World History* 8 (1997): 245.

Anderson, R. Charles. "Do Dragonflies Migrate Across the Western Indian Ocean?" *Journal of Tropical Ecology* 25, no. 4 (2009): 347–358.

Anon. "From Slaughter to Spectacle—Education Inspires Locals to Love Amur Falcon." *BirdLife International*, Jan. 29, 2018. https://www.birdlife.org/worldwide/news/slaughter-spectacle-education-inspires-locals-love-amur-falcon.

Banerjee, Ananda. "The Flight of the Amur Falcon." LiveMINT, Oct. 29, 2013. https://www.livemint.com/Politics/34X8t639wdF1PPhlOuhBlJ/The-flight-of-the-Amur-Falcon.html.

Barpujari, S. K. "Survey Operations in the Naga Hills in the Nineteenth Century and Naga Opposition Towards Survey." *Proceedings of the Indian History Congress* 39 (1978): 660–670.

Baruth, Sanjib. "Confronting Constructionism: Ending India's Naga War." *Journal of Peace Research* 40 (2003): 321–338.

Chaise, Charles. "Nagaland in Transition." *India International Centre Quarterly* 32 (2005): 253–264.

Das, N. K. "Naga Peace Parlays: Sociological Reflections and a Plea for Pragmatism." *Economic and Political Weekly* 46 (2011): 70–77.

Dixon, Andrew, Nyambayar Batbayar, and Gankhuyag Purev-Ochir. "Autumn Migration of an Amur Falcon *Falco amurensis* from Mongolia to the Indian Ocean Tracked by Satellite." *Forktail* 27 (2011): 86–89.

Glancey, Jonathan. *Nagaland*. London: Faber and Faber, 2011.

Kumar, Braj Bihari. *Naga Identity*. New Delhi: Concept Publishing, 2005.

Parr, N., S. Bearhop, D. Douglas, J. Y. Takekawa, D. J. Prosser, S. H. Newman, W. M. Perry, S. Balachandran, M. J. Witt, Y. Hou, Z. Luo, and L. A. Hawkes. "High Altitude Flights by Ruddy Shelduck *Tadorna ferruginea* During Trans-Himalayan Migrations." *Journal of Avian Biology* 48 (2017): 1310–1315.

Sinha, Neha. "A Hunting Community in Nagaland Takes Steps Toward Conservation." *New York Times*, Jan. 3, 2014. https://india.blogs.nytimes.com/2014/01/03/a-hunting-community-in-nagaland-takes-steps-toward-conservation/.

Symes, Craig T., and Stephan Woodborne. "Migratory Connectivity and Conservation of the Amur Falcon *Falco amurensis*: A Stable Isotope Perspective." *Bird Conservation International* 20 (2010): 134–148.

Thomas, John. *Evangelizing the Nation*. London and New York: Routledge, 2016.

Epilogue

342 **"baking":** Rick Thoman, quoted in Susie Cagle, "Baked Alaska," *The Guardian*, July 3, 2019, https://www.theguardian.com/us-news/2019/jul/02/alaska-heat-wildfires-climate-change.

INDEX

Page numbers in *italics* refer to maps.